A

Adult Services
Mead Public Library
Sheboygan, Wisconsin

Borrowers are responsible for all library
materials drawn on their cards and for all
charges accruing on same.

DEMCO

SUSTAINABLE AGRICULTURE SYSTEMS

EDITED BY

J. L. HATFIELD

D. L. KARLEN

LEWIS PUBLISHERS

Boca Raton   Ann Arbor   London   Tokyo

Library of Congress Cataloging-in-Publication Data

Sustainable agriculture systems / edited by J.L. Hatfield, D.L. Karlen.
    p. cm.
   Includes bibliographical references and index.
   ISBN 1-56670-049-3
   1. Sustainable agriculture. 2. Agricultural systems. 3. Agricultural ecology.
 I. Hatfield, Jerry J. II. Karlen, D.L. (Douglas L.)
 S494.5.S86S86   1993
 630 – dc20   93–26659
                                                CIP

© 1994 by CRC Press, Inc.

Lewis Publishers is an imprint of CRC Press
No claim to original U.S. Government works
International Standard Book Number 1-56670-049-3
Library of Congress Card Number 93-26659
Printed in the United States of America 1 2 3 4 5 6 7 8 9 0
Printed on acid-free paper

PREFACE

The concept of sustainable agriculture receives either a strong positive or negative reaction from many individuals. To some, sustainability means a complete switchover to all organic inputs, crop rotations, and low inputs, while to others, sustainability evokes ideas of changing practices which improve efficiency in the use of all resources and increases the profitability of the farming enterprise. There are many reasons for these varied points of view and within a given context, both definitions have merit and value. Our intention and purpose within this volume is to assemble the information on the components which are embodied within the concepts of sustainable agriculture. The role of agriculture in the next century will be to provide food for an ever increasing world population while ensuring that the natural resources, water and soil, are not only conserved, but enhanced.

There is a need to embellish the ideas of land stewardship and create an atmosphere in which conservation and enhancement of resources are the norm rather than the exception. The authors who graciously gave of their time to prepare these chapters are committed to improving our understanding of the components within agricultural systems and then evaluating how these concepts could be incorporated into sustainable systems. These concepts range from water to insect management and include discussions of both the economics and sociology of agricultural enterprises. Our goal is to provide the reader with information about the processes which are involved in sustainable systems in such a way as not to polarize but to direct and stimulate thought toward how we can address the problem as a group rather than individually.

The issues of environmental quality, conservation, diversity, risk, and markets often direct our thinking in developing and adopting systems within a given region. *Alternative Agriculture* provided much of the stimulus for us to consider the work, and we hope that the value to the reader will be in terms of the realization that we in agriculture have an important task ahead of us; a task from which all of the peoples of the world will benefit and enjoy.

J. L. Hatfield
D. L. Karlen
Ames, Iowa

ABOUT THE EDITORS

Jerry L. Hatfield, Ph.D., is the Laboratory Director of the USDA-Agricultural Research Service, National Soil Tilth Laboratory in Ames, Iowa. He was appointed as the first director when the laboratory opened in 1989 and has developed a research program to focus on issues of environmental quality and sustainable agriculture.

Dr. Hatfield received his B.S. degree from Kansas State University in 1971 in Agronomy, a M.S. degree in 1972 in Agronomy from the University of Kentucky, and a Ph.D. in 1975 from Iowa State University in Agricultural Climatology with a minor in Statistics.

From 1975 to 1983 he was a Biometeorologist at the University of California-Davis where he taught undergraduate and graduate courses in biometeorology, microclimatology, micrometeorology, and instrumentation. In 1983 he joined the USDA-ARS Plant Stress and Water Conservation Unit in Lubbock, Texas, to conduct research on the energy exchanges and water use by cropping systems in semiarid agriculture.

Dr. Hatfield is recognized for his research on evapotranspiration and crop water use models and the detection of crop stress. His current research interests center on the evaluation of water use efficiency of different cropping systems in humid agriculture and the evaluation of the role of landscape position on the environmental impact of different soil and crop management practices. He is author of over 190 refereed publications and editor of three volumes of *Advances in Soil Science*. He is a fellow of the American Society of Agronomy and the Soil Science Society of America and currently serves as the editor of *Agronomy Journal*.

Douglas L. Karlen, Ph.D., is a Soil Scientist in the National Soil Tilth Laboratory in Ames, Iowa.

Dr. Karlen received his B.S. degree in Soil Science from the University of Wisconsin in 1973, a M.S. degree in 1975 from Michigan State University in Soil Science, and a Ph.D. in 1978 in Agronomy from Kansas State University.

From 1978 until 1988 he was located in the Coastal Plains Soil and Water Conservation Research Center of the USDA-Agricultural Research Center in Florence, South Carolina. In 1989 he transferred to the National Soil Tilth Laboratory in Ames, Iowa. His research program has centered on the quantitative interactions among soil, water, nutrient, and cultural management practices in different cropping systems and the comparison of alternative and conventional farming systems.

Dr. Karlen is the co-leader of the Leopold Center for Sustainable Agriculture Cropping Systems Issue Team and helps develop interfaces among the research community and the user in sustainable agriculture issues. He is the author of over 130 refereed publications and is a fellow of the American Society of Agronomy and the Crop Science Society of America and is Associate Editor of *Crop Science*.

ACKNOWLEDGMENTS

We owe a large thanks to the authors of each chapter who gave their time and, more importantly, contributed their thoughts and ideas. Without the commitment of dedicated scientists, our knowledge base and information transfer to both present and future generations would not be possible. We are most appreciative of the efforts of Judy Shoen who carefully read and checked each paper to create a high quality work. We do appreciate the willingness of Lewis Publishers to encourage us in this effort and support the distribution of these concepts to the world.

CONTRIBUTORS

John B. Braden
Department of Agricultural
Economics
University of Illinois
Urbana, Illinois 68101

Orvin C. Burnside
Department of Agronomy and
Plant
Genetics
University of Minnesota
St. Paul, Minnesota 55108

Max D. Clegg
Department of Agronomy
University of Nebraska
Lincoln, Nebraska 68583

R. M. Cruse
Department of Agronomy
Iowa State University
Ames, Iowa 50011

Frank Forcella
USDA, ARS, MWA
North Central Soil Conservation
Research Laboratory
Morris, Minnesota 56267

Charles A. Francis
Department of Agronomy
University of Nebraska
Lincoln, Nebraska 68583

Joseph E. Funderburk
North Florida Research and
Education Center
University of Florida
Quincy, Florida 32351

Jerry L. Hatfield
National Soil Tilth Laboratory
USDA/ARS
Ames, Iowa 50011

Leon G. Higley
Department of Entomology
University of Nebraska
Lincoln, Nebraska 68583

Robert H. Hornbaker
Department of Agricultural
Economics
University of Illinois
Urbana, Illinois 61801

Douglas L. Karlen
National Soil Tilth Laboratory
USDA/ARS
Ames, Iowa 50011

Dennis R. Keeney
Leopold Center for Sustainable
Agriculture
Ames, Iowa 50011

K. A. Kohler
National Soil Tilth Laboratory
USDA/ARS
Ames, Iowa 50011

Steve Padgitt
Department of Sociology
Iowa State University
Ames, Iowa 50011

John Pesek
Department of Agronomy
Iowa State University
Ames, Iowa 50011

Peggy Petrzelka
Department of Sociology
Iowa State University
Ames, Iowa 50011

C. A. Robinson
Eastern New Mexico University
Station II
Portales, New Mexico 88130

E. John Sadler
USDA/ARS
Coastal Plains Soil, Water, and
    Plant Research Center
Florence, South Carolina 29502

Andrew N. Sharpley
USDA, ARS, SPA
National Agricultural Water Quality
    Laboratory
Durant, Oklahoma 74702

Neil C. Turner
CSIRO Division of Plant Industry
Private Bag
Wembley, Australia 6014

David C. White
Department of Agricultural
    Economics
University of Illinois
Urbana, Illinois 61801

CONTENTS

SUSTAINABLE AGRICULTURE SYSTEMS

1

Historical Perspective

John Pesek

TABLE OF CONTENTS

SUSTAINABILITY—GLOBAL DIMENSIONS

Agriculture may be defined as the art, practice, or science of crop and livestock production on organized farm units. It emerged as a directed human activity some ten to twelve thousand years ago and is generally recognized as one of the foundations for civilization. It is the dominant industry in many states, and the largest industry in the United States — indeed, of the world. It is indispensable because it provides almost all the food and most of the fiber we consume. The human species cannot exist in large numbers without an adequately productive agriculture.

Lee and DeVore (1968) wrote that of the estimated 80 billion humans who have lived on earth since "Cultural Man" emerged, 90% lived as hunters and gatherers and only 10% under agriculture. They pointed out

1

that the most successful and enduring human adaptation has been to the hunting way of life. However, the earth did not and cannot support many people as hunters. That is why we must make agriculture sustainable. The increasing population of humankind depends on sustaining an increasing agricultural production.

Sustainability of an agriculture that is environmentally benign in relation to world resources, population, and environment is a serious issue — perhaps, along with population, the central issue for the human race. Until relatively recent times, we could move to new locations to produce our food if our land resources were destroyed; but burgeoning global population assures us that if we have not already run out of new places to produce our food and fiber and live, we certainly will and very soon. "Sustainable agriculture is not only worth pursuing, it is inevitable. It seems sustainability, by whatever name, will be the overriding force of agricultural development efforts for the rest of this century and beyond" (Wilken, 1991). Vasey (1992) places this into perspective by asking us to look forward to 10,000 A.D. as we can look back to 10,000 B.C. when agriculture emerged. We have no substitute for agriculture, and it requires resources already being used.

The human species is the only one that can use resources at its disposal to alter the immediate environment in which it lives. Behaving like other organisms, it has reproduced to fill all possible niches where it can exist naturally as well as by use of its intellect. Rather than adjusting to the environment, humans have sought to dominate other species in the ecosystem and to crowd them out to extinction by various manipulations of external, sometimes distant, resources (Hillel, 1991). In doing so, severe instabilities in agroecosystems have developed and threaten the ability of the earth to accommodate the projected human population.

Most of the pressure on the ecosystems started after the emergence of agriculture. Early agriculturalists arose from the landscape and were part of the environment. The indigenous population in the contiguous United States probably was in good balance with the environment — at least, environmental dominance asserted itself if the natural system became violated. We do, indeed, have examples of such assertion prior to the modern era.

Archeological evidence indicates that lower Mesopotamian fields "salted out" under irrigation by about 4000 years ago (Jacobsen and Adams, 1958). Records over about an 1800-year period illustrate what happened. In 3500 B.C., wheat and barley were grown in about equal proportions. Wheat does not tolerate much salt in the soil while barley tolerates salt much better than wheat. By 2500 B.C. only one sixth of the grain was wheat, and by 1700 B.C., less than 2%. Also, in 2400 B.C. wheat yields were 2357 l/ha, only 1460 l/ha by 2100 B.C., and only 897 by 1700 B.C. This was because the small amount of salt in the irrigation water slowly accumulated to seriously damaging levels for wheat. The effect of salt was not understood nor did the farmers know that this might have been prevented by providing adequate drainage and a regular periodic flushing of salt from the soil profile by adding water in excess of that needed to produce the crop. Therefore, Sumerian cities dwindled to villages and were easily overcome by Babylon by the 18th century B.C.

Gardner (1985) addressed this problem of irrigation in dry regions when he wrote, "Successful irrigation schemes in arid regions carry the seeds of their own demise." Not only is water lost from agriculture to increasing populations but "salinity is a classical problem". Nor can modern pesticides, plant nutrients, and industrial wastes be ignored as causes of demise and deterioration of land and water.

The Chinese experience with agriculture has often been cited as an example of sustainability of the land resource. This is generally true where rice was the main food crop, but China underwent severe erosion in other regions before the land surface was stabilized (Bennett, 1939). Bennett also summarized many reports on erosion to indicate serious losses of land in Antioch, Persia, Greece, Rome, and North Africa, all important civilization centers which, like the Sumerians, were overcome by external forces. Rome, Greece, and other empires eventually fell because each, having destroyed its nearby soils, had to extend its food lines so far that they became easy prey to "resentful subjects" (Hillel, 1991).

It is interesting to note, in passing, that the same misfortune did not befall the Nile River valley because it was regularly flooded and the soil replenished and purged of salt accumulation. Its agriculture remained intact for thousands of years until the present (Hillel, 1991). The construction of the Aswan High Dam seems to have engendered some salt problems in unexpected places in Egypt, and one might wonder whether the high technology of this century might not eventually destroy the Nile valley in the next. Irrigation is not alone in causing salinization – the simple wheat-fallow system of farming in the northern Great Plains of North America has caused salt seeps on lower parts of the landscape as a result of more percolation of water higher on the slopes.

In the Western Hemisphere, some city-states in Central America had collapsed before the arrival of Cortez. There is not complete agreement as to why they declined. The fragile tropical ecosystems may have failed after hundreds of years, but evidence also leads some to believe that it was a failure of the culture and the government it created that led to failure (Dunning and Demarest, 1991). In fact, an effort is being made to reconstruct the lost art of tropical agriculture in this ecological zone for potential use in the future.

The history leading to the establishment of the United States and of the two centuries following has been mostly of exploitive behavior of both people and government. Natural resources have been plundered and often were more spoiled than used and we have caused losses of unnumbered species as well as other resources. The expanses of our great forests were cut to make room for farming; others were cut, consumed wastefully, and not reestablished. We have shamelessly allowed our soils to erode and there are millions of acres of formerly cultivated land that has been abandoned because of soil erosion (Pesek, 1980). The National Research Council (NRC) reports that even now we lose an estimated 3 billion tons of soil to erosion each year and, in addition to silt, plant nutrients and agricultural chemicals are finding their ways into surface and ground water (NRC, 1989).

We cannot severely fault our forebears because they, like the people of Mesopotamia, often were not aware of the consequences of their actions, and their personal survival was paramount. Besides, the land, timber, water, and game supply must have looked inexhaustible when viewed as recently as the 18th century by the signers of the Declaration of Independence.

EARLY PERFORMANCE OF SCIENTIFIC AGRICULTURE

Agriculture was largely at the mercy of the elements until 150 years ago, even though usefulness of manure and marl applications and the benefits of leguminous crops were known to the Romans and probably before. The era of scientific agriculture dates from the first publications of Justus von Liebig and James F. W. Johnston in 1840 and 1842, respectively. Their work dealt with the potential of chemistry in agriculture. The discovery of inheritance by Gregor Mendel (published in 1865) completed the basic chemical and biological building blocks for modern day conventional agriculture. It was then up to experiment stations, initially in Europe, and then North America and elsewhere to elaborate practices and crops that increased yields of fields and reduced the uncertainty of production. Scientific agriculture thus stalled the threat of mass starvation as the ultimate population control. It still does, and thereby gives humans time to stabilize population.

Nevertheless, given the present perception and reality of agriculture's operations and effects on the environment, it is not surprising that those outside of agriculture, as well as some of those in it, are becoming concerned about our stewardship as well as about farm workers' safety and the safety of the food we deliver. Some people, rightly or wrongly, feel very strongly that they are unnecessarily exposed to hazards that they do not know about and wonder why we in agriculture have had a tendency quickly to turn to the "technological fix" (often called "chemical fix") to solve problems, rather than to work within the biological and ecological systems first.

Yet, we cannot eschew all new technologies and methods of production and choices of crops to grow. Much of the productivity we enjoy in the United States and worldwide traces to improvements in technology and management and superior crop cultivars. Without these, it is certain that global pressure on food would be regular and profound instead of localized and globally infrequent. The suffering from hunger is still tragic in places like Ethiopia, Somalia, and the Sudan where the food shortage is a problem of distribution from abroad and, locally, due to conflict and failure to adopt technologies that are appropriate and likely adequate to provide food for even denser populations.

We have not always been this confident. Especially following World War II, we were concerned about our ability to "set the fifth plate" at the dinner table of the world. In fact, scientists working on a productive capacity study for the United States, published in 1952 by the United States Department of Agriculture and the land-grant colleges, grappled with the potential effect of emerging technologies in arriving at their productive capacity estimates. These included the use of synthetic chemical fertilizers and pesticides, higher-yielding crop varieties, and improved planting and harvesting machinery.

Thanks, in part, to the fertilizer, seed, and agricultural chemical industries; farm machinery manufacturers; farmers; and the educational and extension systems, these practices have become the conventional agriculture of recent decades and have produced dramatic increases in productivity. Yields per acre of United States crops are among the highest in the world, with fewer farmers feeding more people than ever before. The "technological fixes" have worked to produce more food.

There was little difficulty in getting widespread adoption of these practices once their effectiveness was established. We had everything going for us—most were simple, stood alone and were easy to communicate, the effects were clearly visible, there was a good experimental base, farmers wanted to escape drudgery, there was a growing agrichemical industry, the world needed more food and farmers were determined to supply it, and national agricultural policies encouraged food for export. Best of all, off-farm inputs were involved so at least two people benefited from adoption—hopefully the farmer, if the technology worked, and the supplier selling it. We were so caught up in change that we did not note all the side effects of the new methods and drew some experimentally unverified conclusions.

These thoughts were expressed by Davidson (1989). When he addressed the American Society of Agronomy Administrators' Roundtable he stated that we in the land-grant colleges are confronted with several dilemmas, one of which is

Distrust of non-agricultural groups interested in food quality and safety, natural resources, environmental quality and human resource issues. . . .

The distrust on the part of non-agricultural groups is well justified. With the publication of Rachel Carson's book entitled 'Silent Spring', we, in agriculture, loudly and in unison stated that pesticides did not contaminate the environment — *we now admit that they do*. When confronted with the presence of nitrates in groundwater we responded that it was not possible for nitrates from commercial fertilizer to reach groundwater in excess of 10 parts per million under normal productive agricultural systems — *we now admit they do*. When questioned about the presence of pesticides in food and food quality, we assured the public that if a pesticide was applied in compliance with the label, agricultural products would be free of pesticides — *we now admit they're not*. Certainly the availability of new instrumentation and ability to detect trace amounts of pesticides in water and food have changed the meaning of absolute zero. Although this may be used as an excuse for our belief that agriculture was not a contributor to environmental degradation, the truth is, we were not conducting the research and/or making the appropriate measurements to ensure that this was the case. Very few mass balance studies were conducted involving plant nutrients or pesticides. (Davidson, 1989)

This is a strong indictment!

Had we taken a good hard look at the incidental or concurrent effects of the technology we were studying and developing, we may have avoided some of the mistakes we admit having made and some we seemed to have made. Still, there is a compelling drive to be as efficient as possible in crop production to maintain a competitive edge, and by outward appearances we have succeeded admirably. We spend only a little over a tenth of our take-home pay on food — less than any other country. We have ignored the real cost of our technology at the farm level because we have not had to pay for the consequences (off-site effects), and society at large has not fully determined nor assessed costs of these effects on others and on the environment. After all, the upland farmer does not directly have to pay for dredging silt from the Mississippi River nor does the farmer in north central Iowa have to worry about the nitrate in river water used for drinking in Des Moines. However, even this is beginning to be changed by laws in Europe and the United States.

The conventional agriculture of the 1960s, 1970s, and 1980s emerging from the new production alternatives was characterized by consolidation of land holdings into fewer hands, the use of high capacity machinery, and employment of large amounts of other off-farm production inputs. We were very effective in promoting these changes including the principle of leveraging. Just as some abused borrowing of capital and lost farms, we have evidence that we have been imprudent in the use of other inputs as well, in some cases causing environmental deterioration.

A farm debt crisis in the United States emerged in the 1980s as land values plummeted, increasing the debt-to-asset ratio to a point where many operators could not survive in spite of the efforts of cooperative extension and other information systems. Others were looking for ways to reduce input costs and increase efficiency in order not to add to their debt load and still others were looking for safer farming practices. The stage was set for general consideration of different ways to farm—to search for methods which were not as demanding of capital, especially borrowed capital.

Monitoring records of wells and springs in many states also showed increasing amounts of nitrate and the presence of pesticides previously not observed in ground water, and we had come dangerously close to losing the bald eagle in some ecosystems and other birds of prey were put in jeopardy as a result of agricultural pollution. The fact that analytical methods have become more sensitive and that nitrate often is not at hazardous levels are not adequate defenses as long as too much is being used. Nor is the fact that nitrate is a common component of natural uncontaminated water sources—after all, even rainwater usually contains nitrate. The point is that we should not and, perhaps, need not add to the natural burden.

INCLUDING ECOLOGY IN AGRICULTURAL SYSTEMS

Scientific agriculture and conventional farming tended to drive ecology out of the input-output equation, but a number of activities and events, since 1970 and before, addressed the concern about agroecosystems and how contemporary agriculture changed and managed them. According to the Council for Agricultural Science and Technology (CAST), early in the 20th century, Sir Albert Howard, Director of the Institute of Plant Industry in Lahore, India, first described organic farming as a system, even though necessity had imposed this type of agriculture for thousands of years. His goal was to show how to farm in remote regions where the infrastructure of commerce was poorly developed or nonexistent. More recently, the Rodale Press, under J. I. Rodale, began publishing magazines and other materials dealing with organic farming and gardening in 1942 (CAST, 1980). The concepts for organic farming have also appeared under several different names in Europe and Australia as well as in the United States.

Silent Spring (Carson, 1962), mentioned by Davidson (1989), was the harbinger of a reassessment of agricultural technology application. That book, more than anything else, brought under scrutiny the chemical technology underpinning agriculture and caused many to ask whether there might be a better way. The petroleum fuel crisis of the mid-1970s and the simultaneous global food shortage emphasized the vulnerability of agriculture to nondomestic influences because of the dependence on petroleum for significant fuel and nonfuel agricultural inputs. Momentarily, the direction of agrichemically based agriculture everywhere was challenged, including the continued success of the "green revolution". The Rome World Food Conference in 1974 and the World Food Conference of 1976 in Ames, Iowa were convened to examine, again, the earth's capacity to produce enough food for its population.

A few "independent souls" began creative work on crop protection late in the 19th century, so crop protection concepts have been on the scene for at least a hundred years (Sill, 1982). In early years, these took the form of breeding for resistance of crops to pathogens after the general recognition of Mendel's discoveries at the turn of the century. One of the first successes of major dimensions was the development of rust-resistant wheat in the United States by introducing a resistance factor in wheat brought here from abroad. Numerous other successes in both pathogen resistance and insect and nematode resistance have followed in many crop species. These genetic biological controls have been easy to adopt because adoption usually meant the simple use of a different (resistant) cultivar instead of a failing susceptible one. Still, many crop pests have not succumbed to genetic controls and mutations of some have emerged to attack previously resistant plants. Today, molecular biology and genetics hold promise to provide new tools to confer genetic defenses. Some of these tools will allow a much greater arsenal of countermeasures to make crop plants tougher to attack.

Difficulties with the use of synthetic pesticides brought about by the development of resistance in diverse pests rekindled interest in biological control of pests by all biotic means, including parasites, prey organisms, toxins, antagonism, and antibioses (Sill, 1982). The so-called "natural" controls have probably always operated in nature and still do, but often not as rapidly as needed for commercial protection. The effectiveness of some of these can be enhanced by deliberate infestation or application by growers or by maintaining a favorable environment for the control organisms as is emphasized by organic farmers. Generally, biological control agents are more complex to use than chemical agents, may not operate as quickly, and require a greater degree of management skills in the operator to be successful.

The recorded successes of biological controls, the failure of synthetic pesticides, and the intractability of some pests led to the concept of a multifaceted system of holding pests in check below some established tolerable thresholds. The generally used term to describe this set of procedures is "integrated pest management" or IPM. Significant targeted federal support for research, extension and demonstration programs for IPM started in 1972 (NRC, 1989). The Extension Service of the United States Department of Agriculture (USDA-ES), in cooperation with state extension services, has had an IPM program in place since 1972—a program greatly expanded by congressional action in 1979, which now is administered at the four USDA regional levels. This permits the programs to be better tailored to address local needs both for research and extension, and permits the development of a private independent IPM capacity, further extending its usefulness. Application of IPM methods requires better management skills and a more profound understanding of the interlacing systems involved in crop production. Also, the application of individual IPM practices requires temporal and spatial selectivity. The manager needs to have both real time information and basic knowledge to apply appropriate practices when needed, i.e., the applications need to be site specific.

As early as the 1970s, Robert Rodale offered the concept of "regenerative agriculture" to address both the deterioration of the land base supporting agriculture and the issue of food safety *vis à vis* potential contamination with chemicals. Congress began to explore how it might encourage a reexamination of agricultural research. At the request of the United States Secretary of Agriculture, Bob Bergland, the Director of Science and Education, Anson Bertrand, appointed a study team for organic farming in April, 1979 (USDA, 1980). In the foreword of the resulting report, the Secretary wrote, "Energy shortages, food safety, and environmental concerns have all contributed to the demand for more comprehensive information on organic farming technology." In the preface, the Director wrote, "One of the major challenges to agriculture in this decade will be to develop farming systems that can produce the necessary quantity and quality of food and fiber without adversely affecting our soil resources and the environment."

CAST anticipated the USDA by convening a task force in October 1978 "to provide the needed background . . . to place the issues [related to organic farming and gardening] in perspective." The USDA Report was completed in July, 1980 (USDA, 1980), and the CAST study appeared in October, 1980 (CAST, 1980). With these initiatives and the regenerative agriculture of Robert Rodale in mind, Pesek proposed roles for agronomy in a unified resource conservation thrust in agriculture at the 1979 annual meeting of the American Society of Agronomy (Pesek, 1980).

Much was written about organic farming, and many meetings and discussions followed the USDA and CAST publications. Two of these were a symposium on organic farming conducted by the American Society of Agronomy and the resulting special report, "Organic Farming: Current Technology and Its Role in Sustainable Agriculture" (Bezdicek and Power, 1984). While few, if any, of the authors publishing in these reports found organic farming adequate to meet all food and fiber needs and still protect the soil and environment, these initiatives encouraged scientists, professionals, and farmers to explore alteratives to conventional systems in agricultural production.

Soil erosion, with its associated siltation and nutrient fertilization of water bodies, has been a serious problem in sustainability of agriculture and in the environment (NRC, 1989). This, in spite of a Soil Conservation Service in the USDA which, for over 5 decades, provided state-of-the-art erosion control and water conservation plans for all farmers. These plans were based on cropping systems, configuration of fields for wind and water erosion control, vegetative cover, and mechanical structures. The mechanical structures always were relatively expensive and with a long pay-out period. The emergence of chemically based farming made multiple-crop systems less attractive, because legumes, which provided soil protection and nitrogen for the grian crops, were no longer needed for their nitrogen-supply role. The row crop monoculture was highly profitable, and commodity programs rewarded the practice (NRC, 1989).

This change in agricultural practices left many soils more vulnerable to erosion than ever before because lack of mechanical erosion control and plowing with the moldboard left soil exposed to the elements. One of the first to recognize the damage of plowing was Faulkner (1943), who demonstrated limited crop production without the plow. A crop residue on the surface was shown to be effective in reducing erosion and rate of water runoff. Several technologies had to be developed to permit extensive farming without plowing and they were invented during the past four decades. These included: cultivars resistant to diseases and insects with vigorous emergence under adverse conditions; seed fungicide treatments; herbicides to control weeds; machinery with greater power, capacity, and versatility; and planters and cultivators that functioned in residue cover on the soil surface. These and other innovations permitted cropping that used no tillage, or reduced basic tillage operations and soil surface configuration and preserved the conditions created by surface mulch. This finally placed acceptable erosion control within reach on most prime cropland. The 1990 Farm Bill reflects the usefulness of mulch systems by accepting specified mulch cover as an erosion control measure for highly erodible land.

The need to examine these alternatives led the Board on Agriculture of the National Academy of Sciences, National Research Council to appoint a committee in 1984 to study the role of alternative farming methods in modern production agriculture (NRC, 1989). The Report on Alternative Agriculture published in 1989 was a pivotal contribution which has helped bring the issue of sustainable agriculture to the attention of the general and scientific publics to a degree not previously experienced. Before the report was issued, Congress coined the expression, "low input sustainable agriculture" (LISA), and most writers and investigators commonly use the term "sustainable agriculture" in the United States. The report, with its case studies, provided a visibility and credibility acknowledged by many, including CAST (1990).

INFLUENCE OF POLICY ON AGRICULTURE

The National Research Council Committee concluded that our laws and policies governing agriculture, especially our commodity policies, are among the major culprits (NRC, 1989) interfering with alternative methods of production. The commodity policies, in particular, dominate agricultural behavior at the farm level. Similar conclusions have been drawn by others (Cochrane and Runge, 1992).

About 70% of the nation's cropland is in crops covered by federal commodity programs and some 88% of eligible production is enrolled. The size of a farmer's subsidy is determined by the base acreage and the base yield of land in program crops. The farmer, therefore, is encouraged to produce as much as is possible on each acre and to keep the highest possible acreage of land in program crops. There is evidence that this may have led to overapplication of fertilizers and other chemicals and the bringing of fragile land into cultivation (NRC, 1989).

Just how pervasive government commodity policies are was expressed by John Miller, a Cedar Falls, Iowa farmer, speaking at a conference held by the Leopold Center for Sustainable Agriculture in 1991. He said,

A farmer is always making decisions . . . often . . . decisions on . . . decisions. Government may have more influence on these . . . than either research or education.

Research and education are optional when compared to . . . policy. I can choose to respond to research by adopting it as an . . . innovator, or I can ignore it altogether. . . . education is . . . effective if I choose to be receptive. But failure to pay heed to policy can cause me financial hardship, it might even break me, or it could put me in jail. (Miller, 1991)

Another profound illustration of long-term policy is in a Landsat photograph published in the *Journal of Range Management* by Dormaar and Smoliak (1985) of Agriculture Canada. It shows the border between

Montana and Alberta, clearly reflecting the difference in landscapes resulting from a national policy to produce wheat to the south and a national policy to allow reversion to rangeland and its maintenance and production to the north. According to the authors, land in this semiarid region in Canada once was cultivated as it was in the United States. Reversion to rangeland occurred between 1916 and 1935; the 49th parallel separating Alberta from Montana is as distinct as a fence line between a pasture and a fallow field.

Both policy and technological changes brought about by the land use mix in Iowa since 1939 are shown in Figure 1 (Miller, 1990). Oats declined because mechanical power replaced animal power; soybeans were developed as a new crop that was profitable; much livestock left Iowa to feedlots on the Great Plains, reducing the need for hay; and corn prices and production were kept in check by policy and the agricultural programs. The payment-in-kind (PIK) programs in 1983 (Figure 1) had a particularly profound effect on acreage of corn for 1 year. Many farm programs are deliberately designed to reduce the acreages planted to selected crops.

Other federal policies deal with the cosmetic standards of fruits and vegetables — standards that have little or no bearing on their nutritive value. Often they are achieved with additional and possibly unnecessary

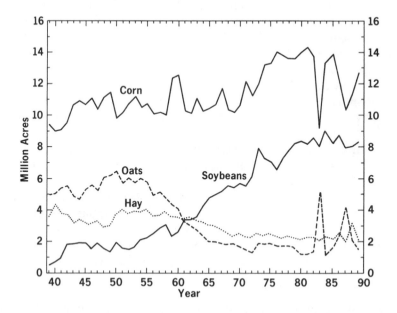

FIGURE 1. Acreage of corn, oats, soybeans, and hay planted each year in Iowa from government reports. (Adapted from Miller, personal communication, 1990).

applications of pesticides that add to the environmental burden. The rules governing the licensing of new materials that may be safer and more effective require procedures that are difficult and costly. It is easier to continue with what has been approved in the past. Even after some materials and practices have been shown to be unsafe or harmful, it is difficult to remove these undesirable materials and end the practices (NRC, 1989).

Both good news and bad news for adapting alternative production methods emerged from case studies by NRC (1989) as well as by USDA (1980). The good news is that alterative agriculture systems and practices do work—they are environmentally beneficial, and, when efficiently managed, can be highly profitable. The bad news is that relatively few farmers are fully benefiting from these systems. A primary reason pointed out earlier is that numerous national policies inhibit their development and widespread adoption. At the same time, research and extension services have not been able to integrate our knowledge into viable alternative agricultural methods. This is due, in part, to the site-specific nature of practices and the temporal requirements of decisions for individual fields. However, progress has been made during the last decade with the deployment of integrated pest management and demonstrations and programs for best farm management practices that are being adopted by farmers.

Support and information for farmers seeking to move toward sustainable agriculture is coming from more than the universities and USDA. The Northwest Foundation provides major funding for sustainable agriculture work in its eight-state area stretching from Minnesota and Iowa to Oregon and Washington. The AERO (Alternative Energy Resources Organization) in Helena is working with Montana State University and other universities to provide networking and publications dealing with sustainability in the Northwest. The Northern Plains Sustainable Agriculture Society in North Dakota provides experience information applicable to the northern Great Plains. The Practical Farmers of Iowa works closely with Iowa State University and the Leopold Center doing on-farm research in sustainable agriculture. The Sustainable Agriculture Society of Nebraska, the Illinois Sustainable Agriculture Society, and the Land Stewardship Project in Minnesota are additional examples of grassroots networking and directories of participating farmers. These initiatives need to be nurtured because farmer networks and directories assist greatly in disseminating application of site- and time-specific options.

Our agriculture exists and functions as it does because of the laws and policies we have established to govern it. If we had different rules, even with similar resources, our agriculture would be different. The Homestead Act of 1862 was an early policy decision affecting agriculture.

Others, also in 1862, were the acts granting land to railroads, and establishing the land-grant colleges and the United States Department of Agriculture. The Hatch Act created the experiment stations in 1887, and the Smith-Lever Act of 1914 established the Cooperative Extension Service, and other acts followed. Many commodity acts passed since 1933 have dominated agricultural production, but more recent acts like those creating the Clean Water initiatives, the Integrated Pest Management program, and Low Input Sustainable Agriculture activities have added environmental considerations to farm policy. More are likely to emerge.

RECENT VIEWS ABOUT SUSTAINABILITY IN AGRICULTURE

No single definition for sustainable agriculture has yet emerged, but, in general, it includes one or more of the following characteristics attributed to "alternative agriculture" (NRC, 1989):

1. Diversification rather than continuous planting of fields to single or only a few annual crops
2. Biological pest control and other innovative methods to reduce pesticide use
3. Disease prevention in livestock rather than routine use of subtherapeutic doses of antibiotics
4. Genetic improvements in crops to resist pests, diseases, and drought and to use nutrients more efficiently

Sustainable agriculture encompasses, but is not limited to, farming systems known as biological, ecologically clean, low-input, organic, and alternative. It is well defined in the legislation establishing the Leopold Center for Sustainable Agriculture. The Iowa Groundwater Protection Act of 1987 says, "[Sustainable agriculture is] the appropriate use of crop and livestock systems and agricultural inputs supporting those activities which maintain economic and social viability while preserving the high productivity and quality of Iowa's land." (Iowa General Assembly, 1987.)

All definitions tend to include the elements the Leopold Center uses as the "day-to-day" definition: "[Sustainable agriculture is] farming systems that are environmentally sound, profitable, productive, and maintain the social fabric of the rural community." (Keeney, 1991).

Hamilton (1990), of the Agricultural Law Center at Drake University, wrote: "The theory of sustainable agriculture is fairly simple—the development of policies and practices that ensure our nation's ability to produce the food and fiber we need without degrading our natural resources, while preserving the economic health of farmers and agricultural businesses and the social values contributed by the agricultural community to U.S. society. The potential of the concept to serve as a new way of looking at agriculture and analyzing the impact and value of decision-making is significant."

An admittedly value-laden workable definition is given by Harwood (1990), "[Sustainable agriculture is] an agriculture that can evolve indefinitely toward greater human utility, greater efficiency of resource use, and a balance with the environment that is favorable both to humans and to most other species." He points out that this generic definition permits specification of details by countries and of time frames. Harwood cited a definition that states the ideal of the Technical Advisory Committee, Consultative Group of International Agricultural Research, to wit, "Sustainable agriculture should involve the successful management of resources for agriculture to satisfy changing human needs while maintaining or enhancing the natural resource base and avoiding environmental degradation."

Legislatures in other states, as well as Iowa, have placed their confidence in an informed farm and agribusiness population, and the Agricultural Water Quality Incentive Act of 1990 suggests that the United States Congress might provide incentives for ground water protection in targeted areas, pending funds. Iowa has led in providing for new research and demonstration efforts in order to learn and illustrate use of enlightened technology that is friendlier to the environment.

As a result of these efforts, we are learning and demonstrating that careful soil testing and fertilizer applications reduce potential contamination of both surface water and ground water and, at the same time, lead to greater profits because of their reduced use. We are learning also, that scouting for pests and careful management of their control is providing for savings in farm operations and should lead to improvements in water quality and to lessen the potential of unwanted residues in food crops. It is evident that farmers and others will respond very positively to these kinds of initiatives and that our agriculture will become more efficient by using better information for more careful use of technological inputs.

Agriculture is a strongly regulated social activity and, as pointed out earlier, functions as it does because of the laws and policies established to govern it. Different rules, even with similar natural resources, would change agriculture, and the allocation of production resources would be adjusted. Legislation, custom, and necessity under the law have driven agriculture in the direction of strong commercialization and consolidation. Likewise, legislation can cause changes in the direction of sustainability to an extent not being achieved at present.

The National Research Council Report on Alternative Agriculture urged that the Congress restructure federal commodity programs to remove disincentives for the adoption of alternative agricultural techniques (NRC, 1989). It specifically recommended that if the existing commodity programs are retained, they should be reviewed at least to remove penalties for adopting crop rotations or planting alternate crops, and that farmers should be free to decide for themselves how best to produce or deliver an allotted amount of a commodity for a given production period. The first part of this idea has shown up in the 1990 Farm Bill as the "triple base" provision to give more flexibility. However, this was not welcomed with open arms by all.

Even with such changes to commodity programs, however, alternative farming will not be an immediate solution for all our problems. Alternative farming typically requires more information, more and better trained labor, and more diverse management skills per unit of production than conventional farming. This is because diversification into each additional crop and into each additional animal species requires additional and different skills to be successful. It takes better production management and a different kind of labor resource to handle a diversified farm like some of those studied (NRC, 1989; USDA, 1980) that may have beef cows, finishing cattle, swine, corn, oats, soybeans, and hay than it does to operate a similar-sized farm growing only corn and soybeans.

Unfortunately, the information farmers need has not been generally available from traditional sources such as the United States Department of Agriculture, the Cooperative Extension agents, or agricultural supplies dealers. As a result, most alternative systems and other changes have been developed and implemented by innovative farmers. It is up to us to change this, and, as pointed out before, we are; and the networks of farmers evaluating more sustainable practices are as well. We do have the technology to distribute real time information to every farmer and we are delivering more of it this way. However, we also know that "believing is first seeing", so demonstrations are still important and effective.

Most farmers incorporate new practices gradually, learning as they go. Therefore, they will not immediately abandon all of the practices and technologies with which they already feel comfortable—nor does the National Research Council Committee suggest that they do. Yet, more and more people are seriously exploring the possibilities of adopting sustainable agricultural systems, and if the rules of the game are gradually rewritten to encourage these, then there will be as much impetus to make these changes as there was for adopting the chemical technologies that emerged after World War II and have become the conventional agriculture of recent decades. Sustainable agriculture does and must have a future—it is either that, or our species as it has evolved socially has no future.

Hamilton (1990), had this to say:

The relation of sustainable agriculture to the multitude of environmental, social, and economic issues associated with modern farming practices makes the debate over the issue one of the most significant in the history of U.S. farm policy.

Sustainable agriculture may provide the nation with a mechanism for protecting our environment from pollution by agricultural practices in a method that minimizes regulation and emphasizes research, education and sound economic decisions to promote alternative production practices. The concept has great potential for our nation. Sustainable agriculture may remove the tension from the debate between the farm sector and environmentalists, it may restore and protect consumer confidence in the quality of the nation's food supply, it may justify continued federal spending on the farm sector at a time when federal price support expenditures are under fire, and most importantly, it may provide farmers with the opportunity to rightfully claim the title of *land steward* [emphasis by author] to which they aspire.

Saying it differently, David Masumoto, a California farmer, is of the opinion that agriculture has been estranged from the rest of the population; still, he wrote: "A courtship has blossomed between farmers and environmental concerns. If given time it can lead to marriage. But we need to start thinking of farmers as part of our ecological landscape." (Masumoto, 1990).

CONCLUSIONS

The general public will not let agriculture forget its responsibility for these sustainable practices, the safety of their food and the importance of the environment and the landscape. We are overdue in adopting new policies — replacing the old with those that are better and safer for farmers, healthier for consumers, more benign to the environment and sustainable. After all, we will depend on agriculture for food into the distant future. Even if we do not look forward any farther than we look backward to the beginning of agriculture, we are speaking not of decades or centuries but of thousands of years.

At the same time, the general public, more and more, has come to the realization that there is a cost associated with a wholesome environment, including a pleasing landscape, contamination-free ground water, useful rivers and lakes, and clean air. The debate now is not whether the desirable benefits of a sustainable agriculture will be required, but rather how they will be achieved. This applies both to the execution and how the costs will be borne. There already is a major public investment in agriculture, and the public is seeking to learn how to get the most of what it wants for the price paid.

Most recent experiences in technology transfer in selected regions and research on environmentally benign agricultural practices show that efficiency of production *vis-á-vis* external inputs can be improved and profits to farmers increased. They also show that a more sophisticated management of the agroecosystem is required when emphasis is on optimizing the basic biological systems involved instead of trying to dominate natural systems with a broad range of synthetic external agents. This management need represents major new opportunities in formal education in schools and universities and informal education through extension and other outreach programs. It also holds an opportunity for "crop production advisors" as a professional activity and for new research thrusts directed toward sustainable practices. The forthcoming chapters address various aspects of establishing and maintaining a sustainable agriculture or ways to move in that direction.

REFERENCES

Bennett, H. H. 1939. *Soil Conservation*. McGraw-Hill. New York.

Bezdicek, D. F. and J. F. Power. 1984. Organic Farming: Current Technology and Its Role in a Sustainable Agriculture. Spec. Rep. No. 46, American Society of Agronomy, Crop Science Society, and Soil Science Society of America, Madison, WI.

Carson, R. L. 1962. *Silent Spring*. Houghton Mifflin. Boston.

Cochrane, W. W. and C. F. Runge. 1992. *Reforming Farm Policy Toward a National Agenda*. Iowa State University Press, Ames.

Council for Agricultural Science and Technology (CAST). 1980. Organic and Conventional Farming Compared. Rep. No. 84. Ames, IA.

Council for Agricultural Science and Technology (CAST). 1990. Alternative Agriculture: Scientists' Review. Spec. Publ. No. 16. Ames, IA.

Davidson, J. M. 1989. Anticipating future research needs in response to an expanding clientele. 27th Annual Agronomic Administrator's [sic] Roundtable. Personal communication.

Dormaar, J. F. and S. Smoliak. 1985. Recovery of vegetative cover and soil organic matter during revegetation of abandoned farmland in a semiarid climate. *J. Range Manage.* 38:487–491.

Dunning, N. P. and A. A. Demarest. 1991. Sustainable agriculture systems in the Petexbatan, Pasion, and Peten regions of Guatemala: perspectives from contemporary ecology and ancient settlement. University of Cincinnati and Vanderbilt University. Personal communication.

Faulkner, E. 1943. *Plowman's Folly*. Grossett and Dunlap. New York.

Gardner, W. R. 1985. Arid lands, today and tomorrow. In: *Proc. Int. Res. Dev. Conf.*, Tucson, AZ. October 20 to 25, 1985. pp. 167–172.

Hamilton, N. D. 1990. Sustainable agriculture: the role of the attorney. *Environ. Law Rep.* 20:10021–10036.

Harwood, R. R. 1990. A history of sustainable agriculture. In: *Sustainable Agricultural Systems*, C. A. Edwards, R. Lal, P. Madden, R. M. Miller, and G. House, Eds. Soil and Water Conservation Society. Ankeny, IA. pp. 3-19.

Hillel, D. J. 1991. *Out of the Earth: Civilization and the Life of the Soil*. The Free Press. New York.

Iowa General Assembly. 1987. Groundwater Protection Act of 1987. Code of Iowa.

Jacobsen, T. and R. M. Adams. 1958. Salt and silt in ancient Mesopotamian agriculture. *Science*. 128:1251-1258.

Keeney, D. 1991. Flexible dealers will prosper in sustainable agriculture. *Custom Applicators*. January:18-19.

Lee, R. B. and I. DeVore. 1968. Problems in the study of hunters and gatherers. In: *Man the Hunter*. R. B. Lee and I. DeVore, Eds. Aldine, Chicago. pp. 3-12.

Masumoto, D. 1990. Farming's place on ecological agenda. *Des Moines Register*, May 7, 1990. p. 10A.

Miller, G. A. 1990. Historical acreages of crops in Iowa. Personal communication, 1991.

Miller, J. 1991. Policies of little assurance. Leopold Center Conference, Ames, IA. February 1991. pp. 1-4. Personal communication.

National Research Council (NRC). 1989. *Report on Alternative Agriculture*. National Academy Press. Washington, D.C.

Pesek, J. 1980. Unified resource conservation in agriculture: roles for agronomy. *Agron. J.* 72:1-4.

Sill, W. H. Jr. 1982. *Plant Protection: An Integrated Interdisciplinary Approach*. Iowa State University Press. Ames.

United States Department of Agriculture (USDA). 1980. Report and Recommendation on Organic Farming. U.S. Government Printing Office. Washington, D.C.

Vasey, D. E. 1992. *An Ecological History of Agriculture, 10,000 B.C.-A.D. 10,000*. Iowa State University Press. Ames.

Wilken, G. C. 1991. Sustainable agriculture is the solution, but what is the problem? Occas. Pap. No. 14. BIFADEC, [U.S.] Agency for International Development. Washington, D.C.

2

Water Relationships in a Sustainable Agriculture System

E. John Sadler and Neil C. Turner

TABLE OF CONTENTS

INTRODUCTION

Sustainable agriculture requires management of the land so that production and productivity are enhanced while sustaining a healthy ecological balance within the agricultural ecosystem (Ruttan, 1990). One of the key inputs to any agricultural system is water. Because of the strong link between photosynthesis and transpiration at the level of the individual leaf, crop and pasture production are usually highly correlated with water use (Fischer and Tumer, 1978). For agricultural systems to be both productive and sustainable in the long term, management of the water resource is required to ensure that sufficient water is available for plant growth and excess water is not allowed to contribute to land degradation (Fillery and Gregory, 1991).

Water is not only important because it contributes to plant growth, but also because it is a transporting agent for dissolved materials, nutrients, chemicals, and solids. Although its ability to transport nutrients, pesticides, and phytohormones is fundamental to plant growth and protection, excess water can lead to pesticides, salts, and nutrients entering the ground water or surface water and to soil particles being moved downslope, resulting in soil erosion and land degradation.

As a framework for our discussion of water in a sustainable agricultural system, we will use the hydrologic cycle because sustainability impacts all aspects of the cycle. We can then discuss these impacts on its individual components. The hydrologic cycle is depicted in Figure 1. The primary input to the cycle is precipitation, which is generally considered to be unaffected by the agricultural system, but which varies enormously both spatially and temporally. In agricultural systems, precipitation is

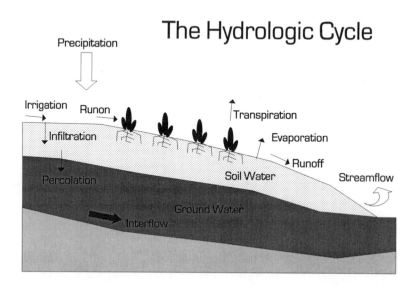

FIGURE 1. A diagrammatic representation of the hydrologic cycle. Terms defined are used in the text.

sometimes supplemented by irrigation or by runon from runoff higher in the landscape. Losses from the system include evaporation from the soil, transpiration from the crop, surface runoff, interflow along an impermeable subsoil layer, and deep percolation. The balance among these fluxes constitutes the soil water storage term, so that the capacity of the soil to store water is a candidate for management in a sustainable system.

In looking at water in sustainable agricultural systems, we draw heavily on modern research findings. However, agricultural systems have changed dramatically in the past half century with energy inputs from petrochemical sources increasing dramatically. On a world scale, water inputs through irrigation have also increased significantly and are likely to continue to increase until and beyond the turn of the century (Alexandratos, 1988). It is, therefore, not clear how sustainable some of the recent agricultural systems are likely to be in the long term. Therefore, in this chapter we will also look at some ancient agricultural systems and their sustainability. Some of these ancient agricultural systems were sustainable over millennia, and an understanding of water use in these systems may be instructive to modern scientists.

PRECIPITATION

The Food and Agriculture Organization of the United Nations estimates that 600 million hectares of potentially arable land is unused because of limited water (Alexandratos, 1988). That area is almost equivalent to the area currently utilized in arable production. In many parts of the world, crop production is limited by rainfall, and production varies markedly from year to year depending on water availability. With precipitation being such a major limitation to crop production its efficient use is an important consideration in any sustainable system. While little can be done to alter the amount of precipitation that is received, current scenarios suggest that global climate change resulting from the increase in greenhouse gases may result in marked regional changes in precipitation in the next half century (Pittock, 1988). Increasing cyclonic activity is predicted to move the subtropical regions polewards, and warmer temperatures are predicted to result in the arid, semiarid, and temperate regions moving closer to the poles. The role of global climate change cannot be discussed in detail in this chapter, but it needs to be recognized that the predicted changes resulting from global warming may have significant effects on water use and the sustainability of future agricultural systems (Adams et al., 1990).

Precipitation at a particular location can be supplemented with water arising from rainfall elsewhere, either by irrigation or by runon farming practices. Alternatively, the rainfall in one season may be supplemented from storage of prior rainfall. Irrigation from deep aquifers utilizes water accumulated over many years or even centuries. A much more immediate use of rainfall storage occurs when water is stored in one season for use in the subsequent season. This process, fallowing, is widely practiced in dryland agriculture where the seasonal rainfall is limited.

Supplementing Precipitation—Irrigation and Runon Practice

We first discuss larger-scale rainfall harvesting, such as is done where mountain rainfall and snowmelt, or simply large catchment areas, provide water that can be used by producers at lower levels. Areas proximal to mountain ranges (e.g., coastal Peru [Browman, 1983; Ortloff, 1988]) or to major rivers (e.g., the Nile valley and the Dujiang Weir system on the Minjiang River in China) have had extensive development of irrigation based on such practices. The irrigation water may be utilized many kilometers from its collection and river waters may be reused several times. Sustainability of these practices is enhanced by the low cost of the water supply and the natural head that drives the water delivery system. Questions of salinity (Ponting, 1990), luxury consumption, municipal competition for water, and other considerations may decrease the sustainability of these practices, as is already evident in contemporary California (Reisner, 1986). However, irrigation has been used for centuries and, in many places, has allowed food production in areas that otherwise would not support large populations (Ponting, 1990).

The Tigris and Euphrates River valleys in southern Mesopotamia provide the earliest known example both of formal irrigated culture and of decline due to salinization. Agriculture in the area dates to about 5000 B.C., but probably was limited to riverbanks and small-scale irrigation. By 3000 B.C., an irrigation canal over 15 km in length had been built, indicating larger-scale irrigation. By 2400 B.C., a second large and important canal further increased the scope of irrigation (Minc and Vandermeer, 1990). However, salinization had begun to occur within 1500 years of the beginning of irrigation. From 3500 B.C. on, farmers responded to salinization by shifting from wheat to barley, which is more tolerant to salt. During the period from 3500 to 1700 B.C., the area planted to wheat dropped from about equal to that for barley to none at all. Records indicate that crop yields remained high until about 2400 B.C., by 2100 B.C., they had dropped 42%, and by 1700 B.C., yields had dropped by 65% (Ponting, 1990). Similar problems of salinization after irrigation are widely distributed throughout the world, though few have such a historical perspective.

However, not all irrigation schemes have led to salinization and land degradation. In Arizona, ancestors of the Navajo used steep slopes of the

Black Mesa Mountains to funnel water toward small fields at the bases. There, they grew corn, squash, melons, and even fruit trees. Although yields were low relative to modern standards, the practice was successful and continues today (Seery, 1990).

Irrigation from deep aquifers is a more recent phenomenon because it depends on energy, particularly fossil fuels, to pump the water to the surface. However, its sustainability does not only depend on the sustainability of the water supply. Water in deep aquifers has often collected over many years or even centuries and, if not replenished at the rate of utilization, will ultimately be a finite resource. Irrigation from the Ogallala and associated high plains aquifers by U.S. Great Plains farmers has had to be restricted to prevent it from either being quickly depleted or at least dropping too deep for economical pumping. Since pumping began, the water levels in the Ogallala have decreased more than 3 m in 29% of the aquifer, by more than 15 m in 6.9%, and by more than 30 m in 1.4% of the aquifer's 450,000 km^2 (Gutentag et al., 1984). These drops have occurred even though the volume of water in the aquifer has decreased by only 5% (Weeks and Gutentag, 1984).

An interesting form of supplemental rainfall in areas of very low rainfall (100 to 150 mm annual rainfall) was developed by the Nabateans about 2000 years ago in the Negev Desert (Evenari et al., 1982; Hillel, 1982). Rainfall on surrounding barren hills was collected and channeled to the terraced valley floors, which had deep loessal soils. Recent reconstruction of these ancient runoff/runon farming systems indicate that a crop can be grown on a single rainfall event of 20 mm, and fruit trees and other perennial plants can survive and produce in regions in which the annual rainfall is too low to allow normal horticultural production (Evenari et al., 1982). The Nabatean civilization lasted for about 700 years in the Negev Desert, but gradually declined as trading routes and religious and civilian empires in the region changed (Evenari et. al., 1982; Hillel, 1982). The successful reconstruction of these ancient systems in modern Israel is testimony to the fact that the runoff/runon farming system is sustainable and that the decline of agriculture did not arise from land degradation.

Runon farming in dry environments is still practiced in modern agriculture, albeit on a different scale. Evenari et. al. (1982) describe a microcatchment system used in the Negev Desert, in Afghanistan, and in parts of Africa. There, fruit trees are planted in a depression 0.9-m deep and 4-m square at the corner of a microcatchment of 250 m^2, and the rainfall from the entire catchment is channeled to that corner. Depending on rainfall, the microcatchment required for an individual fruit tree will vary, but the principle is that water is not lost but channeled to the plant roots. Conservation bench terraces (Zingg and Hauser, 1959; Jones, 1975, 1981) are also a small-scale application of water harvesting for use elsewhere. Here, a slightly sloping (1 to 2%) area is modified to provide a watershed area that contributes runoff to a level bench area about half as large. The level area is continuously cropped, and the watershed area is in some rotation that includes fallow. The conservation bench terraces control runoff water, prevent water erosion, and contribute water to the level bench (Zingg and Hauser, 1959; Hauser, 1968). However, the success of the level bench system depends on soil type. Positive results were obtained on a Pullman clay loam soil, but on an Amarillo fine sandy loam with a higher infiltration rate and lower water holding capacity, similar terraces provided no yield benefit for either sorghum or cotton (Armbrust and Welch, 1966). Lower water holding capacities and higher infiltration rates resulted in the soil water reservoir being filled before runoff arrived from the contributing watershed, and the impounded water was lost to deep percolation.

While runoff from high in the landscape can be beneficially utilized for crop production lower in the landscape, the management of runoff water is not always easy nor well controlled. As a consequence, water tables exist within a meter or two of the surface in many parts of the world. Where there are shallow water tables, root depth may not be adequate for sustained crop growth. Historically, one procedure that has been used to overcome this has been the use of raised beds and canals.

The Yucatan peninsula was farmed by this technique during the period from 200 B.C. to 850 A.D. Evidence of raised fields, terraces, and possibly irrigation of normally dry areas has been found in Pulltrouser Swamp in southern Quintana Roo, Mexico (Turner and Harrison, 1981; Chen, 1987). In the study area, over 600 ha of channelized or raised fields exist, indicating that extensive effort had been invested to alter the natural landscape. The soils in the area were 0.5- to 1.0-m deep and overlaid weathered limestone. Channels from 2- to 10-m wide were cut up to 1-m deep into the weathered limestone. The soil extracted to make the channels was used to raise the level of the soil surface between the channels. In most cases, the original topsoil was removed and replaced above the fill. Such extensive labor inputs, estimated to be about 12,000 worker-years for the 600-ha site, plus suggestions of mulching and mucking, are indicative of high-output agriculture. The Mexican chinampas, which are the modern equivalent, support about 19 people per hectare. If 70% of the total area were raised fields and 30% canals, then about 8000 people could have been supported by the 600 ha of the Pulltrouser Swamp site. Such intensive agricultural development coincided with the rise in population in the Classic Maya. However, why this civilization rapidly collapsed is not clear. Ponting (1990) argued that the population increase outstripped rural production, causing increased pressure on marginal lands to provide not only food but also timber for construction and fuel. Certainly there is evidence of increased infant and maternal mortality and signs of nutrient deficiency from about 800 A.D. However, the breakdown of civilization could have arisen as much from internal strife as from land degradation.

Modern research in similar environments emphasizes the importance of controlling the water table by using a combination of drainage and irrigation. Lowland agriculture in the Atlantic coastal plain and Mississippi Delta of the U.S. has been enhanced by water table management using simple or advanced water control structures such as flashboard risers and automated, water-filled fabric dams (Doty et al., 1984b, 1985, 1987). Both act to retain the water resource (about 1000 mm annual rainfall) without leaving the land subject to flooding such as occurs in undrained land. Conventional drainage without water table control causes overdrainage and, ironically, in many cases, results in water shortage between rainfall events (Doty et al., 1984a).

Supplementing Precipitation—Season-to-Season Supplementation

Storing rain during fallow for use in a later cropping season has been used in most semiarid regions of the world to supplement the growing-season rainfall. The primary objective in most fallowing is to store water during the noncrop season for use in the following season, but the efficiency of such storage is, on average, low. Mathews and Army (1960) reported precipitation storage efficiencies from harvest to seeding of 12 to 40% for annual spring wheat and 10 to 28% for annual winter wheat. Similar values for wheat-fallow systems were 10 to 25% for spring wheat and 6 to 22% for winter wheat. These values for the U.S. are similar to ones obtained in Australia (Ridge 1986). Studies show that during periods without rain, water evaporates from the top 0.20 to 0.30 m of the soil profile, while subsoil water is only slowly subject to evaporative loss. The storage efficiency will vary with soil type, with frequency and quantity of rainfall in individual storms, and with the rate of evaporation. Studies in South Australia indicate that the best soil moisture storage after fallow occurs in fine-textured soils or soils with a clay subsoil and a minimum storage capacity of 125 mm in the top 1.2 m of soil. Coarse-textured soils with a moisture storage capacity of 50 to 85 mm in the top 1.2 m gain little from fallowing (Schultz, 1971; French, 1978a; Tennant, 1980). Water from frequent small storms that do not allow penetration of the rainfall below 20 cm will be evaporated away, while infrequent heavy storms allow deeper water penetration and less evaporative loss. Storage efficiencies also tend to increase as one moves north in the U.S. Great Plains, a consequence of reduced evaporative demand. Lavake and Wiese (1979) documented the influence on storage during fallow when different tillage and weed control practices were used. Storage efficiency decreased from

18 to 14% as the delay between weed emergence and tillage increased from 14 to 24 d after emergence. Clearly, control of weeds during fallow is important if fallowing is to be practiced efficiently.

Cereal yields in Australia are usually higher as a result of fallowing (French, 1978b; Tennant, 1980; Ridge, 1986) and yields in the southern Great Plains (Unger, 1983) are 40 to 50% higher for fallow-rotation crops than continuous crops. However, as yields are not doubled, the total production from continuous cropping is higher (as are costs of harvesting greater areas). Nevertheless, the risk of low crop yields is higher where continuous cropping is practiced than where a fallow is included in the rotation. For the central and northern Great Plains, where the storage efficiencies are higher and risk of crop failure (yields below 1000 kg ha^{-1}) is greater, Smika (1970) reported that fallow-wheat yields averaged more than three times the continuous-crop wheat yields. The fallow system had no complete crop failures, whereas the continuous wheat system had 6 out of 27 years with yields less than 100 kg ha^{-1}, and 15 out of 27 years with yields less than 1000 kg ha^{-1}. Tennant (1980) showed a similar increase in the probabilities of low crop yields in marginal areas of Western Australia with continuous cropping compared with fallowing.

There are, however, some drawbacks to fallowing that raise questions regarding its long-term sustainability. The necessity to control weeds may lead to excessive cultivation and the exposure of the soil to wind and water erosion. This risk, together with low storage efficiencies and development of suitable annual medic and clover species, has led to its demise in southern Australia (Tennant, 1980; Ridge, 1986). Also, the lower total production or periods of low production may reduce the organic matter in the soil, with detrimental effects on the water holding capacity, erosion susceptibility, and fertility of the soil (Jennings et al., 1990). However, the use of herbicides to control weeds can reduce erosion risks and reduce the detrimental effects of cultivation on soil organic matter.

We conclude that systems that supplement rainfall can be sustainable in the long term, but for long-term sustainability, the balance of water income to water outflow of the soil/crop system must be maintained. Irrigation management to prevent overwatering, soils impermeable to deep percolation, and crops that utilize water efficiently all contribute to long-term stability. Likewise, runon farming systems need to be carefully balanced to utilize all incoming rainfall without creating excesses in wet years or shortages and crop failure in dry years.

Infiltration

Unless a system of runoff/runon farming as described above is to be practiced, it is important in any sustainable agricultural system for the water to quickly penetrate the soil and for runoff to be minimized. The infiltration of water into the soil will initially be affected by the soil slope and characteristics at the soil surface. However, subsurface characteristics that prevent water draining to deeper in the soil profile can also ultimately affect infiltration, so modifications to increase infiltration both at and below the soil surface will be discussed.

Anecdotal evidence of southwestern U.S. Indians disturbing crusts during periods between rains to increase infiltration probably constitutes an early example of soil surface modification. With conventional tillage (complete soil surface tillage), examples of techniques for rainfall capture and runoff prevention include contour farming, strip cropping, graded terraces, bench terraces, furrow-diking or tied ridges, basin tillage, basin listing, and microbasins (Lyle and Dixon, 1977; Jones and Clark, 1987), even though the initial objective for some of these may have been to reduce erosion (Unger et al., 1988; Jones et al., 1985).

Other approaches to increase infiltration retain residue on the surface by means of reduced tillage or conservation tillage practices (Laflen et al., 1978; West et al., 1991; Steiner, 1992). Mills et al. (1988) examined the probabilities of rainfall retention for several conventional and conservation tillage systems and showed that the median rainfall retention from six conservation tillage systems was 9% higher than that from four conventional tillage systems. Improvement in rainfall retention was attributed to a buildup of surface residue and improved soil tilth in the surface horizon, especially the protection of the surface from raindrop impact (Laflen et al., 1978).

Surface residues increase the ponding depth, which increases infiltration via the direct effect of increasing the time of ponding. However, a second, less direct effect of surface residue is its effect on the potential infiltration rate at the soil surface. Residues intercept rainfall, absorbing and dissipating impact energy. This transfer of energy reduces degradation of soil aggregates at the surface, which, in turn, reduces the sealing of the surface against infiltration. Residues also shade the soil surface, which reduces the surface temperature during periods of high evaporative demand.

Increased residue on the soil surface has also been associated with increased organic matter content of soils (Karlen et al., 1989) or altered distribution in the profile (Unger, 1991), thereby affecting infiltration. Typically, management to improve organic matter by changes in tillage practice causes relatively subtle effects. Therefore, long-term (~ 5- to 10-year) studies are normally required to distinguish differences, although Wood et al. (1991) were able to detect changes in 4 years after initiation of a no-till system. Blevins et al. (1977) found higher organic matter in the surface 50-mm layer of a no-till system after 5 years of continuous maize, while Juo and Lal (1979) observed elevated organic carbon in the upper 100 mm of a tropical Alfisol under no-tillage after 6 years of continuous maize. The increased organic matter may affect infiltration in several ways. For example, organic matter can increase the water holding capacity of the soil, decrease its bulk density, increase its aggregate stability, and increase its cation exchange capacity (Hargrove et al., 1982; Klavidko et al., 1986; Bruce et al., 1987).

Little information exists to separate the effects of surface residues on rainfall capture from the effects of organic matter on infiltration. West et al. (1991) compared runoff and soil loss for conventional and no-till grain sorghum at three Piedmont sites. Under simulated rainfall at 60 mm for 1 hour, the runoff from the conventional-tilled sorghum was 10 mm, compared to 1 mm from no-till sorghum when the previous year's residue was left on the surface. When it was removed, the values were 19 and 5 mm, respectively. Thus, there appear to be benefits in terms of reducing runoff from both the retention of residues at the surface and the improvement of organic matter content and consequent aggregate stability of the soil (West et al., 1991).

Increased porosity of the soil and, in particular, increased macroporosity increase the infiltration of water into soils. As the southwestern U.S. Indians mentioned at the beginning of this section demonstrated, tillage is one of the primary methods employed to accomplish an increase in porosity. Mukhtar et al. (1985) reported infiltration rates varying from 1 to 30 min, depending on whether the soil was plowed in various ways or not tilled. The Paraplow®* (The Tye Co., Lockney, TX), which fractures the subsoil, induced infiltration rates about twice as high as the other treatments. While the primary objective of subsoiling is to create zones acceptable for rooting in lower horizons (Campbell et al. 1974; Doty and Reicosky, 1978), clearly the operations also increase infiltration. There are, however, other methods of increasing macroporosity that do not rely on tillage. For example, burrows of earthworms and other soil fauna provide channels for water infiltration.

Use of tillage to increase the infiltration of rainfall into the soil in the short term may have long-term negative effects. If loosening of the upper soil to increase infiltration comes at the expense of compaction below the tillage zone (Busscher et al., 1986), this can impede infiltration to lower storage zones and can induce a perched water table above the tillage pan. This, in turn, can increase the chance of waterlogging, thereby increasing runoff and decreasing infiltration. Indeed, cultivation may be detrimental to earthworm activity and lead to poorer infiltration and degradation (Abbott et al., 1979; Parker, 1989). Thus, systems that rely less heavily on tillage to increase infiltration have the greatest chance for long-term

---

*Tradenames are provided for the benefit of the reader, and use does not imply endorsement of the product by the USDA or CSIRO.

sustainability. Hence, there is interest in the use of earthworms and other soil macrofauna to increase the soil macroporosity and, thus, infiltration of rainfall.

SOIL WATER EVAPORATION

In mediterranean climatic regions, water loss by direct evaporation from the soil during the cropping season can represent 30 to 50% of the annual rainfall. This represents a significant loss of water that is potentially available for crop growth. Recent estimates of soil water evaporation under a crop indicate that the rate of soil water evaporation was 1 to 1.6 mm/d in crops of lupin with a leaf area index of 0.5 to 2.5 and a total crop evapotranspiration of 1.8 to 3.4 mm/d (Greenwood et al., 1992). These estimates compare favorably with other estimates of soil water evaporation, which suggest that beneath wheat and barley crops growing in mediterranean climatic zones, the soil water evaporation totals 60 to 120 mm per growing season in regions in which the rainfall is 200 to 500 mm per growing season (French and Schultz, 1984; Shepherd et al., 1987; Perry, 1987; Cooper et al., 1987a,b).

Decreasing water loss by soil water evaporation provides the potential for improved crop production on the same rainfall input, thereby maintaining sustainability as defined by Ruttan (1990). Selection of cultivars of wheat with rapid early growth (Turner and Nicolas, 1987; Whan et al.,1991; Regan et al., 1992) or use of higher plant densities and/or earlier dates of planting to increase early growth (Turner et al., 1987; Turner et al., 1993; Greenwood et al., 1992) are mechanisms for reducing soil water evaporation and increasing crop water use.

Residues and mulches provide a physical barrier to diffusion of water vapor from the soil surface and thus also act to conserve water for later use by the plant (Rosenberg et al., 1983; Steiner, 1992). Mulching has been practiced at least since the ancient Chinese and Romans put pebbles on the soil surface near plants (Unger, 1983). Layers of straw (Unger and Parker, 1976), gravel (Unger, 1971a,b), plastic (Willis et al., 1963), or other materials placed on or just below the surface have been proposed to reduce the rate of soil water evaporation. The addition of external materials, such as gravel or plastic, on the soil surface is generally, but not always, restricted to ornamentals, vegetables, or other high-value crops. The most economical mulch for large-area application is plant residue (standing or flattened) or dust created by tillage.

Willis et al. (1963) found that a plastic-covered ridge between maize rows suppressed soil water evaporation, resulting in higher water use in dry years and relatively higher yields in both wet and dry years. This gave a water use efficiency for grain 44 and 96% higher than the conventionally tilled control. Griffin et al. (1966) evaluated the effects of plastic film, asphalt film, and asphalt-covered paper mulches on grain sorghum yield and soil moisture. Under irrigated conditions, all these mulches reduced soil water evaporation and improved water use efficiency, even though yields were higher on the unmulched check. Unger (1971a,b) evaluated the effect of a 25-mm layer of gravel as a surface mulch on growth and water use of a hybrid forage sorghum. The gravel improved the water use efficiency at the first harvest, but not at subsequent harvests when the ground cover by the forage was greater.

When plant residues were used as a mulch, Unger and Parker (1976) found that wheat straw was twice as effective as sorghum residue and four times as effective as cotton residue. For any given material, increasing the amount of residue decreased the evaporation from the soil. Although one cannot partition the effects of decreased evaporation and increased infiltration, both Greb et al. (1967) and Unger (1978) found increased soil water storage levels of wheat residue during fallow. However, where there are frequent small showers, the mulch will intercept the rainfall and evaporation from the mulch will be as high as from the soil. A. P. Hamblin and D. Tennant (personal communication) showed that it required more than 4000 kg ha^{-1} of straw and heavy rainfall events before mulching gave increased soil water storage.

Irrigation placement will also have important effects on water use. Use of subsurface and microirrigation methods provides water to the plant without wetting the soil surface, thereby reducing soil water evaporation and allowing less water to be applied without reduced plant water use or yield reduction (Batchelor et al., 1990; Turner, 1990a).

A problem that has been observed in many regions of the world (Jamison, 1945; Letey et al., 1975; Wallis et al., 1990) and is causing increasing concern in southern Australia is the increased water repellency of the sandy-surfaced soils common in agricultural areas (Bond, 1964; Wetherby, 1984; McGhie and Tipping, 1990). The problem appears to be increasing with the increasing productivity of the region. The water repellency results in poor establishment of crops and pastures (Roberts, 1966; Osborn et al., 1967; Bond, 1972) and greater risk of wind and water erosion (Osborn et al., 1964; McGhie, 1980). One benefit of the increased water repellency is that if the soils are ridged at seeding, subsequent rainfall runs into the furrows under which the seed is sown. Recent studies suggest that with the ridging of nonwetting soils, soil water evaporation is reduced when the crops are small due to the smaller degree of evaporation from the dry soil ridges (Yang et al., 1993).

The reduction in soil water evaporation provides a mechanism for greater water use by agriculturally important crops provided they can capture the water not lost by soil water evaporation. There is, therefore, the potential to make the system more sustainable by using water more efficiently in crop and pasture production.

MODIFICATION OF TRANSPIRATION

Transpiration for a given crop has been shown to be almost always directly proportional to dry matter production (de Wit, 1958; Fischer and Turner, 1978; Tanner and Sinclair, 1983), so reduced transpiration is generally not desirable for crop productivity and may be detrimental to sustainable agricultural practice. Indeed, the requirement in most agricultural systems is to match crop evapotranspiration with rainfall and to maximize the water use by the crop rather than losses by soil water evaporation and nonproductive losses of water via transpiration of weeds and barren plants. Matching crop water use to rainfall and irrigation supply is necessary for sustainable agricultural production.

The recent history in parts of southern Australia of clearing native vegetation for agriculture provides an example of lack of consideration of water use by crops and pastures leading to unsustainable production systems and land degradation. Replacement of deep-rooted perennial vegetation by shallow-rooted pastures and crops has led to rising water tables, waterlogging, and secondary salinization lower in the landscape (Figure 2), particularly in regions of southwestern Australia, in which the rainfall is less than 1100 mm and where cyclic salt occurs in the profile (Schofield et al., 1988). Figure 2 shows a generalized picture of the impact of clearing in regions of southwestern Australia where the annual rainfall is about 650 to 700 mm. Replacement of deep-rooted perennial native vegetation by annual pastures has resulted in the annual evapotranspiration falling by about 60 to 100 mm. This water percolates to the saline ground water table, causing it to rise, and resulting in waterlogging and salinization of lower parts of the landscape. While the reduction in evapotranspiration may increase the streamflow by about 60 mm, the increase in total soluble salts in the streamflow from 100 mg l^{-1} to 5500 mg l^{-1} reduces the potability of the water. The data in Figure 2 are taken from studies in Western Australia. Greenwood et al. (1981, 1985) and Sharma et al. (1991) showed that the water use by pastures was considerably lower than water use by native vegetation or pines, leading to an additional 20 to 50 mm of rainfall percolating beneath the root zone to ground water annually (Nulsen and Baxter, 1982). Replacement of part of the catchment to evergreen trees is currently being employed to reverse land degradation from secondary salinization and to reduce the salinity of streams and water catchments in the region (Schofield et al., 1989; Schofield and Scott, 1991).

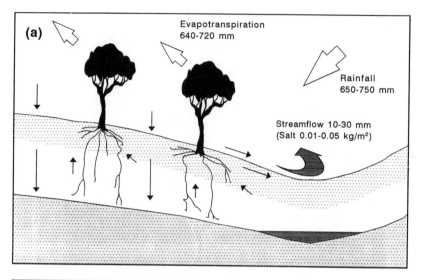

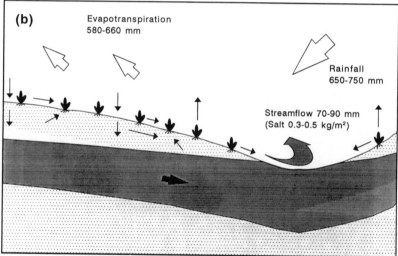

FIGURE 2.   The effect of land clearing on the water balance of a forested catchment in a region where the annual average rainfall is 650 to 750 mm. In (a) the vegetation utilizes most of the incoming rainfall resulting in a streamflow of 10 to 30 mm/year and with a low salt concentration. In (b) the vegetation uses less of the annual rainfall resulting in more streamflow with a greater annual output of salt from that stored in the soil profile.

An alternative or complementary solution to the problem is to increase the water use by the crops and pastures in such catchments. Early sowing, use of higher planting densities, use of fertilizers to stimulate early growth, and selection of cultivars with high early growth are all methods that can increase crop water use early in the season leading to potentially higher yields in water-limited environments (Turner and Nicolas, 1987; Shepherd et al., 1987; Whan et al., 1991; Turner et al., 1993) and decrease the flow to the water table, thereby reducing the potential for land degradation (Greenwood et al., 1991).

SOIL STORAGE INCREASERS

Management to increase soil water storage has been discussed above in that all reductions of losses result in increased soil water storage and, therefore, increased water available for evapotranspiration. However, some management techniques act to increase the capacity of the soil to store water, regardless of whether there is more water to store. Storage capacity here is meant to be capacity to store water that will later be available to plants.

One traditional method to increase storage capacity has been to increase the volume of soil suitable for rooting. Subsoiling (Campbell et al., 1974; Doty and Reicosky, 1978), deep tillage (Karlen et al., 1992), the use of gypsum to produce stable aggregates (Blackwell et al., 1991a,b), and breeding programs that select for stronger rooting cultivars (Turner and Nicholas, 1987; Kasperbauer and Busscher, 1991) all act in this manner. Where such methods increase water penetration, root growth, and water use by the crop, they will have a major impact on sustainability.

A second method that acts to increase capacity of storage does so not by increasing volume, but increasing the capacity to store water in a unit volume of soil. Addition of organic matter or other soil amendments should increase the available water holding capacity. This concept receives enthusiastic support in the popular press (e.g., Kendall, 1988). However, the effects may be subtle. Anderson et al. (1990) were able to detect increases in saturated hydraulic conductivity after 100 years of manure additions, but the effect on water retention was not consistent across cropping systems. Unger and Fulton (1990) compared soil properties among a conventional tillage system and two no-till systems that had been in place for 6 and 8 years, respectively. Organic matter concentrations at the 0.04- to 0.07-m depth were, surprisingly, lower in the no-till than in the conventional system. This was attributed to mixing of surface organic matter in the conventional system, whereas the no-till systems had organic matter distributed more toward the surface. Mean weight diameter of water-stable aggregates presents mixed results, with the conventional system the same as the younger no-till system, and the older no-till system higher than both. For the 0.04- to 0.07-m depth, the water retention curve for both no-till systems was higher than the conventional at saturation, and higher at 4.89 kPa tension for the older no-till system. None of the intermediate tensions and none of the values for the 0.14- to 0.17-m depth were different. Bulk density at the 0.04- to 0.07-m depth was lower for the no-till systems and unchanged from the conventional system for the 0.14- to 0.17-m depth.

Mixing the surface and subsurface soils has been attempted for soils in which a sandy surface layer overlies a clayey subsoil (Miller and Aarstad, 1972; Campbell et al. 1988), in the hope that the retention curve for the composite soil might be superior to that for the sandy surface layer. On a southeastern U.S. typic Paleudult, the result was a sand-clay mix that cemented during dry periods (W. J. Busscher, personal communication). Results for deeper soils suggest that sustained improvements can be made. Chaudhary et al. (1985) found that mixing the surface 0.45-m layer of a coarse-textured soil reduced bulk density, reduced penetration resistance at the 0.2- to 0.4-m depth, and increased maize yields 70 to 350%, even though there were no root-limiting layers in the soil prior to mixing. Measurements made 21 years after mixing a Pullman clay loam to 0.9- or 1.5-m depth (Eck, 1986) indicated that mixing still affected infiltration rate and bulk density. Yield responses were mixed, and Eck (1986) concluded that the benefits did not exceed the cost of mixing.

PERCOLATION REDUCERS

Compaction of the subsoil and placement of relatively impervious materials at depth are methods used to reduce percolation (e.g., Erickson et al., 1968; Robertson et. al., 1973). This may increase sustainability in situations where deep drainage is a problem, such as with irrigated rice cultivation in the New South Wales region of Australia. There, deep percolation of water from rice growing has caused saline water to flow back into the Murray River downstream from the irrigation district. Compaction of the subsoil, as is common in flooded rice culture in Asia, may help to reduce deep percolation (De Datta and Kerim 1974; Saroch and Thakur, 1991) and to increase sustainability.

However, use of artificial barriers to water movement through the soil has been less successful. Willis et al. (1963) tested plastic buried at the 0.2-m depth as well as testing the surface plastic mulches discussed above. Neither a buried ridge nor slanted plastic surface performed as well as the ridged surface with 90% plastic cover in terms of crop growth and water use.

Soil and crop management to reduce the likelihood of wet soil conditions can also reduce the risk of deep percolation or runoff on intersoil layers. Establishment of deeper-rooting crops can reduce percolation by lowering the water content in the subsoil (Unger et al., 1988) and by allowing water to penetrate the clay subsoil in duplex soils in Australia. However, the success of management practices to decrease percolation is strongly dependent on soil type. For instance, conservation bench terraces on a fine sandy loam in Woodward, OK (Armbrust and Welch, 1966) had more percolation than similar terraces on a clay loam in Bushland, TX (Unger, 1983). The lower water holding capacity and higher rainfall in Woodward combined to increase the loss to percolation.

Management of irrigation to avoid overwatering is an important method of reducing deep percolation in many soils. Irrigators frequently overwater, particularly where water is cheap and plentiful and the benefits from irrigation are large. This has led to the problems of salinization and land degradation from irrigation discussed above. Regulated deficit irrigation can not only lead to less deep percolation of water and hence less land degradation, but also can improve crop and pasture production (Turner, 1990a,b) as well as horticultural production and management (Chalmers et al., 1981; Mitchell et al., 1989; English et al., 1990).

OPPORTUNISTIC CROPPING

There remains one class of management tools that can be used to increase sustainability, primarily by conserving water or by increasing beneficial use of water. These techniques share the characteristic that the farmer must adapt not only to soil characteristics and average climates, but must remain sufficiently flexible to adapt his farming enterprise to utilize unusual rainfall events. Bond et al. (1964) reported that yields could be increased in Bushland, TX, by reducing row width from 1.01 to 0.51 m (increasing population from about 45,000 to 90,000 plants ha^{-1}) only if the initial soil water content (0 to 2 m) was above 150 mm. Unger et al. (1988) reported that opportunistic farmers could evaluate soil water conditions at wheat harvest, and plan accordingly. If sufficient moisture existed, they should plant sorghum in a double-crop sequence. If not, they could defer the decision until it is time to plant wheat. With sufficient autumn rains, they could then plant wheat. If not, they should defer planting until the next summer. Flexibility to implement such decisions may not be available to all farmers because of residual herbicides, farm programs, or restrictions by landlords or lenders. In spite of these potential limitations, the potential appears high for increasing total production per unit water.

Another, similar technique is termed "response farming" (Stewart, J. I., 1988). Here, the probable seasonal rainfall is inferred from the timing of the onset of monsoon rains. With an early onset, decisions are made that make use of the likely higher rainfall, and for late onset, decisions are made that increase the chance of surviving the likely lower rainfall. A final example of dynamic management is limited-irrigation dryland farming (Stewart et al., 1983; Stewart, B. A., 1988). Here, the extent of irrigation in a furrow-irrigated field depends upon the rainfall; if more rain falls, the irrigated area is extended because less irrigation is needed per unit land area. This extension can be achieved by adjusting seeding densities and fertility down the field. Alternate furrows are diked or used for irrigation. The upper one half of the field is irrigated normally. The lower one half uses a lower planting density, and the first half of this (the 3rd quarter of the field) is irrigated by tailwater from the upper half. The lower quarter of the field is normally dryland, but if rain follows an irrigation, the runoff from the upper area will reach this zone. The system maximizes production per unit of water for systems with fixed water supplies, whether fixed by physical or regulatory limitations, and reduces the risk of waterlogging and loss of excess water. The fact that crops can withstand extraction of 50 to 60% of the water in the root zone without detrimental effects on yield (Ritchie, 1981; Turner et al., 1986; Turner 1990a) can be used to irrigate larger areas where water supply is limited (Hearn and Constable, 1981, 1984).

The dynamic management techniques listed above accomplish several things relative to water use that pertain directly to sustainability. First, they decrease the chance of crop failure (as can a rotation with fallow), yet they take advantage of the better years, in which rainfall is sufficient for continuous or even double cropping. Moreover, in good years they do not waste the rain that happens to fall during a fallow period.

SUMMARY

We have tried to develop the hypothesis that for agriculture to be sustainable and productive, crop water use must balance water available from rainfall, irrigation, and soil storage. Both crop and soil management strategies are available to enable this balance to be achieved. Where farm management systems have not taken water use into consideration and imbalances have occurred, land degradation has often ensued by soil erosion, waterlogging, and salinization.

Moving toward sustainability is an important goal for agriculture. Historical examples suggest that agricultural systems can be sustainable in the long term if water is managed wisely. History also shows that sustainable production can decline even in systems that initially appeared sustainable. Agriculture will need to be responsive to declines in sustainability in order to avoid land and water degradation. Recent examples of increased fertilizer and pesticide use resulting in contamination of ground water, streams, and lakes indicate areas where new technologies and farming systems need to be developed. Likewise, the clearing of native vegetation that leads to secondary salinization and waterlogging lower in the landscape is being reversed by strategic use of evergreen trees, perennial pastures, and better management of the annual crops and pastures to increase water use. Thus we conclude that management of water to match supply with demand is important for sustainable agricultural production. Nevertheless, subtle changes in climate and increased climatic variability may make currently sustainable practices unsustainable in the future. Being able to predict and adapt to these changes will be an important component of future crop production.

REFERENCES

Abbott, I., C. A. Parker, and I. D. Sills. 1979. Changes in abundance of large soil animals and physical properties of soils following cultivation. *Aust. J. Soil Res.* 17:343–353.

Adams, R. M., C. Rosenzweig, R. M. Peart, J. T. Ritchie, B. A. McCarl, J. D. Glyer, R. B. Curry, J. W. Jones, K. J. Boote, and L. H. Allen, Jr. Global climate change and U.S. agriculture. *Nature.* 345(6272):219–224.

Alexandratos, N. 1988. *World Agriculture: Toward 2000—An FAO Study.* Food and Agriculture Organization, Rome.

Anderson, S. H., C. J. Gantzer, and J. R. Brown. 1990. Soil physical properties after 100 years of continuous cultivation. *J. Soil Water Conserv.* 45(1):117–121.

Armbrust, D. V. and N. H. Welch. 1966. Evaluation of Zingg conservation bench terraces on Amarillo fine sandy loam soil. *J. Soil Water Conserv.* 21:224–226.

Batchelor, C. K., G. C. Soopramanien, J. P. Bell, R. Nagamuth, and M. G. Hodnett. 1990. Importance of irrigation regime, dripline placement, and row spacing in the drip irrigation of sugar cane. *Agric. Water Manage.* 17:59–73.

Blackwell, P. S., N. S. Jagawardane, T. W. Green, J. T. Wood, J. Blackwell, and H. J. Beatty. 1991a. Subsoil macropore space of a transitional red-brown earth after either deep tillage, gypsum, or both. I. Physical effects and short term changes. *Aus. J. Soil Res.* 29:123–140.

Blackwell, P. S., N. S. Jagawardane, T. W. Green, J. T. Wood, J. Blackwell, and H. J. Beatty. 1991b. Subsoil macropore space of a transitional red-brown earth after either deep tillage, gypsum, or both: II. Chemical effects and long-term changes. *Aus. J. Soil Res.* 29:141–154.

Blevins, R. L., G. W. Thomas, and P. L. Cornelius. 1977. Influence of no-tillage and nitrogen fertilization on certain soil properties after 5 years of continuous corn. *Agron. J.* 69:383–386.

Bond, J. J., T. J. Army, and O. R. Lehman. 1964. Row spacing, plant populations, and moisture supply as factors in dryland grain sorghum production. *Agron. J.* 56:3-6.

Bond, R. D. 1964. The influence of microflora on the physical properties of soil: II. Field studies on water repellant sands. *Aus. J. Soil Res.* 2:123-131.

Bond, R. D. 1972. Germination and yield of barley when grown in a water-repellant sand. *Agron. J.* 64:402-403.

Browman, D. L. 1983. Tectonic movement and agrarian collapse in prehispanic Peru. *Nature.* 302:568-569.

Bruce, R. R., S. R. Wilkinson, and G. W. Langdale. 1987. Legume effects on soil erosion and productivity. In: *The Role of Legumes in Conservation Tillage Systems.* J. F. Power, Ed. Soil Conservation Society of America, Ankeny, IA. pp. 127-138.

Busscher, W. J., R. E. Sojka, and C. W. Doty. 1986. Residual effects of tillage in Coastal Plain soils. *Soil Sci.* 141(2):144-148.

Campbell, R. B., D. C. Reicosky, and C. W. Doty. 1974. Physical properties and tillage of paleudults in the southeastern coastal plains. *J. Soil Water Conserv.* 29:5.

Campbell, R. B., W. J. Busscher, O. W. Beale, and R. E. Sojka. 1988. Soil profile modification and cotton production. Proc. Beltwide Cotton Prod. Res. Conf., January 3 to 8, New Orleans, LA. National Cotton Council of America, Memphis, TN.

Chalmers, D. J., P. D. Mitchell, and L. van Heek. 1981. Control of peach tree growth and productivity by regulated water supply, tree density, and summer pruning. *J. Am. Soc. Hortic. Sci.* 106:307-312.

Chaudhary, M. R., P. R. Gajri, S. S. Prihar, and R. Khera. 1985. Effect of deep tillage on soil physical properties and maize yields on coarse-textured soils. *Soil Tillage Res.* 6:31-44.

Chen, A. 1987. Unraveling another Mayan mystery. *Discover.* June:40-49.

Cooper, P. J. M., P. J. Gregory, J. D. H. Keatinge, and S. C. Brown. 1987a. Effects of fertilizer variety and location on barley production under rainfed conditions in Northern Syria. II. Soil water dynamics and crop water use. *Field Crops Res.* 16:67-84.

Cooper, P. J. M., P. J. Gregory, D. Tully, and H. C. Harris. 1987b. Improving water use efficiency of annual crops in the rainfed farming systems of west Asia and north Africa. *Exp. Agric.* 23:113-158.

De Datta, S. K. and M. S. Kerim. 1974. Water and nitrogen economy of rainfed rice as affected by soil puddling. *Soil Sci. Soc. Am. Proc.* 38:515-518.

de Wit, C. T. 1958. Transpiration and Crop Yields. Agric. Res. Rep. No. 646. Wageningen, The Netherlands.

Doty, C. W., R. B. Campbell, and D. C. Reicosky. 1975. Crop response to chiseling and irrigation in soils with a compact A2 horizon. *Trans. ASAE.* 18(4):668-672.

Doty, C. W., J. E. Parsons, A. W. Badr, A. Nassehzadeh-Tabrizi, and R. W. Skaggs. 1985. Water table control for water resource projects on sandy soils. *J. Soil Water Conserv.* 40(4):360-364.

Doty, C. W., J. E. Parsons, A. Nassehzadeh-Tabrizi, R. W. Skaggs, and A. W. Badr. 1984a. Stream water levels affect field water tables and corn yields. *Trans. ASAE.* 27(5):1300-1306.

Doty, C. W., J. E. Parsons, and R. W. Skaggs. 1987. Irrigation water supplied by stream water level control. *Trans. ASAE.* 30(4):1065-1070.

Doty, C. W. and D. C. Reicosky. 1978. Chiseling to minimize the effects of drought. *Trans. ASAE.* 21(3):495-499.

Doty, C. W., W. B. Thayer, and R. G. Jessup. 1984b. Automated fabric dam aids water research project. In: *Proc. Int. Conf. Geomembranes 2C 127,* Denver, CO. p. 132.

Eck, H. V. 1986. Profile modification and irrigation effects on yield and water use of wheat. *J. Soil Sci. Soc. Am.* 50:724-729.

English, M. J., J. T. Musick, and V. V. Murty. 1990. Deficit irrigation. In: *Management of Farm Irrigation Systems.* G. J. Hoffman, T. A. Howell, and K. H. Solomon, Eds. American Society of Agricultural Engineers, St. Joseph, MI. p. 631-663.

Erickson, A. E., C. M. Hansen, and A. J. M. Smucker. 1968. The influence of subsurface asphalt barriers on the water properties and the productivity of sandy soil. In: *Trans. 9th Int. Congr. Soil Sci.* Elsevier, New York. pp. 331-337.

Evenari, M., L. Shanan, and N. Tadmor. 1982. *The Negev — The Challenge of a Desert.* Harvard University Press. Cambridge, MA.

Fillery, I. R. and P. J. Gregory. 1991. Defining research goals and priorities for sustainable dryland farming. In: *The Nature and Dynamics of Dryland Farming Systems — An Analysis of Dryland Agriculture in Australia.* V. R. Squires and P. Tow, Eds. Sydney University Press, Sydney, Australia. p. 162-168.

Fischer, R. A. and N. C. Turner. 1978. Plant productivity in the arid and semiarid zones. *Annu. Rev. Plant Physiol.* 29:277-317.

French, R. J. 1978a. The effect of fallowing on the yield of wheat. I. The effect of soil water storage and nitrate supply. *Aust. J. Agric. Res.* 29:653-666.

French, R. J. 1978b. The effect of fallowing on the yield of wheat. II. The effect on grain yield. *Aus. J. Agric. Res.* 29:666-684.

French, R. J. and J. E. Schultz. 1984. Water use efficiency of wheat in a mediterranean-type environment. I. The relation between yield, water use, and climate. *Aust. J. Agric. Res.* 35:743-764.

Greb, B. W., D. E. Smika, and A. L. Black. 1967. Effect of straw mulch rates on soil water storage during summer fallow in the Great Plains. *Soil Sci. Soc. Am. Proc.* 31:556-559.

Greenwood, E. A. N., N. J. Beresford, and J. R. Bartle. 1981. Evaporation from vegetation in landscapes developing secondary salinity using the ventilated chamber technique. III. Evaporation from a *Pinus radiata* tree and surrounding pasture in an agroforestry plantation. *J. Hydrol.* 50:155-166.

Greenwood, E. A. N., L. Klein, J. D. Beresford, and G. D. Watson. 1985. Differences in annual evaporation between grazed pasture and *Eucalyptus* species in plantations on a saline farm catchment. *J. Hydrol.* 78:261-278.

Greenwood, E. A. N., N. C. Turner, E.-D. Schulze, G. D. Watson, and N. R. Venn. 1992. Groundwater management through increased water use by lupin crops. *J. Hydrol.* 134:1-11.

Griffin, R. H., B. J. Ott, and J. F. Stone. 1966. Effect of water management and surface applied barriers on yield and moisture utilization of grain sorghum in the southern great plains. *Agron. J.* 58:449-452.

Gutentag, E. D., F. J. Heimes, N. C. Krothe, R. R. Luckey, and J. B. Weeks. 1984. Geohydrology of the high plains aquifer in parts of Colorado, Kansas, Nebraska, New Mexico, Oklahoma, South Dakota, Texas, and Wyoming. *U.S. Geol. Surv. Prof. Pap.*, No. 1400-B.

Hargrove, W. L., J. T. Reid, J. T. Touchton, and R. N. Gallaher. 1982. Influence of tillage practices on the fertility status of an acid soil double-cropped to wheat and soybeans. *Agron. J.* 74:684-687.

Hauser, V. L. 1968. Conservation bench terraces in Texas. *Trans. ASAE.* 11:385-386, 392.

Hearn, A. B. and G. A. Constable. 1981. Irrigation for crops in a semi-humid environment. I. Stress day analysis for soybeans and an economic evaluation of strategies. *Irrig. Sci.* 3:1-15.

Hearn, A. B. and G. A. Constable. 1984. Irrigation for crops in a semi-humid environment: VII. Evaluation of irrigation strategies for cotton. *Irrig. Sci.* 5:75-94.

Hillel, D. 1982. *Negev — Land, Water, and Life in a Desert Environment.* Praeger, New York.

Jamison, V. C. 1945. The penetration of irrigation and rain water into sandy soils of central Florida. *Soil Sci. Soc. Am. Proc.* 10:25-29.

Jennings, M. D., B. C. Miller, D. F. Bezdicek, and D. Granatstein. 1990. Sustainability of dryland cropping in the Palouse: An historical view. *J. Soil Water Conserv.* 45(1):75-80.

Jones, O. R. 1975. Yields and water-use efficiencies of dryland winter wheat and grain sorghum production systems in the southern high plains. *Soil Science Soc. of America Proc.* 39:98-103.

Jones, O. R. 1981. Land forming effects on dryland sorghum production in the southern Great Plains. *J. Soil Sci. Soc. Am.* 45:606-611.

Jones, O. R. and R. N. Clark. 1987. Effects of furrow dikes on water conservation and dryland crop yields. *J. Soil Sci. Soc. Am.* 51(5):1307-1314.

Jones, O. R., H. V. Eck, S. J. Smith, G. A. Coleman, and V. L. Hauser. 1985. Runoff, soil, and nutrient losses from rangeland and dry-farmed cropland in the Southern High Plains. *J. Soil Water Conserv.* 40(1):161-164.

Juo, A. S. and R. Lal. 1979. Nutrient profile in a tropical Alfisol under conventional and no-till systems. *Soil Sci.* 127(3):168-173.

Karlen, D. L., W. R. Berti, P. G. Hunt, and T. A. Matheny. 1989. Soil-test values after eight years of tillage research on a Norfolk loamy sand. *Commun. Soil Sci. Plant Anal.* 20:1413-1426.

Karlen, D. L., W. J. Busscher, S. A. Hale, R. B. Dodd, E. E. Strickland, and T. H. Garner. 1992. Conservation tillage implement effects on energy use and soil strength, *Trans. ASAE.* 34(5):1967-1972.

Kasperbauer, M. J. and W. J. Busscher. 1991. Genotypic differences in cotton root penetration of a compacted subsoil layer. *Crop Sci.* 31:1376-1378.

Kendall, D. 1988. Good soil eases drought worries. *The New Farm.* 10(7):44-47.

Kladivko, E. J., D. R. Griffith, and J. V. Mannering. 1986. Conservation tillage effects on soil properties and yield of corn and soya bean in Indiana. *Soil Tillage Res.* 8:277-287.

Laflen, J. M., J. L. Baker, R. O. Hartwig, W. F. Buchele, and H. P. Johnson. 1978. Soil and water loss from conservation tillage systems. *Trans. ASAE.* 21(5):881–885.

Lavake, D. E. and A. F. Weise. 1979. Influence of weed growth and tillage interval during fallow on water storage, soil nitrates, and yield. *J. Soil Sci. Soc. Am.* 43:565–569.

Letey, J., J. F. Osborn, and N. Valoras. 1975. Soil water repellancy and the use of nonionic surfactants. Tech. Rep. 154, California Water Resource Center. University of California, Davis, CA.

Lyle, W. M. and D. R. Dixon. 1977. Basin tillage for rainfall retention. *Trans. ASAE.* 20(6):1013–1017, 1021.

Mathews, O. R. and T. J. Army. 1960. Moisture storage on fallowed wheatland in the Great Plains. *Soil Sci. Soc. Am. Proc.* 24:414–418.

McGhie, D. A. 1980. The contribution of the Mallet Hill surface to runoff and erosion in the Narrogin region of Western Australia. *Aust. J. Soil Res.* 18:299–307.

McGhie, D. and P. Tipping. 1990. *The Alleviation of Water Repellance in Soils.* Western Australian Department of Agriculture, South Perth.

Miller, D. E. and J. S. Aarstad. 1972. Effect of deep plowing on the physical characteristics of Hezel soil. *Wash. Agric. Exp. Stn. Circ.*, No. 556. Washington State University, Pullman, WA.

Mills, W. C., A. W. Thomas, and G. W. Langdale. 1988. Rainfall retention probabilities computed for different cropping-tillage systems. *Agric. Water Manage.* 15:61–71.

Minc, L. D. and J. H. Vandermeer. 1990. Origin and spread of agriculture. In: *Agroecology.* C. R. Carroll, J. H. Vandermeer, and P. Rosset, Eds., McGraw-Hill, New York. p. 65–111.

Mitchell, P. D., B. van den Ende, P. H. Jerie, and D. J. Chalmers. 1989. Responses of 'Bartlett' pear to withholding irrigation, regulated deficit irrigation, and tree spacing. *J. Am. Soc. Hortic. Sci.* 114:15–19.

Mukhtar, S., J. L. Baker, R. Horton, and D. C. Erbach. 1985. Soil water infiltration as affected by the use of the Paraplow. *Trans. ASAE.* 28(6):1811–1816.

Nulsen, R. A. and I. N. Baxter. 1982. The potential of agronomic manipulation for controlling salinity in Western Australia. *J. Aust. Inst. Agric. Sci.* 48:222–226.

Ortloff, C. R. 1988. Canal builders of Pre-Inca Peru. *Sci. Am.* December:100–107.

Osborn, J. F., J. Letey, L. F. DeBano, and E. Terry. 1967. Seed germination and establishment as affected by non-wettable soil and wetting agents. *Ecology.* 48:494–497.

Osborn, J. F., R. E. Pelishek, J. S. Krammes, and J. Letey. 1964. Soil wettability as a factor in erodibility. *Soil Sci. Soc. Am. Proc.* 28:294–295.

Parker, C. A. 1989. Soil biota and plants in the rehabilitation of degraded agricultural soils. In: *Animals in Primary Succession—The Role of Fauna in Reclaimed Lands.* J. D. Majer, Ed. Cambridge University Press, Cambridge, UK. pp. 423–438.

Perry, M. W. 1987. Water use efficiency of non-irrigated field crops. In: Proc. the 4th Aust. Agron. Conf., Melbourne. August. pp. 83–99.

Pittock, A. B. 1988. Actual and anticipated changes in Australia's climate. In: *Greenhouse – Planning for Climate Change*. G. I. Pearman, Ed. E. J. Brill, Leiden. pp. 35–51.

Ponting, C. 1990. Historical perspectives on sustainable development. *Environment.* 32(9):4–9, 31–33.

Regan, K. L., K. H. M. Siddique, N. C. Turner, and B. R. Whan. 1992. Potential for increasing early vigour and total biomass in spring wheat. II. Characteristics associated with early vigour. *Aust. J. Agric. Res.* 43(3):541–553.

Reicosky, D. C., R. B. Campbell, and C. W. Doty. 1976. Corn plant water stress as influenced by chiseling, irrigation, and water table depth. *Agron. J.* 68:499–503.

Reisner, M. 1986. *Cadillac Desert: The American West and Its Disappearing Water*. Viking Penguin, New York.

Ridge, P. E. 1986. A review of long fallows for dryland wheat production in southern Australia. *J. Aust. Inst. Agric. Sci.* 52:37–44.

Ritchie, J. T. 1981. Water dynamics in the soil-plant-atmosphere system. *Plant Soil.* 58:81–96.

Roberts, F. J. 1966. The effects of sand type and fine particle amendments on the emergence and growth of subterranean clover with particular reference to water relations. *Aust. J. Agric. Res.* 17:657–672.

Robertson, W. K., L. C. Hammond, G. K. Saxena, and H. W. Lundy. 1973. Influence of water management through irrigation and a subsurface asphalt layer on seasonal growth and nutrient uptake of corn. *Agron. J.* 65:866–870.

Rosenberg, N. J., B. L. Blad, and S. B. Verma. 1983. *Microclimate: The Biological Environment*. John Wiley & Sons, New York.

Ruttan, V. W. 1990. 'Alternative Agriculture' sustainability is not enough. In: *'Alternative Agriculture' Scientists' Review*. Spec. Pub. No. 16. Council for Agricultural Science and Technology, Ames, IA. pp. 130–134.

Saroch, K. and R. C. Thakur. 1991. Effect of puddling (wet tillage) on rice yield and physico-chemical properties of soil. *Soil Tillage Res.* 21:147–157.

Schofield, N. J. and P. R. Scott. 1991. Planting trees to control salinity. *West Aust. J. Agric.* 32:3–10.

Schofield, N. J., I. C. Loh, P. R. Scott, J. R. Bartle, P. Ritson, R. W. Bell, H. Borg, B. Anson, and R. Moore. 1989. Vegetation strategies to reduce stream salinities of water resource catchments in south-west Western Australia. Rep. WS 33, Water Authority of Western Australia. Leederville, Western Australia.

Schofield, N. J., J. K. Ruprecht, and I. C. Loh. 1988. The impact of agricultural development on the salinity of surface water resources of southwest Western Australia. Rep. WS 27, Water Authority of Western Australia. Leederville, Western Australia.

Schultz, J. E. 1971. Soil water changes under fallow crop treatments in relation to soil type, rainfall, and yield of wheat. *Aust. J. Exp. Agric. Anim. Husbandry.* 6:425–431.

Seery, D. 1990. Navajos grow corn as their elders did. *Soil Water Conserv. News.* 11(3):12.

Sharma, M. L., R. J. Barron, and A. B. Craig. 1991. Land use effects on groundwater recharge to an unconfined aquifer. Div. Rep. 91/1, CSIRO Division of Water Resources. Wembley, Western Australia.

Shepherd, K. D., P. J. M. Cooper, A. Y. Allan, D. S. H. Drennan, and J. D. H. Keatinge. 1987. Growth, water use, and yield of barley in Mediterranean-type environments. *J. Agric. Sci.* 108:365-378.

Smika. D. E. 1970. Summer fallow for dryland winter wheat in the semiarid Great Plains. *Agron. J.* 62(1):15-17.

Steiner, J. L. 1992. Crop residue effects on water conservation. In: *Managing Agricultural Residues.* Unger, P. W., Ed. CRC Press, Boca Raton, FL.

Stewart, B. A. 1988. Dryland farming: the North American experience. In: *Challenges in Dryland Agriculture: a Global Perspective.* P. W. Unger, W. R. Jordan, T. V. Sneed, and R. W. Jensen, Eds. Proc. Int. Conf. Dryland Farming. August 15 to 19, Amarillo/Bushland, TX. pp. 54-59.

Stewart, B. A., J. T. Musick, and D. A. Dusek. 1983. Yield and water use efficiency of grain sorghum in a limited irrigated-dryland farming system. *Agron. J.* 75:629-634.

Stewart, J. I. 1988. Risk analysis and response farming. In: *Challenges in Dryland Agriculture: a Global Perspective.* P. W. Unger, W. R. Jordan, T. V. Sneed, and R. W. Jensen, Eds. Proc. Int. Conf. Dryland Farming. August 15 to 19, Amarillo/Bushland, TX. pp. 322-324.

Tanner, C. B. and T. R. Sinclair. 1983. Efficient water use in crop production: research or re-search? In: *Limitations to Efficient Water Use in Crop Production.* H. M. Taylor, W. R. Jordan, and T. R. Sinclair, Eds. Agron. Monogr. American Society of Agronomy, Madison, WI. pp. 1-27.

Tennant, D. 1980. Effect of fallowing on cereal yields. *West Aust. J. Agric.* 21:38-41.

Turner, B. L. and P. D. Harrison. 1981. Prehistoric raised-field agriculture in the Maya lowlands. *Science.* 213:399-405.

Turner, N. C. 1990a. Plant water relations and irrigation management. *Agric. Water Manage.* 17:59-73.

Turner, N. C. 1990b. The benefits of water deficits. In: *Proc. Int. Congr. Plant Physiol.* S. K. Sinha, P. V. Sane, S. C. Bargava, and P. K. Agrawal, Eds. Society for Plant Physiology and Biochemistry, New Delhi. pp. 806-815.

Turner, N. C., A. B. Hearn, J. E. Begg, and G. A. Constable. 1986. Cotton (*Gossypium hirsutum* L): physiological responses to water deficits and their relationships to yield. *Field Crops Res.* 14:153-170.

Turner, N. C. and M. E. Nicolas. 1987. Drought resistance of wheat for light-textured soils in a mediterranean climate. In: *Drought Tolerance in Winter Cereals.* J. P. Srivastava, E. Porceddu, E. Acevedo, and S. Varma, Eds. John Wiley, Chichester, U.K. pp. 203-216.

Turner, N. C., P. Prasertak, and T. L. Setter. 1994. Influence of plant spacing and planting density on wheat subjected to post-anthesis water deficits. *Crop Sci.* In press.

Turner, N. C., D. Tennant, A. P. Hamblin, I. E. Henson, C. R. Jensen, and M. W. Perry. 1987. Influence of nitrogen on the yield water use and water deficits of wheat grown in a mediterranean climate. *Proc. 4th Aust. Agron. Conf.* Melbourne. August. 254 pp.

Unger, P. W. 1968. Soil organic matter and nitrogen changes during 24 years of dryland wheat tillage and cropping practices. *Soil Sci. Soc. Am. Proc.* 32:427–429.

Unger, P. W. 1971a. Soil profile gravel layers. I. Effect on water storage, distribution, and evaporation. *Soil Sci. Soc. Am. Proc.* 35:631–634.

Unger, P. W. 1971b. Soil profile gravel layers. II. Effect on growth and water use by a hybrid forage sorghum. *Soil Sci. Soc. Am. Proc.* 35:980–983.

Unger, P. W. 1978. Straw-mulch rate effect on soil water storage and sorghum yield. *Soil Sci. Soc. of Am. J.* 42:486–491.

Unger, P. W. 1983. Water conservation: Southern Great Plains. In: *Dryland Agriculture.* Agron. Monogr. American Society of Agronomy, Madison, WI. pp. 35–55.

Unger, P. W. 1991. Organic matter, nutrient, and pH distribution in no- and conventional-tillage semiarid soils. *Agron. J.* 83:186–189.

Unger, P. W. and L. J. Fulton. 1990. Conventional- and no-tillage effects on upper root zone soil conditions. *Soil Tillage Res.* 16:337–344.

Unger, P. W., O. R. Jones, and J. L. Steiner. 1988. Principles of crop and soil management procedures for maximizing production per unit rainfall. In: *Drought Research Priorities for the Dryland Tropics.* F. R. Bidinger and C. Johansen, Eds. International Crops Research Institute for the Semi-Arid Tropics (ICRISAT) Center, Patancheru, India.

Unger, P. W. and J. J. Parker. 1976. Evaporation reduction from soil with wheat, sorghum, and cotton residues. *Soil Sci. Soc. Am. Proc.* 40:938–942.

Wallis, M. G., D. J. Horn, and K. W. McAuliffe. 1990. A study of water repellancy and its amelioration in a yellow-brown sand. I. Severity of water repellancy and the effects on wetting and abrasion. *N. Z. J. Agric. Res.* 33:139–144.

Weeks, J. B. and E. D. Gutentag. 1984. High plains regional aquifer— Geohydrology. In: *Proc. Ogallala Aquifer Symp. II*, Lubbock, TX. p. 6.

West, L. T., W. P. Miller, G. W. Langdale, R. R. Bruce, J. M. Laflen, and A. W. Thomas. 1991. Cropping system effects on interrill soil loss in the Georgia Piedmont. *J. Soil Sci. Soc. Am.* 55:460–466.

Wetherby, K. G. 1984. The Extent and Significance of Water Repellant Sands on Eyre Peninsula. Tech. Rep. No. 47, Department of Agriculture, South Australia.

Whan, B. R., G. P. Carlton, and W. K. Anderson. 1991. Potential for increasing early vigour and total biomass in spring wheat. I. Identification of genetic improvements. *Aust. J. Agric. Res.* 42:347–361.

Willis, W. O., H. J. Haas, and J. S. Robins. 1963. Moisture conservation by surface or subsurface barriers and soil configuration under semiarid conditions. *Soil Sci. Soc. Am. Proc.* 25:577–580.

Wood, C. W., D. G. Westfall, and G. A. Peterson. 1991. Soil carbon and nitrogen changes on initiation of no-till cropping systems. *J. Soil Sci. Soc. Am.* 55:470–476.

Yang, B. J., P. S. Blackwell, and D. F. Nicholson. 1993. Numerical simulation of heat movement and water loss from furrow-sown water repellant sand. *Aust. J. Soil Res.* In press.

Zingg, A. W. and V. L. Hauser. 1959. Terrace benching to save potential runoff for semiarid land. *Agron. J.* 51:289–292.

3

Management Strategies for Sustainable Soil Fertility

Douglas L. Karlen and Andrew N. Sharpley

TABLE OF CONTENTS

INTRODUCTION

Sustainable soil fertility is defined for this chapter using a mass-balance concept. Nonanthropogenic and anthropogenic factors, processes, or strategies that sustain the fertility of a soil, are discussed with respect to their ability to supply nutrients essential to plant growth without creating conditions that degrade the environment. Our objectives are to examine factors which affect soil and to present management strategies for sustaining fertility without adverse environmental or economic impacts.

MASS-BALANCE APPROACH TO SUSTAINABLE
SOIL FERTILITY

Achieving sustainable soil fertility with a mass-balance approach requires an understanding of both the anthropogenic (controllable) and nonanthropogenic (uncontrollable) factors which affect soil fertility (Figure 1). The five basic soil-forming factors (Jenny, 1941) (parent material, climate [including moisture and temperature effects], macro- and microorganisms, topography, and time) can be considered the nonanthropogenic factors which interact to create an inherent fertility level in every soil. This unique characteristic reflects physical, chemical, biological, and morphological effects created by interactions among naturally occurring and relatively uncontrollable factors. A sixth soil-forming factor, occasionally included in the basic model for describing a soil (SSSA, 1987), is humankind. We consider this to be our anthropogenic factor, since activities of humankind are generally controllable.

Soil fertility is defined as the ability of a soil to supply nutrients essential to plant growth (SSSA, 1987). Therefore, to achieve sustainable soil fertility, our goal must be to develop and implement optimum techniques and management practices that supply appropriate amounts of all

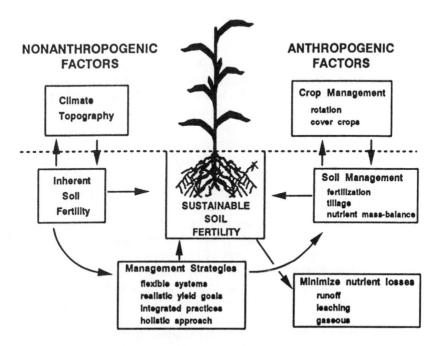

FIGURE 1. Management strategy considerations for sustainable soil fertility.

essential plant nutrients at optimum times for producing our food and fiber resources. Achieving a balance between nutrient inputs and outputs is essential to prevent excess accumulations which can be transferred to nontarget locations, such as surface or ground water resources, or be wasteful of natural and economic resources within the farming enterprise.

Nonanthropogenic Factors

Parent material influences inherent soil fertility primarily by determining the predominant mineral content and particle size which influence soil texture and the type and rate of chemical reactions that occur. The deposition method (residual, or transported by gravity, ice, water, or wind) primarily affects soil texture and landscape topography. Climate influences inherent soil fertility because precipitation, temperature, and the amount of erosive wind and water action determine weathering rates of parent materials. Amount and distribution of precipitation and evapotranspiration, as well as temperature regimes, are important nonanthropogenic factors because they influence biological and chemical reaction rates including solution, hydration, mineralization, immobilization, and leaching. Climate also influences the kind and quantity of vegetation found throughout various landscapes and the amount of organic material added to the soil each year.

Interest in understanding nonanthropogenic climatic effects on inherent soil fertility from the perspective and scale of seasonal weather patterns (local microclimate) has recently increased because of major efforts to quantitatively understand the impact of farming practices on leaching of residual nitrates and pesticides in ground water resources (Hallberg, 1989a,b; Keeney, 1989). This is important because mechanisms resulting in transport of water and indigenous soil chemicals, including nitrate, are natural processes occurring throughout the soil profile.

For example, in the southeastern Coastal Plains, soils are generally highly weathered and have dense, eluviated (E) horizons, low available water retention characteristics (Campbell et al., 1974), and low nutrient retention because of the predominance of 1:1 clay minerals and low organic matter levels. These soil physical and chemical characteristics, combined with a seasonal microclimate that provides excess water for leaching, often results in movement of both anions and cations through the soil profile. Karlen et al. (1984) demonstrated the effects of these nonanthropogenic factors when, after applying K at higher than normal rates for 4 years, they found no effect on extractable K concentrations in the Ap or E horizons, even though values were increased by more than 40% in the subsoil. This observation for Coastal Plain soils was in direct contrast with previous information for midwestern soils (Hanway, 1964), and showed that cation stratification with reduced tillage is less permanent on sandy Coastal Plain soils than in midwestern loam and silt loam soils. Therefore, when developing sustainable soil fertility programs for different regions, soil type and seasonal microclimate must be considered.

Microclimate and nonanthropogenic soil characteristics influence inherent soil fertility by affecting microbial activity and associated respiration, mineralization, and denitrification processes. Doran et al. (1990) and Linn and Doran (1984a,b) demonstrated that for a wide range of soils aerobic microbial activity increased in a linear relationship with increased water content between 30 and 60% water-filled pore space (WFPS). Between 60 and 70% WFPS, aerobic microbial activity declined, presumably because the additional water presented a barrier for oxygen diffusion to and waste products away from soil microorganisms. They also demonstrated that given an adequate available C and NO_3-N supply, very little denitrification occurred at 60% WFPS. However, denitrification increased exponentially at WFPS exceeding 70 to 75% WFPS.

The biota type and quantity are nonanthropogenic factors that determine the kind and amount of organic materials that are returned to the soil. In natural grasslands, over 2.5 Mg ha^{-1} of dry matter can be added to the soil each year since roots and total root systems may contain more than 12 Mg ha^{-1} of dry matter as compared with the production of only 2 to 5 Mg ha^{-1} of above-ground material (Russell, 1973). The biota type influences spatial organic material distribution in the soil, i.e., grasses differ from trees in that they translocate much of the organic matter they synthesize into the soil or root systems, whereas the primary contribution from trees is through leaf drop on the soil surface. Vegetation also influences many microbial processes, including nitrogen mineralization, fixation, and immobilization, as well as organic matter and crop residue decomposition processes that influence nutrient cycling by providing energy sources to drive those processes.

Topography influences water infiltration, soil temperature, and erosion. Soils on sloping land (especially those having medium or fine texture) have a high probability of losing water as runoff. Soils located on level or nearly level landscape positions or in depressions usually have higher water content than those on slopes. Sloping topographies are also subject to more water erosion than more level areas with similar land cover. Slope orientation and elevation are topographical factors that influence local microclimate and thus influence the soil-forming processes. Formation time, in conjunction with intensity of temperature and rainfall, determines the degree of soil development which has a major influence on inherent soil fertility and nutrient reactions.

Anthropogenic Factors

We consider humankind to be our anthropogenic factor because over relatively short time intervals, controllable activities generally determine whether the existing condition, created by either nonanthropogenic factors or prior anthropogenic activities, will be lowered, sustained, or improved.

Unique combinations that result from overlapping of nonanthropogenic and anthropogenic factors give rise to different soils (Arnold, 1983), each with an inherent and current soil fertility status. Decisions regarding fertilizer rates, tillage practices, water management, and crop selection are among the anthropogenic factors that determine the relative sustainability of a soil.

Several studies of agricultural production system effects on soil P cycling dynamics have shown a differential behavior of inorganic and organic P forms (Agboola and Oko, 1976; Harrison, 1978; Sharpley, 1985a; Tiessen et al., 1983). In production systems where no fertilizer P is added, there is a general inorganic P consistency, although available levels will gradually decline. The net P loss from the system through removal in the harvested crop is primarily accounted for by a decrease in soil organic P content. For example, 60 years (1913 to 1973) of cotton (*Gossypium hirsutum* L.) growth on the Mississippi Delta soil, Dundee silt loam, with no reported fertilizer P applications, resulted in no appreciable effect on inorganic P content (Sharpley and Smith, 1983). However, organic P content of cultivated (93 mg kg^{-1}) compared to virgin analogue (223-mg kg^{-1}) surface soil (0 to 15 cm) decreased. Mineralization of organic P replenished the inorganic pool and thus provided adequate amounts of plant available P.

More detail on the processes involved has been provided by sequential chemical fractionation of soil organic P (Hedley et al., 1982; Bowman and Cole, 1978b). Temporal changes in amounts of labile organic P extracted by NaHCO$_3$, indicate the importance of this fraction to crop growth (Figure 2) (Sharpley, 1985a; O'Halloran et al., 1985). Figure 2 also illustrates the importance of silt and clay fractions on the physical stabilization of root exudates and microbial products, which enter the soil system as soluble or colloidal matter. Obviously, this mineralization process and its interaction with soil physical properties is affected by microclimate, with amounts of organic P mineralized annually varying by three orders of magnitude from temperate (0.3 kg P ha^{-1} year^{-1}) to tropical soils (396 kg P ha^{-1} year^{-1}) (Stewart and Sharpley, 1987).

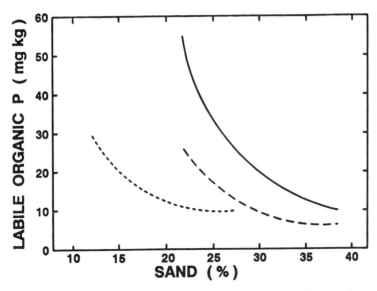

FIGURE 2.   Labile organic P content (bicarbonate extractable) as a function of percent sand in a Canadian Mollisol at the end of a second wheat crop of a fallow-wheat rotation (_____), at the end of the fallow year on the same plots (– –) (from O'Halloran et al., 1985), and for several Southern Plains soils (----) (from Sharpley, 1985a).

Interactions among nonanthropogenic and anthropogenic factors discussed in this section demonstrate a few of the physical, chemical, and biological factors which must be considered and emphasize that careful long-term soil and crop management are critical for developing sustainable soil fertility programs. For this reason, no single management strategy can be developed and implemented across all soils and regions of the U.S. or the world. Rather, the selection process must consider the type of soil for which a sustainable fertility status is being developed and what types of microclimate constraints are typically going to be placed upon the soil and plant system. Consequently, there is a need for developing computer simulation models to evaluate the relative long-term effects of alternative management strategies on the sustainability of soil fertility. Linking soil management and water quality models will facilitate evaluation of relative effects of both soil and crop management strategies. This will make possible more site specific management strategies which utilize available water and nutrient resources more efficiently and thus have no adverse effects on water quality.

SOIL MANAGEMENT STRATEGIES FOR SUSTAINABLE SOIL FERTILITY

Fertilization Strategies

The goal for a sustainable soil fertility program is to efficiently use all of the nutrients applied to a soil for food and fiber production without accumulating excess amounts that can be lost to environmentally sensitive areas such as surface and/or ground water resources. Nitrogen and P are the plant nutrients which are of primary concern from an environmental perspective. Nitrogen, because at concentrations exceeding 10 mg l^{-1} NO_3-N it has been attributed to causing methemoglobin (Hallberg, 1989a), and P because it can stimulate the growth of algae in lakes and rivers (Schindler, 1977; Vollenweider and Kerekes, 1980). Potassium, the third primary plant nutrient which is often applied to build and maintain soil fertility, has not been implicated in any deleterious environmental effects, although in some environments concentrations high enough to cause salinity problems have been identified.

In noncultivated or "virgin" soils the amounts and rates at which plant nutrients are supplied to plants is controlled by weathering of soil minerals and nutrient cycling through the vegetation. When the soils are farmed to continually provide food and feed, nonanthropogenic soil mineral sources are gradually depleted and must be supplemented by application of fertilizers or manures to maintain the mass-balance process of sustainable soil fertility (Cooke, 1984).

Even though nutrients removed by crop uptake, chemical and biological fixation, and loss to the environment can be replaced by fertilizer applications, there are more subtle changes in the inherent soil fertility which must be considered in developing sustainable soil fertility management strategies. For example, Sharpley and Smith (1983) and Smith and Young (1975) investigated the effects of up to 72 years of cultivation and fertilizer application by quantifying relative amounts, distribution, and forms of C, N, and P in eight soils, representing different U.S. cropping areas. By comparing cultivated soils with virgin analogues, they showed that on an average basis, organic C and total N concentrations in cultivated surface horizons (0 to 15 cm) decreased 42 and 35%, respectively. Total and inorganic P contents increased 25 and 118%, respectively, suggesting a positive mass balance, but surface soil organic P content decreased an average of 43% with cultivation and fertilizer P application. Stable organic P forms constituted the major portion of this decrease, with labile organic P maintained at a fairly constant level. Clearly, cultivation of these soils decreased inherent or sustainable soil fertility. For P, changes in stable organic P forms may indicate the sustainability of different agricultural production systems.

A wealth of information is available on fertilizer N and P management strategies which can be used to maximize the efficiency with which those plant nutrients are used by crops. This includes information on timing, placement, sources, and mixtures of inorganic and organic amendments to fertilizer materials (Halvorson et al., 1989; Peterson and Frye, 1989; Power and Broadbent, 1989; Randall et al., 1985; Schepers and Fox, 1989; Sharpley et al., 1992). For additional details about specific practices, the reader is referred to those articles. Our focus will be to identify techniques to assess the effectiveness of those different management strategies toward achieving a sustainable soil fertility by using a nutrient mass balance approach.

When developing sustainable fertilizer management strategies, it is important to know the nutrient balance that must be maintained. Typical amounts of N, P, and K accumulation and removal by four major U.S. crops are presented in Table 1. Values such as these are useful for developing a nutrient budget or balance sheet showing nutrient "exports" and "imports" to a farming system. Plant nutrients are exported when plant material or animal products are moved from the farm and imported with animal feeds, off-farm waste products, or commercial fertilizers.

To achieve sustainable soil fertility without excess accumulation and possible contamination of surface or ground water resources, the mass balance between imports and exports must approach zero. If the balance is negative, there may be a gradual depletion of nutrients and declines in productivity. This scenario can occur for any of the essential plant nutrients, although because of the amount required, it most likely will begin with the macronutrients N, P, and K. Some N can be supplied through fixation by including legumes in the crop rotation, but importation of these nutrients in either organic or inorganic forms must generally occur to build and maintain a sustainable soil fertility.

A positive mass balance occurs when there is a net gain of nutrients within the farming enterprise. This occurs mainly on farms with

Table 1. Nutrient Accumulation in the Aboveground Portion and Removal in the Harvested Portion of Several Crops

Crop	Yield (Mg ha^{-1})	Accumulation (kg ha^{-1})			Removal (kg ha^{-1})		
		N	P	K	N	P	K
Corn	10.0	238	44	198	134	34	43
Soybean	4.0	353	28	191	269	23	78
Wheat	4.0	140	20	128	86	14	17
Alfalfa	13.4	377	43	335	377	43	335

Adapted from Potash and Phosphate Institute, 1990.

relatively large numbers of livestock, especially if there is a high importation of feed (Wallingford et al., 1975). These situations often contribute to increased nutrient loss to surface water and ground water resources.

Nutrient importation into a farming enterprise does not indicate lack of stability. In fact, with an increase in nutrient input intensity, amounts leaving the system will generally increase. The critical need for implementing sustainable soil fertility production systems is to develop methods to minimize nonpoint N and P losses in runoff, drainage, and groundwater recharge.

To ensure nutrient balance, farming practices which promote nutrient recycling should be encouraged. On-farm nutrient sources such as animal manures, livestock bedding, and plant residues should be considered first since they have been demonstrated to be economically and environmentally responsible (Powers et al., 1974). Although high transportation costs have generally limited the practice to date, recycling off-farm nutrient sources such as sewage sludge, leaf litter, and manufacturing by-products for their fertilizer value should be encouraged as methods for developing sustainable soil fertility programs.

Strategies for Balancing Nutrient Inputs and Outputs

Several strategies can be used to balance nutrient inputs and outputs in farming enterprises. These include the use of soil tests and plant analyses to monitor the accuracy of the budget, as well as systems modeling to project long-term impacts of various management decisions. Important considerations for each of these strategies are discussed below.

Soil Testing

Soil testing is one of the best management strategies for monitoring effects of anthropogenic practices being used to build or maintain a sustainable soil fertility. Melsted and Peck (1973) stated that soil testing is an accurate and indispensable tool essential for assessing soil fertility. They summarized their soil testing objectives to be an accurate determination of the available nutrient status of soils, a clear indication to the farmer of the seriousness of any deficiency or excess that may exist for various crops, a basis for making fertilizer recommendations, and a mechanism facilitating an economic evaluation of the fertilizer recommendations. Their objectives recognized that soil test results per se must include both the factual data, reported in a finite manner, and interpretations based upon crop sequence, anticipated weather conditions, managerial ability of the farmer, and several other judgmental factors.

Basic Soil Testing Practices. Soil testing programs generally consist of four phases: sample collection, extraction and analysis, interpretation, and use. Sample collection is critical because a very small amount of material is used to represent a large area. To obtain a sample that will best represent a specific area, soil-forming factors, landscape position, and time of sampling should all be considered in developing the sampling plan. Local consultants or cooperative extension service agents can generally provide guidance in sample collection techniques for routine soil test measurements such as pH, NO_3-N, P, K, Ca, Mg, S, and micronutrients. The procedures used for testing, correlation, calibration, and interpretation of soil tests for predicting crop response to fertilizer is well documented (Brown, 1987), but outside the scope of this chapter. Use of these tests is encouraged as a basis from which to establish and maintain a sustainable soil fertility practice. A research need for establishing these practices is to develop new calibrations that can accurately predict what will happen if soil-test levels have become extremely high because of prior management and to determine how much time will elapse before additional fertilizer will again be needed.

Nitrogen Soil Tests. Routine procedures for measuring the soil N status have developed slowly because 95 to 97% of the N in soil is present in very complex organic compounds that are not directly available to plants (Dahnke and Vasey, 1983). Nitrogen from such sources slowly becomes available to plants through microbial decomposition, but N soil tests are complicated because conversion rates for this organic matter are dependent upon the type of organic material (C:N ratio), soil temperature, water content, aeration status, pH, and other factors which influence the microbial population. Furthermore, nitrate, the final and primary plant available product of this conversion is subject to loss through leaching, denitrification, and immobilization by microorganisms.

Several soil tests to measure potential N mineralization in soils have been proposed (Keeney and Bremner, 1966; Jenkinson, 1968; Dahnke and Vasey, 1973; Fox and Piekielek, 1978; Stanford, 1982; Keeney, 1982; and Meisinger, 1984), but few have been successful in the U.S. Corn Belt where substantial amounts of fertilizer N are applied annually (Blackmer et al., 1989). In nonirrigated, semiarid regions, where leaching losses are negligible, measurements of residual soil nitrate have been used for some time to adjust annual fertilization rates (Dahnke and Vasey, 1973). Autumn soil sampling and nitrate testing programs are common in these areas, but those programs are not satisfactory in areas of high autumn and winter precipitation because of the potential for leaching.

Magdoff et al. (1984) proposed a soil test for available N that is gaining acceptance in the eastern U.S. Their procedure is based on concentrations of nitrate found in the surface 30 cm of soil when corn (*Zea Mays* L.) plants are 15 to 30 cm tall. This "late-spring" procedure has been adapted and tested in Iowa (Blackmer et al., 1989) and found to have good potential for improving N management in the northern Corn Belt.

Nitrate soil tests may seem to be of recent origin, but Dahnke and Vasey (1973) pointed out that the literature shows how workers in the early 1900s were using soil nitrate concentrations to predict crop yields. Regardless of who is credited with the concept, use of an appropriate soil nitrate test to adjust fertilizer applications is a management strategy that should be adopted for sustainable soil fertility.

Phosphorus Soil Test Methods. Soil test procedures to estimate plant available P include a variety of chemical extraction methods, which are closely related to plant growth and uptake of P under certain climatic and edaphic conditions (Fixen and Grove, 1990). Alternative procedures including use of anion exchange resin, isotopic ^{32}P, buffer capacities, adsorption-desorption relationships, and quantity-intensity curves have been developed to predict soil solution levels of P for environmental reasons. None of those procedures have found widespread or routine application for predicting plant response. Therefore, dilute solution extraction techniques used by soil test laboratories to predict P availabilities have changed little over the last 25 years. According to Fixen and Grove (1990), the most commonly used extraction methods are Bray I (Bray and Kurtz, 1945), Mehlich III (Mehlich, 1984), and Olsen (Olsen et al., 1954).

Despite this lack of change in methodology, recent advances in soil characterization, sampling, and fertilizer application techniques have made the soil P test and interpretation the weakest part of the P recommendation process. Thus, further test development is necessary, particularly in assessing the sustainability of soil P fertility for different management strategies, which will rely more heavily on precise fertilizer placement, incorporation of crop residues, and utilization of residual inorganic and organic P. These developments should involve zonal sampling of P-stratified soils, assessing organic and microbial P contents, and use of nonconventional procedures.

One test development is zonal sampling. Due to the immobility of P in the soil profile, fertilizer placement is generally more critical for P than N. Thus, the precise subsurface placement or banding of fertilizer P will be of particular importance in developing sustainable soil P fertility in reduced tillage systems, with minimal surface soil disturbance. A greater crop yield response to surface- or subsurface-banded application of fertilizer P has been measured, compared to broadcast or mixing (Alston, 1980; Bailey and Grant, 1989; Lammond, 1987; Welch et al., 1966). Holford (1989) found that fertilizer P effectiveness, as measured by yield response in a wheat-clover rotation, was greater for shallow banding at 5 cm depth compared to banding at 15 cm and broadcast application for six soils having a 100-fold variation in P sorptivity. However, the almost equal effect of mixing P throughout the soil regardless of P sorptivity suggests the important factor in maximizing fertilizer effectiveness is its positional availability in the root zone, rather than reduction of chemical immobility by concentration bands (Holford, 1989).

Soil test P procedures may also need refining for soils common to the western U.S., which contain a lime-enriched layer (Westermann, 1992). Cultivation of such soils may expose subsoil of high indigenous $CaCO_3$ content, initially capable of fixing large amounts of P. In a greenhouse study of the effect of lime on P availability in a calcareous soil (Portneuf silt loam) to sudangrass (*Sorghum vulgare sudanese*) and potatoes (*Solanum tuberosum* L.) in Idaho, Westermann (1992) concluded that the critical soil test P concentration (Olsen P), should increase as $CaCO_3$ content increases, in order to recommend adequate fertilizer P rates to sustain, particularly, potato production in these calcareous soils.

Clearly, zonal concentrations of P in soil will present sampling problems to determining subsequent fertilizer P requirements (Havlin, 1991). For example, if location of the P band is known, what portion of samples should be collected on and off the band; and, if the band location is not known, is a random sampling strategy adequate? Collection of 15 (Ward and Liekam, 1986; Shapiro, 1988) to 30 random samples (Hooker, 1976) has been reported to adequately reflect P availability in fields where P bands exist. Kitchen et al. (1990) found that only slightly increased soil test P variability was obtained by collecting two samples. One sample was taken randomly and one at 50% of the band spacing distance from the first and perpendicular to the band direction.

Another test assessed organic and microbial phosphorus contents. Mineralization of soil organic P can be an important source of plant-available P in some temperate soils and in some soils under reduced tillage (Harrison, 1987; McLaughlin et al., 1988a). Organic P mineralization in several unfertilized and P-fertilized soils in the Southern Plains was quantified by Sharpley (1985a) as the decrease in soil organic P content during the period of maximum crop growth (spring and early summer). As soil organic P may be formed by plant residue incorporation during this period, the value of net organic P mineralization will underestimate the actual value. Averaged for each soil type, amounts of organic P mineralized ranged from 15 to 33 kg P ha^{-1} $year^{-1}$ (Figure 3). Mineralization was not completely inhibited by fertilizer P application of 20 to 28 kg P ha^{-1} $year^{-1}$, with similar amounts of P contributed by both sources (Figure 3). Amounts of organic P mineralized in these soils are similar to other temperate soils, ranging from 0.3 to 40 kg P ha^{-1} $year^{-1}$. They were lower than for soils from the tropics where amounts from 6 to 396 kg P ha^{-1} $year^{-1}$ have been measured, because the distinct wet and dry

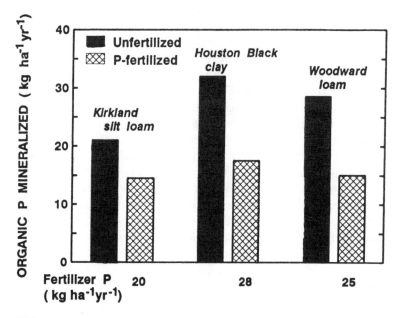

FIGURE 3. Annual net mineralization of organic P in unfertilized and P-fertilized soils from the Southern Plains.

seasons and higher soil temperatures can enhance rapid organic P mineralization (Stewart and Sharpley, 1987).

Consequently, it is possible that soil P fertility tests may be improved in certain situations by giving credit to mineralizable organic P as well as inorganic P. Several studies have reported that potential soil P supply, as reflected by crop yields, was more closely estimated by including extractable organic P (Abbott, 1978; Adepetu and Corey, 1976; Bowman and Cole, 1978a; Daughtrey et al., 1973). Bowman and Cole (1978a) used a modification of the Olsen method (Olsen et al., 1954), which measured the total P extracted by the reagent. Where other soil P test methods are recommended, a similar adaptation may be used. As conditions of organic P extraction may not replicate the dynamic field condition influencing organic P mineralization, caution must be used in relating amounts of extractable organic P to expected crop response.

Since fumigation-extraction techniques were developed to measure soil microbial biomass P (Brookes et al., 1982; Hedley and Stewart, 1982), its importance in P cycling has been recognized (McLaughlin et al., 1988b; Stewart and Tiessen, 1987). For example, McLaughlin and Alston (1986) demonstrated that microbial biomass can assimilate a similar proportion of fertilizer P to that taken up by wheat. In a study of P cycling through soil microbial biomass, Brookes et al. (1984) measured annual P fluxes of 5 kg P ha^{-1} year^{-1} in soils under continuous wheat and 23 kg P ha^{-1} year^{-1} under permanent grass in England. Although biomass P flux under continuous wheat was less than P uptake of 20 kg P ha^{-1} year^{-1} by the crop, annual P flux in the grassland soils was much greater than uptake of 12 kg P ha^{-1} year^{-1} by the grass.

Not all this biomass P is available for plant uptake each year; some will be directly transferred to subsequent microbes and some released to the soil solution, which can then be fixed by soil material or taken up by microbes or plants (Brookes et al., 1984). However, rate and extent of microbial P flux emphasize the importance of microbial P in controlling organic P transformation dynamics and thereby, sustainable soil fertility.

There are also some nonconventional procedures used in testing soil. Generally, crops only recover a small portion of fertilizer P applied, with the remaining P either becoming fixed in unavailable forms or available as residual P to succeeding crops (McLaughlin et al., 1988a; Sharpley, 1986). Field studies using ^{32}P labeled fertilizer demonstrated that annual uptake from 30 kg of applied P ha^{-1} by winter wheat decreased from 23 to 9% over a 4-year period for several soils in the Southern Plains (Sharpley, 1986). In this study, initial Bray-I P levels (7 mg kg^{-1}) increased eightfold. However, with excessive fertilizer P applications, residual soil P can accumulate to levels in excess of 300 mg kg^{-1} (as Bray I P) (Yerokum and Christenson, 1990; Pierzynski et al., 1990b).

Many studies have shown that accumulated residual P can sustain crop production for 7 to 13 years without fertilizer P application (Brams, 1973; Halvorson and Black, 1985; Novias and Kampráth, 1978; Read et al., 1973; Spratt et al., 1980; Yost et al., 1981). However, with an increase in residual P levels and potential formation of less soluble P-rich particles with time (Adepoju et al., 1982; Fixen and Ludwick, 1982; Pierzynski et al., 1990a,c), it is possible that current soil P test extractants and their interpretation may not adequately represent residual P availability. Additional sampling problems may occur in determining the residual availability of banded fertilizer P, particularly as the residual value of banded P is greater than broadcast P (Sander et al., 1990; Eghball et al., 1990). Accurately quantifying residual P availability is essential for developing sustainable soil P fertility systems, which will depend on the efficient utilization of this P pool.

Nonconventional soil test methods may have application to determine residual P availability. Such methods involve using resin accumulators (Cooperband and Logan, 1991; Yang et al., 1991) or Fe and Al oxide-impregnated filter paper strips (Menon et al., 1989a,b) as a sink for plant available soil P. Strip P was more closely related to both dry matter yield and P uptake by maize (*Zea mays* L.) than was Bray I extractable P for four soils ranging in pH from 4.5 to 8.2 (Menon et al., 1989a,b,c).

Sharpley (1991) reported that Fe-Al oxide-impregnated paper strips removed primarily physically bound P (anion exchange resin P) from 203 soils representing all soil orders. Resin P described soil P availability as determined by crop P uptake, A value, crop yield, and relative yields over a wider range of properties than other soil P test methods (e.g., Olsen, Bray I, and Mehlich III) (Fixen and Grove, 1990; Sibbesen, 1978, 1983). In addition, P extracted by paper strips embedded in soil columns was closely related to Bray I P for acid and Olsen P for alkaline and calcareous soils (Menon et al., 1990). The potential use of paper strips as a nondestructive method to measure *in situ* soil P availability may be of value in estimating banded residual P availability.

As strip P was closely related to different P tests for soils on which the use of the test is recommended and with resin P, it is possible that the paper strip may extract amounts of P closely related to plant availability for soils ranging widely in physical and chemical properties. The close correlation between strip P, soil test P, and resin P does not in itself justify adoption of the procedure to quantify P uptake. However, it emphasizes the potentially wide applicability of these nonconventional "P-sink" methods to estimate plant available soil P, including residual P, and suggests further evaluation of the methods is warranted.

Plant Analyses

The use of plant analyses is an effective management strategy for a sustainable soil fertility program because it provides a direct measure of nutrient concentrations and balance within the plant. Principles and procedures used for plant analyses have evolved over many years and changed as knowledge increased about each element that is essential for a plant to complete its life cycle (Munson and Nelson, 1990). As such, use of plant analyses has become an integral part of most agronomic research and a tool for crop consultants and fertilizer dealers to monitor production fields.

Principles for Basic Plant Analyses. The basis for plant analyses is a determination of the concentration of an element or compound such as chlorophyll that is essential for a plant to survive and reproduce. These measurements can be made using whole plants or an individual part which has been shown to be responsive to the medium in which the plant is being grown and proportional to the growth or yield of the plant. The results of these measurements can be interpreted using several techniques including critical levels or sufficiency ranges, a diagnosis and recommendation integrated system (DRIS), total amounts and rates of accumulation, and nutrient use efficiencies (Munson and Nelson, 1990).

Critical Values. Critical values were generally established by graphing crop yield and elemental concentration data for essential nutrients within a plant (Munson and Nelson, 1990). This technique has been used by many people (Chapman, 1966), but has also been criticized because plant dry matter yield differences result in dilution effects, sufficiency ranges are often quite broad, and these values are not sensitive to small treatment differences imposed to optimize nutrient applications.

DRIS Approach. The DRIS concept was introduced by Beaufils (1961) for analysis of rubber trees (*Hevea brasiliensis* L.) in Vietnam to determine (1) the concentrations and interrelations of essential elements in plants, and (2) the soil nutrient levels, conditions, and cultural practices associated with optimum plant growth and the highest attainable yields. The basis for this approach is that with optimum concentrations of one or more elements, yield levels will be low, medium, or high depending upon one or more factors which also influence plant growth. These factors include: availability of other nutrients, soil characteristics, water availability, temperature, plant population, weeds, insects, diseases, and cultural practices. The DRIS approach allows the user to determine a relative ranking from the most to the least deficient nutrient. It is also less sensitive to dry matter dilution than critical values because nutrient ratios rather than concentrations are used for interpretation (Sumner, 1977). Originally, a common data base for interpretation of DRIS norms was thought to be feasible, but studies with corn, soybean (*Glycine max* L. Merr.), alfalfa (*Medicago sativa* L.), wheat (*Triticum aestivum* L.), and potato (*Solanum tuberosum* L.) have indicated that local or regional norms provide more accurate deficiency diagnosis than those developed in other regions (Escano et al., 1981; Beverly et al., 1986; Walworth et al., 1986; Amundson and Koehler, 1987; MacKay et al., 1987). For more detailed information about the DRIS approach, the reader is referred to the recent review of plant analysis principles and practices by Munson and Nelson (1990).

Total Nutrient Accumulation. Use of plant analyses to determine amounts and rates of nutrient uptake is a useful management strategy for sustainable soil fertility. The information can be used to anticipate when high plant nutrient concentrations must be made available for rapid uptake and assimilation and when they would be more vulnerable to loss through processes such as leaching. Karlen et al. (1987b,c; 1988) determined amounts and rates of nutrient accumulation for high yielding corn, and Karlen and Sadler (1989, 1990) determined amounts and rates for soybean at different yield levels throughout the U.S. and for wheat in the southeastern Coastal Plain. They used compound cubic polynomials (splining) to describe plant growth and nutrient accumulation data measured through plant analyses. The equations computed to describe this information were then differentiated to show daily rates of dry matter and nutrient accumulation for these crops. This approach identified periods of intraseasonal variation in accumulation which could be used as a sustainable soil fertility management tool to schedule fertilizer applications to provide a minimum nutrient stress environment during both vegetative and reproductive plant growth stages.

The splining approach was also used by Karlen and Doran (1991) to evaluate dry matter and nutrient accumulation for soybean and corn grown using alternative management practices on the Richard and Sharon Thompson farm near Boone, Iowa. Those analyses showed that there were no unique plant or development patterns associated with using alternative management practices for production of those crops.

Nutrient Use Efficiencies. Use of plant analyses data to determine relative nutrient use efficiencies for various crop and soil management practices is one of the most important applications of the information relative to sustainable soil fertility management. Measuring total dry matter and nutrient concentrations provides the needed information to compute total uptake in the entire plant or harvested portion. By dividing these values by the amount of fertilizer applied or the amount of nutrient available in the soil, nutrient use efficiency (NUE) values can be determined. These efficiency values can then be used to determine the recovery of applied fertilizer and the uptake of residual nutrients. One application for this type of plant analysis information is to compute percent fertilizer recovery with isotopic sources such as ^{15}N (Broadbent and Carlton, 1978; Hauck and Bremner, 1976; Schepers et al., 1989).

Munson and Nelson (1990) defined NUE values as the internal efficiency with which crops convert nutrients taken up into grain yield. Their definition can also be used to evaluate sustainability of different management practices, but the methods of calculating NUE values can not be compared. Munson and Nelson (1990) also emphasized that recovery of one nutrient from the soil or fertilizer depends upon availability of other nutrients as well as other plant growth factors.

Emerging Plant Analyses Technologies

Chlorophyll Monitoring. Several new plant analysis techniques are being evaluated and may provide effective management strategies for sustainable soil fertility. One technique that has shown very promising results for evaluating the N status of plants is the use of a Minolta®* SPAD-502

---

*Mention of trademark, proprietary product, or vendor does not constitute a guarantee or warranty of the product by the USDA and does not imply its approval to the exclusion of other products or vendors that may also be suitable.

chlorophyll meter (Schepers et al., 1990b). This instrument was used for an evaluation of several corn hybrids grown at different N fertilization rates. Results using this nondestructive technique were very similar to those obtained by punch sampling leaves and subsequently analyzing the samples (Schepers et al., 1990a).

Basal Stem Nitrate Monitoring. Use of basal stem nitrate N measurements at either the V6 growth stage or at physiological maturity provides another management strategy for developing a sustainable N management program. The early-season stem nitrate test has been studied (Rauschkolb et al., 1974; Iverson et al., 1985; McClenahan and Killorn, 1988; Schepers et al., 1990a), but critical concentrations vary significantly between 0.9 to 1.8% on a dry weight basis. Plant growth stage is also critical for early season stem nitrate analysis because the concentrations decrease as the corn plant develops (Hanway, 1962). If the sampling can be delayed, there is an increased chance that N deficiencies will be detected. However, if deficiencies do occur, there will be less time to respond with sidedress fertilizer applications. Another use for stem nitrate analyses is to evaluate effectiveness of fertilization rates that were used during the growing season. Binford et al. (1990) stated that if corn stalk nitrate N concentrations exceeded 1.8 g kg^{-1} during the first 2 weeks after physiologic maturity, excess N was available for the crop. Although this "postmortem" analysis has no benefit for the current crop year, it may be useful for developing more sustainable strategies for subsequent years.

CROP MANAGEMENT FOR SUSTAINABLE SOIL FERTILITY

Crop selection and sequencing is one of the easiest anthropogenic factors to utilize in the development of sustainable soil fertility. Crop rotation is not new, but its advantages will probably be exploited more in the future. For example, corn following soybean often shows a yield advantage of 10 to 15%, with yields during periods of drought being 25 to 30% higher with rotation than for continuous corn production (Wikner, 1990; Karlen et al., 1991). Crop rotation aids in weed and disease control, disrupts insect cycles, and frequently provides opportunities to distribute economic risk and to better utilize labor resources within the farm enterprise (Wikner, 1990).

Growing cover crops during an interim between cash or grain crops is another crop selection practice which may be useful for developing a more sustainable soil fertility program. Although discussion and use of cover crops may seem to some as "reinventing the wheel", Pieters and McKee (1938) pointed out that compared to green-manuring practices, which were utilized by the Greeks more than 300 years B.C., cover cropping is of relatively recent origin. A better perspective when considering cover cropping as a sustainable soil fertility strategy would be to assume that, some 53 years later, we are just a mere 2% further along the same timeline.

Rotation Management

Crops managed in a rotation should be selected in a sequence where complementary root systems fully exploit available water and nutrients. For example, the inclusion of deep-rooted legumes in a rotation may necessitate the need to allow the water-depleted soil profile to recharge before the next crop (Grecu et al., 1988; Hoyt and Leitch, 1983). The sequence with which crops are grown in a rotation also influences N movement through the soil profile and ultimately into groundwater resources (Carter et al., 1991; Carter and Berg, 1991). Legumes, including soybean and alfalfa which do not require supplemental N inputs, can effectively use or "scavenge" residual N remaining in the soil from previous crops (Johnson et al., 1975; Mathers et al., 1975; Muir et al., 1976; Olson et al., 1970; Stewart et al., 1968). To illustrate this benefit, Sharpley et al. (1992) overlayed hypothetical root development for several cropping sequences on typical N-leaching patterns (Figure 4). Clearly, inclusion of alfalfa in the rotation increased sustainability with regard to N recovery when compared to continuous corn.

Alfalfa has been shown to develop roots to depths greater than 5.5 m, and research has shown that nitrate can be utilized by the crop from any depth where soil solution is extracted by its roots. Mathers et al. (1975) reported that alfalfa removed nitrate from the soil profile at a depth of 1.8 m during the first year of establishment and to a depth of 3.6 m during the second and third years of stand development. In a study of rotation effects on N movement in a silt loam soil, Olson et al. (1970) found that the concentration of nitrate in the soil solution was inversely proportional to the amount of time oats (*Avena sativa* L.), meadow, or alfalfa were grown in sequence with corn. Depending upon the rotation being measured, they found soil solution nitrate levels at a depth of 1.2 to 1.5 m to be 34 to 82% lower than at the same depths beneath continuous corn. They reported that the amount of reduction was directly proportional to the number of years in oats, meadow, or alfalfa production, and attributed this to combined recovery of nitrate by shallow-rooted oat crops followed by deep-rooted alfalfa crops.

Olson et al. (1970) reported that N applications which were not used by crops grown on Plainfield sand were often leached below rooting depths in less than 1 year. They concluded that alfalfa following corn or other crops receiving relatively high rates of N on sandy soils would not be able

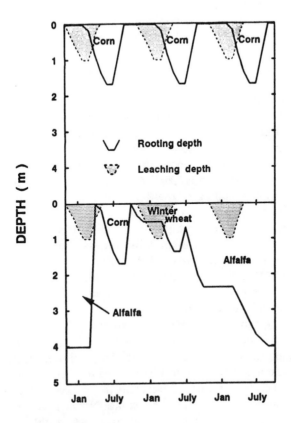

FIGURE 4. Typical root development of continuous corn and corn-winter wheat-alfalfa rotation in relation to soil drainage over a 3-year period.

to recover nitrate which passed through the profile. Therefore, removal of subsoil nitrates by deep-rooted legumes such as alfalfa will probably be more effective on medium and heavy textured soils than on sands.

Soybean can also effectively scavenge residual soil N (Johnson et al., 1975; Havlin et al., 1990; and Karlen et al., 1991), but Jackson et al. (1987) reported that in Wisconsin they are not as effective as alfalfa because of their shallower rooting depth.

Although a wealth of information documents the benefits for N use efficiency of careful selection and sequencing of crops in a rotation, less information is available for P. It is possible that appropriate selection and use of a crop with a higher affinity for P may reduce soil P stratification and increase P use efficiency, particularly if the nonharvested portion of the crop is returned to the soil. Selection of crops which can more efficiently utilize residual soil inorganic and organic P, while being economically viable to the farmer, will enhance the sustainability of soil P fertility.

The N fixation potential of legumes such as alfalfa and soybean must also be considered when including them in crop rotations to develop or maintain a sustainable soil fertility program. In the southeastern Coastal Plains, where maturity groups VII and VIII soybean are grown, Hunt et al. (1985) found that net N return to the soil from soybean ranged from 14 to 123 kg ha^{-1} year^{-1}, depending more upon seasonal rainfall than upon tillage system or inoculation treatment. In an arid environment, where furrow irrigation is essential for crop production, Carter and Berg (1990) found that up to 300 kg NO_3-N ha^{-1} could be formed in the soil during the season following the killing of alfalfa. For U.S. Corn Belt states, a soybean N credit of 20 g kg^{-1} seed yield (1 lb/bu) is recommended, with upper limits depending upon individual state recommendations (Bundy et al., 1990; Killorn, 1988). For alfalfa, Bundy et al. (1990) recommend using N credits ranging from 78 to 156 kg ha^{-1} (70 to 140 lb/ac), depending upon the stand density when the crop is killed or incorporated. Nitrogen credits for other forage crops, green manures, and leguminous vegetable crops have also been determined (Table 2) and can significantly reduce annual expenditures for N fertilizers (Table 3). Research is needed to determine under which crop rotation strategies P credits may also be used to reduce fertilizer P requirements. Using these N and P credits to replace fertilizer inputs and varying crop sequences to periodically scavenge the soil profile of residual N and P are two management strategies for developing a sustainable soil fertility program through crop selection, particularly for N.

Cover Crop Management

Cover crops can contribute to a more sustainable soil fertility status by protecting the soil surface from the impact of raindrops, minimizing runoff, and preventing erosion (Pieters and McKee, 1938); improving soil structure and tilth characteristics including aggregation, porosity, bulk density, and permeability (Rogers and Giddens, 1957); fixing atmospheric N; and preventing leaching of N, K, and possibly other nutrients by incorporating them into their biomass. Cover crops add organic matter to the soil and by providing a readily available carbon source, they can cause an increase in microbial activity which subsequently increases aggregation. One example of this showed aggregation increases of 30% in response to cover crops plus woodchips (Ram and Zwerman, 1960).

Table 2.   Legume-Derived N Fertilizer Replacement Value for Selected Crops Grown in Wisconsin

Legume Crop	N Credit	Exceptions
Alfalfa (1st year)[a] (*Medicago sativa* L.)	45 kg ha^{-1} + 1.1 kg ha^{-1} for every percent legume in the forage stand 100% stand = 157 kg ha^{-1} 60% stand = 112 kg ha^{-1} 30% stand = 78 kg ha^{-1}	Use 50% of the normal credit if the crop was harvested after September 10 in the prior year
Red clover (*Trifolium pratense* L.) or Birdsfoot trefoil (*Lotus corniculatus* L.)	N Credit is 80% of alfalfa 100% stand = 126 kg ha^{-1} 60% stand = 90 kg ha^{-1} 30% stand = 62 kg ha^{-1}	Use 50% of the normal credit if the crop was harvested after September 10 in the prior year
Green manure crops Sweet clover (*Melilotus alba* L.)	90 to 134 kg ha^{-1}	Use 22 kg N ha^{-1} credit if field has less than 15 cm of growth before tillage
Alfalfa	67 to 121 kg ha^{-1}	
Red clover	56 to 90 kg ha^{-1}	
Soybean	0.2 kg kg^{-1} harvested seed with a maximum credit of 45 kg ha^{-1}	No credit on sandy soils
Leguminous vegetable crops Pea (*Pisum sativum* L.)	22 kg ha^{-1}	No credit on sandy soils
Snap beans (*Phaseolus vulgaris* L.)	22 kg ha^{-1}	No credit on sandy soils
Lima beans (*Phaseolus limensis* L.)	22 kg ha^{-1}	No credit on sandy soils

[a]A second year credit of 34 kg N ha^{-1} can be taken for any "good" legume sod.
Adapted from Bundy et al., 1990.

If legumes are used as cover crops, they can provide N through fixation for subsequent crops (Smith et al., 1987). Cover crops also provide residual effects, as demonstrated by McCracken et al. (1989) in Kentucky. They reported that growing hairy vetch (*Vicia villosa* L. Roth) as a winter cover crop for 10 years increased nitrogen uptake by a subsequent corn crop by an average of 20.4 kg ha^{-1} and resulted in higher grain yields. For more northern locations, however, Karlen (1990) reported

Table 3. Dollar Value of N Fertilizer Credit Following Selected Legumes Grown in Wisconsin

Legume Crop	First Year N Fertilizer Replacement Credit (kg ha^{-1})	Value ha^{-1} at N Fertilizer Cost of		
		$0.33 kg^{-1}	$0.44 kg^{-1}	$0.55 kg^{-1}
Alfalfa				
100% stand	157	$51.81	$69.08	$86.35
60% stand	121	$39.93	$53.24	$66.55
30% stand	78	$25.74	$34.32	$42.90
Red clover				
100% stand	123	$40.59	$54.12	$67.65
60% stand	90	$29.70	$39.60	$49.50
30% stand	62	$20.46	$27.28	$34.10
Red clover Green manure	90	$29.70	$39.60	$49.50
Soybeans (2.7 Mg ha^{-1} yield)	45	$14.85	$19.80	$24.75
Peas, snapbeans, or lima beans	22	$ 7.26	$ 9.68	$12.10

Adapted from Bundy et al., 1990.

that a major conservation tillage research need was to develop cropping strategies and management schemes that make cover crops more compatible with common crop rotations. Improving tolerance of legume cover crop germplasm to more northern conditions was also identified by Power (1987) as a major research need for the Midwest.

Farmer interest in and research emphasis for incorporating cover crops into production systems decreased dramatically during the 1960s. This occurred because fertilizer N became more available at low cost and could be substituted for green manure and cover crop additions of N (Melsted, 1954). Herbicide technology also developed rapidly during that period, making it easy to control perennial and annual weed species (Moody, 1961; Triplett, 1966).

Cover crops provide an effective method for improving water quality because they accumulate and retain plant nutrients as well as reduce soil erosion. Ram et al. (1960) stated that effectiveness of cover crops with regard to canopy protection depended upon stand density, soil coverage, total cover, average height, rate and period of growth, plant spacing, and harvest method. Collectively these can be termed "method of management". For nutrient retention and/or biocycling, these same aerial characteristics, plus the extent and vigor with which the cover crop root systems explore the soil, are also important.

Several state, federal, and private research scientists are beginning to reevaluate cover crops. The focus of many studies is to determine optimum cover crop management practices so that competition for available soil water and yield-limiting factors are minimized. Research scientists at the Rodale Research Institute (1990) have provided leadership for much of the renewed effort with on-farm studies of alternate cover crop species or mixes (Hofstetter, 1988). One farmer-research cooperator involved with cover crop studies in the Midwest is Mr. Richard Thompson. Among the visitors who come to see his on-farm research program, discussions focusing on his tests and experiences with cover crops result in the most questions and greatest interest (R. L. Thompson, Boone, Iowa, personal communication, 1990).

Through cooperative efforts, USDA-ARS scientists have established on-farm cover crop evaluations with farmers including Mr. Thompson. Initial results of the investigations have suggested that tower net organic matter mineralization and/or greater denitrification losses before planting apparently reduced soil N availability and created early-season N deficiencies in corn grown using ridge tillage in fields with cover crops (Doran et al., 1989; Karlen and Doran, 1991).

Stand establishment is one of the most difficult challenges for incorporating cover crops into current crop rotations in the Midwest. During the 1950s and 1960s when interseeding and forage establishment were more prevalent, practices using tools such as a cultipacker seeder were developed to interseed directly over corn (Jackobs and Gossett, 1956; Peterson, 1956). Those systems were developed, however, when plant populations for corn were much lower and when standard row width was at least 1 m instead of current widths of 0.75 m or less. The combined effects of higher plant density and narrower rows have contributed to higher corn grain yields with current practices. However, with nearly total light interception (Karlen et al., 1987a), it is futile to plant a cover crop until the grain crop is nearly mature. For much of the Midwest, this results in minimal growing time for establishing cover crops using current germplasm resources (Power, 1987),

On-farm research trials have shown that for success in the northern Corn Belt, cover crops should be overseeded when soybean leaves begin to turn yellow, but before they drop from the plant (R. L. Thompson, personal communication, 1990). Another study initiated in 1990 by Soil Conservation Service (SCS) agronomists in Missouri was designed to grow no-till soybean after killing 60-cm strips of annual or perennial cover crops around each row and suppressing cover crop growth between rows with various herbicide combinations. The preliminary observations from this study indicate that competition for water by the cover crop and other weeds may be difficult to overcome (Steve Bruckerhoff, USDA-SCS, personal communication, 1990).

Winter rye (*Secale cereale* L.) and Italian ryegrass (*Lolium multiflorum* Lam.) have both been shown to be very effective cover crops for reducing leaching losses of residual nitrogen (Brinsfield et al., 1988; Staver and Brinsfield, 1990; Martinez and Guiraud, 1990). A disadvantage of a winter rye cover crop is that it can be very aggressive during the following spring and deplete available soil water supplies and thus stress the primary crop. This has been reported for sandy soils in the southeastern U.S. (Campbell et al., 1984) and observed in on-farm experiments in Iowa (R. L. Thompson, personal communication, 1990). A current project funded in part by the Aldo Leopold Center for Sustainable Agriculture is evaluating the use of oats (*Avena sativa* L.) instead of rye as a cover crop, since it grows rapidly during autumn, but freezes in winter and therefore does not deplete soil water reserves in spring. Oats are also very economical, generally available, and provide an excellent companion crop for hairy vetch when broadcast into scenescing soybean in Iowa (S. J. Corak, personal communication, 1991).

These studies suggest that use of cover crops would be an effective management strategy for developing a more sustainable soil fertility status and to reduce soil erosion losses, especially following a soybean crop. In the southern U.S., cover crops such as crimson clover (*Trifolium incarnatum* L.) are being successfully used in many locations (Frye et al., 1988), but limited germplasm selections currently prevents recommendations for the northern Corn Belt. Currently, interseeding a mixture of oats and hairy vetch appears to be a promising combination for this region, but much more research is needed to improve this management strategy. Efforts should also be directed toward combining the use of cover crops with conservation tillage practices such as ridge-tillage.

MINIMIZING NUTRIENT LOSSES

Minimizing nutrient losses from agricultural production systems is a primary consideration for developing sustainable soil fertility systems. Although nutrient removal in crop harvest and animal production will occur regardless of management practices, losses via runoff or leaching can degrade environmental quality and must be minimized. Recycling of manure as a nutrient source is advantageous for minimizing environmental impact, but it is covered in the section Fertilizer Management.

Renewed interest in implementing management practices that minimize nutrient loss from farmland may be partially attributed to the increased public perception of the role of agriculture in nonpoint source pollution (USEPA, 1984). Because of easier identification and control of point sources of pollution, agricultural sources now account for a larger share of all discharges than they did a decade ago.

Runoff Losses

Loss of N and P in runoff from agricultural land occurs in soluble and particulate forms. Particulate nutrients encompass all solid phase forms, which include nutrients sorbed by soil particles and organic matter suspended during runoff. While soluble P and N are generally available for immediate biological uptake (Peters, 1981; Walton and Lee, 1972), particulate N and P can provide a long-term source of these elements to aquatic biota (Bjork, 1972; Carignan and Kalff, 1980; Sharpley et al., 1991a). In the past, most studies have measured management effects on only soluble and total nutrient loss in runoff. However, measurement of particulate N and P bioavailability is needed to more accurately estimate the impact of agricultural production systems on the biological productivity of surface waters. This is of particular importance for particulate P, because despite recycling, P is often the growth-limiting element, due to the ability of blue green algae to fix atmospheric N.

Particulate nutrients may account for up to 90% of the total loss in runoff from cultivated land (Schuman et al., 1973a,b; Sharpley et al., 1987). With a reduction in soil erosion, as from conservation tillage and grassed systems, nutrient loss in runoff is generally dominated by the soluble form (Barisas et al., 1978; Johnson et al., 1979; Laflen and Tabatabai, 1984; Yoo et al., 1988). Thus, reducing nutrient losses in runoff will primarily depend on erosion control and fertilizer management.

Erosion Control

Soil losses can be reduced by minimizing the forces of raindrop impact, through surface soil stabilization or by maintaining a vegetative cover in the form of a growing crop or crop residue. Soil loss can also be reduced by weakening the erosive forces of runoff by reducing slope, diverting runoff, or otherwise reducing runoff velocity by contour plowing, terracing, and channeling.

Implementing conservation tillage practices has decreased total N and P loss in runoff by increasing soil vegetative cover (Langdale et al., 1985; McDowell and McGregor, 1984). From an intensive monitoring program of 28 watersheds in Oklahoma and Texas, it is clear that reduced and no-till management of both sorghum and wheat reduced N and P loss in runoff compared to conventional tillage (Table 4). The loss of bioavailable particulate P, as determined by NaOH extraction (Sharpley et al., 1991a), was also lower from conservation compared to conventional practices, although tillage management had no consistent effect on soluble N and P loss (Table 4). However, no-till management of sorghum and, to a lesser degree, wheat reduced nutrient losses in runoff to levels similar to those from unfertilized native grass (Table 4).

Conservation tillage for corn in Mississippi reduced total N losses fivefold and P losses ninefold compared to conventional tillage (McDowell and McGregor, 1984). Although total quantities lost were low, an eightfold increase in soluble P loss was measured for the no-till compared to conventional till practice. Fertilizer application to no-till corn reduced particulate nutrient transport in runoff, probably because of increased vegetative cover resulting from fertilization.

A review of recent literature clearly shows that inclusion of a cover crop in several management systems consistently decreased runoff, erosion, and amounts of N and P transported (Table 5). Further, from additional data provided by Klausner et al. (1974), it is apparent that the relative effect of cover crops in reducing nutrient transport was a function of soil fertility. The percent reduction in runoff, nitrate N, and soluble P transport from corn and wheat with and without a ryegrass and ryegrass-alfalfa cover crop, respectively, was greater for a high compared to a low fertility management system. By maintaining higher soil fertility, the relative effectiveness of cover crops in reducing nitrate N and soluble P transport was enhanced an average 8-fold for corn and 48-fold for wheat. It is clear that judicious fertilizer use can enhance cover crop growth and thereby increase nutrient retention in the soil-crop system. This may be brought about via increased cover and aggregation of surface soil by shoots and roots, respectively.

A more detailed explanation of cover crop effect on P transport may be obtained from an analysis of runoff from several management

Table 4. Effect of Tillage Practice on Soil, N, and P Loss in Runoff from Sorghum and Wheat in the Southern Plains

Management	Runoff	Soil	Nitrate N Loss	Total N	Soluble P	Bioavailable Particulate P[a]	Total P
	cm				$kg\ ha^{-1}\ year^{-1}$		
Sorghum							
No till	3.13	280	0.35	1.11	0.08	0.05	0.28
Reduced till	2.12	520	0.41	1.40	0.04	0.12	0.37
Conventional till	12.11	16,150	0.62	1.34	0.24	1.56	4.03
Wheat							
No till	7.73	540	1.52	5.12	0.53	0.13	0.98
Reduced till	6.10	800	1.92	4.59	0.10	0.10	0.59
Conventional till	10.08	8,470	1.74	20.19	0.21	0.61	3.96
Native grass	9.2	43	1.38	1.11	0.12	0.05	0.20

[a]Determined by NaOH extraction. Soluble P plus bioavailable particulate P is total bioavailable P in runoff.
Adapted from Sharpley et al. (1991b), and Smith et al. (1991).

Table 5. Effect of Cover Crops on Soil, N, and P Loss in Runoff for Several Management Systems

Crop	Cover	Location[a]	Fertilizer (kg ha⁻¹ year⁻¹)		Runoff (cm)	Soil Loss	Nitrate N (kg ha⁻¹ year⁻¹)	Total N	Soluble P (kg ha⁻¹ year⁻¹)	Total P
			N	P						
Corn	None	MD[1]	67	47	0.40	262	0.36	0.95	0.02	0.14
Corn	Barley		67	47	0.08	33	0.05	0.12	0.01	0.01
Corn	None	GA[2]	—	20	15.85	3663	—	—	0.28	4.08
Corn	Winter rye		—	50	9.65	938	—	—	0.30	1.39
Cotton	None	AL[3]	101	0	9.10	1067	1.40	3.13	0.31	0.44
Cotton	Winter wheat		101	0	3.47	260	0.56	0.88	0.16	0.20
Soybean	None	MO[4]	15	12	23.10	1439	3.36	—	0.46	—
Soybean	Common chickweed		15	12	13.25	233	0.77	—	0.17	—
Soybean	Canada bluegrass		15	12	14.20	93	0.88	—	0.43	—
Soybean	Downy brome		15	12	11.55	118	0.84	—	0.27	—
Wheat	None	NY[5]	241	64	17.3	—	1.14	—	0.32	—
Wheat	Ryegrass-alfalfa		241	64	7.4	—	0.93	—	0.17	—

[a]Reference for each study location is: 1, Angle et al., 1984; 2, Langdale et al., 1985; 3, Yoo et al., 1988; 4, Zhu et al., 1989; and 5, Klausner et al., 1974.

systems in the Southern Plains by Sharpley and Smith (1991). Cover crops reduced the loss of total P up to 90% and bioavailable P up to 80%. Bioavailable P is estimated by NaOH extraction of runoff (Sharpley et al., 1991a) and represents P in soluble and particulate forms that is potentially available for algae uptake. However, bioavailable P represented a greater portion of total P transported from systems including cover crops (Sharpley and Smith, 1991; Table 5). This may result from the P leaching from the cover crop or residue and preferential transport of clay-sized material, with a higher P content than coarser silt-sized material. Further, a difference in nutrient loss from alfalfa, cotton, and wheat watersheds at Chickasha, Oklahoma was also observed (Table 6). During the period summarized (1973 to 1974), no fertilizer was applied to any of the watersheds. Nitrate N and soluble P concentrations from alfalfa were greater than from cotton and wheat (Table 6). In fact, soluble P concentrations were greater than the critical value of 0.01 mg l^{-1}, above which Sawyer (1947) and, more recently, Vollenweider and Kerekes (1980) proposed that accelerated eutrophication would occur. Muir et al. (1973) also found a relationship between soluble P concentrations in major streams of Nebraska and legume growth statewide. It was suggested that soluble P concentrations in the Platte River, Nebraska may reflect P leached from alfalfa residues carried in runoff during months when the crop is dormant (Muir et al., 1973).

Clearly, cover crops can reduce soil and nutrient loss in runoff. However, the bioavailability of N and P transported in runoff may increase with cover crops as a function of soil and crop management. This emphasizes the need to consider these factors in developing flexible agricultural production systems that sustain soil fertility while reducing water quality impacts. In some production systems, the cover crop is killed prior to maturity to reduce water and light competition, with the subsequent crop either harvested or plowed into the soil. Research is needed to determine if N and P availability from the cover crop residue differs

Table 6. Mean Annual Soil Loss and Flow-Weighted N and P Concentration of Runoff from Alfalfa, Cotton, and Wheat Watersheds at Chickasha, OK during 1973 and 1974[a]

	Units	Alfalfa	Cotton	Wheat
Runoff	cm	14.00	11.07	18.00
Soil loss	kg ha^{-1} year^{-1}	300	3800	1900
Soluble P	mg l^{-1}	0.81	0.36	0.26
Total P	mg l^{-1}	1.77	2.68	1.59
Nitrate N	mg l^{-1}	1.57	0.73	0.80
Total N	mg l^{-1}	3.01	3.45	2.52

[a]No fertilizer N or P was applied to the watershed during the study period.

from that of soil N and P prior to uptake by the cover crop. Several studies have shown an increase in soil N and P availability following decomposition of incorporated crop residues (Sharpley and Smith, 1989; Smith and Sharpley, 1990). The extent to which the crop residue is increased will influence positional availability of N and P to subsequent crops and, thereby, sustainable soil fertility.

Although it is difficult to distinguish between losses of fertilizer and native soil nutrients without the use of expensive and hazardous radiotracers, in the case of ^{32}P, the loss of fertilizer in runoff is generally less than 5% of that applied (Ryden et al., 1973; Sharpley and Menzel, 1987). Such losses may not be of economic concern to a farmer even though they must be addressed in developing a sustainable fertility system that is environmentally sound. It should be recognized that an increase in bioavailability of nutrients transported in runoff as a result of conservation tillage may not bring about as great a reduction in the trophic state of a water body as may be expected from inspection of total loads only. Because such water quality impact studies are time and labor intensive, the use of computer models will be invaluable in developing and evaluating alternative management strategies that are sustainable. The use of such models will be discussed in the section Model Evaluation of Sustainable Soil Fertility Systems.

Fertilizer Management

Management of fertilizer P is generally more important than that of fertilizer N in reducing potential losses in surface runoff because nitrate N is mobile in the soil profile and algae growing in surface waters can fix N. Fertilizer P management should minimize stratification of nonmobile P in surface soil (0 to 5 cm), where interaction with runoff water via the physiochemical processes of soil particle detachment and P desorption control P loss (Sharpley, 1985b,c). This may be brought about by the careful placement of appropriate fertilizer amounts to optimize both positional and chemical availability in the soil.

In conservation tillage systems with broadcast fertilizer and manure application, it may be necessary to periodically plow no-till soils to redistribute surface soil accumulations of N and P (Cruse et al., 1983; Rehm, 1991; Triplett and Van Doren, 1969). Without the redistribution of stratified P, Rehm (1991) suggested that conservation tillage systems in the Midwest may need higher fertilizer P rates for optimum crop production compared to conventional tillage systems. Subsurface placement of fertilizer and manure during planting should also be encouraged on soils and management systems suited to this practice. Credit for on-farm nutrient sources and soil N and P forms that may become available during the growing season may also reduce fertilizer application rates.

Thus, soil test procedures should account for mineralizable organic N and P in soil and crop residue material and residual inorganic forms (Stewart and Sharpley, 1987; Sharpley et al., 1992).

Clearly, judicious management of fertilizer N and P can directly and indirectly reduce losses in runoff. Indirect benefits may be achieved through increased nutrient availability and thus, crop yields, which may reduce erosion by increased cover and aggregation of surface soil by shoot and roots, respectively.

Leaching Losses

In some sandy soils or where excessive rates of P have been applied via manure, sludge, or other fertilizer sources, P leaching to depths of several feet has been documented (Ellis and Olson, 1986; King et al., 1990). However, P fixation by amorphous Al and Fe oxides, Ca compounds, and organic matter of clay-sized material, generally limits P movement in soil to no more than a few inches. Consequently, the primary concern for most sustainable fertility systems is managing N to minimize leaching losses. In theory, applying fertilizer N to provide optimum nitrate N levels in the root zone during periods of rapid crop uptake will maximize N uptake and thereby minimize potential N leaching. In reality, climate, soil properties, and vegetation type will often limit achieving this goal.

Leaching of water through the soil profile can occur whenever precipitation exceeds potential evapotranspiration. The quantity and frequency of leaching is determined by the amount, type, and seasonal distribution of precipitation. Vegetation influences leaching losses by decreasing the amount of water and nutrients available to move through the soil profile. Soil structure and texture influence permeability (Cassel and Vasey, 1974) and available water holding capacity (Cassel and Nielson, 1986). These conditions have been used to group soils in hydrologic groups (Musgrave, 1955) and leaching classes (Kissel et al., 1982). This information was incorporated into climatic and leaching potentials by Smith and Cassel (1991) to develop a method to determine whether a N-leaching problem exists on a given farm. Use of such potentials will help identify soils and production systems susceptible to leaching, and facilitate implementing corrective measures to minimize such losses.

Management strategies to minimize N leaching loss will require a better synchronization between N availability and crop N requirements by considering the rate, type, timing, and placement of fertilizer applications. Currently, rates of fertilizer application are generally governed by yield goals, general recommendations for the area, and experience or tradition. Use of N soil tests may gradually increase, although limitations of that approach have already been discussed. Slow-release N fertilizers such as sulfur-coated urea and urea formaldehyde compounds have been evaluated under different climatic and management conditions, but their cost has restricted widespread use. Also, continued N release after crop harvest may increase leaching losses.

Timing fertilizer applications to coincide with crop uptake needs is essential and split applications can enhance crop N utilization. However, practical restrictions of crop height and economic limitations of split applications must also be considered in developing sustainable systems.

Use of green manure or cover crops to fix atmospheric N and scavenge excess available N in the soil profile have been studied extensively and implemented in various production systems (Frye and Blevins, 1989; Smith et al., 1987; Wagger and Mengel, 1988). Cover crops can reduce N leaching losses by up to 90% (Meisinger et al., 1991), but several questions as to cover crop type and fate of N in the residue remain unanswered. Under certain conditions, rye appears most suited to reducing N leaching (Meisinger et al., 1991; Smith et al., 1990), but it can also reduce the amount of water available for the primary crop in some areas (Campbell et al., 1984). Growing alfalfa has reduced subsoil nitrate N accumulation (Mathers et al., 1975) because of its potential for developing roots to depths in excess of 5 m.

Gaseous Losses

Gaseous N losses include ammonia N volatilization and denitrification. The ammonia N losses are more common with broadcast than incorporated N (Smith et al., 1990). Management strategies to minimize ammonia N volatilization from ammonia or urea-based fertilizers include cultivating or irrigating after broadcast application or by using subsurface placement. Foliar application and uptake of ammonia fertilizers by the crop canopy can reduce volatile losses (Denmead et al., 1976). However, the time of application for most efficient utilization may vary with crop type. Use of ammonium salts or urea N in fertigation systems should occur when wind speed and air temperature are at a minimum to limit volatilization losses. Irrigation water pH should also be considered when developing practices to minimize volatilization losses. More detailed fertilizer management information is beyond the scope of this chapter, but it can be obtained from Freney and Black (1988), Freney et al. (1983), and Terman (1979).

Volatile denitrification products, primarily N_2 and N_2O, are evolved whenever anaerobic sites develop within the soil profile, and when sufficient C, as supplied by soil organic matter, plant materials, and manure, is available. Other soil factors that influence denitrification are soil pH, moisture, and temperature. Management of these factors may help reduce the incidence of soil conditions favoring denitrification and thereby, minimize N losses. This is mainly achieved by managing to optimize soil aeration and drainage following N application and during periods of crop growth.

There remains uncertainty as to the seasonal and annual combination of denitrification to N losses because of experimental difficulties (Rolston et al., 1979; Ryden et al., 1979). Seasonal dynamics of gaseous fluxes are influenced by variations in rainfall and irrigation, plant growth, and substrate production via organic matter mineralization. These are further compounded by fertilizer application but even so, they are generally low. Gilliam et al. (1985) approximated annual N_2 loss from cropland to be about 15 kg N ha^{-1} year^{-1} and N_2O losses to range from 2- to 10-kg N ha^{-1} year^{-1}. Losses from denitrification are frequently estimated as 10 to 15% of fertilizer N applied (Fried et al., 1976). Appreciably higher losses can occur following manure addition, where increased rates of microbial respiration stimulated by the large additions of readily available C can induce localized anaerobic conditions. Summarizing several studies, King (1990) reported that up to 706 kg N ha^{-1} may be lost by denitrification following application of organic materials to soil. Because of the beneficial effects of organic matter addition on soil structure, these high losses may be minimized by synchronizing manure application and crop N uptake or use of a cover crop to conserve N for the subsequent main crop.

In addition to soil aeration, minimizing high soil water content following N application may reduce denitrification losses. In irrigated soils, gaseous N losses may also be high immediately following water application (Rolston et al., 1980). These losses may be minimized by careful timing of the irrigation cycles and fertilizer applications. For example, Rolston et al. (1982) measured denitrification loss (N_2 plus N_2O) of 4.1 kg N ha^{-1} over a 50-day period following a N fertilizer application of 300 kg N ha^{-1}, when irrigation water was applied 3 times a week to a Yolo loam soil under field conditions. The incorporation of barley straw into the top 10 cm of soil, 2 months prior to fertilization, increased denitrification losses to 14.9 kg N ha^{-1} (Rolston et al., 1982). When the same amount of water in each irrigation cycle was applied less frequently (once over 2 weeks) denitrification N losses were reduced 66 and 54% with and without barley straw residues, respectively, for Yolo loam. However, the less frequent irrigation increased nitrate N mobility in the soil and thereby, potential leaching losses (Rolston et al., 1982). A balance between denitrification and leaching losses should be considered in developing sustainable management criteria. Rolston et al. (1982), concluded from their study that increased denitrification with small frequent irrigations may be more than compensated for by less leaching and more plant uptake of N.

Few attempts have been made to develop a nutrient mass balance for several management systems, considering N and P inputs and losses in crop harvest, runoff leaching, and volatilization (King, 1990). This has resulted from analytical difficulties in measuring some avenues of loss and economic constraints in measuring all losses at a given location. Although such mass balances will be site specific, information on the relative magnitude of each avenue of loss will facilitate identifying management strategies that will target the most important losses. Because of large spatial and temporal variations, in addition to experimental difficulties, it is possible that models simulating soil nutrient cycling may be used to identify the relative sustainability of agricultural production systems.

STRATEGIES FOR IMPROVED MANAGEMENT

Achieving a sustainable soil fertility status requires an integrated soil and crop management approach. Use of realistic crop yield goals, development of "full-service" soil fertility and crop nutrition services, and use of simulation modeling are anthropogenic factors which may be very useful strategies for achieving this goal.

Realistic Yield Goals

Establishing realistic crop yield goals is essential for sustainable soil fertility because many fertilization programs, including N rates, are often based upon yield goals. If these goals are too high, nutrients such as N will probably be applied in excess of crop needs and result in reduced recovery, decreased profitability or return on fertilizer investment, and an increased potential for N loss to ground water resources. If yield goals are too low, crop nutrient needs may be underestimated leading to loss of yield, quality, and profit.

Crop yield goals must be both realistic and achievable, but this is not typical in many locations. Schepers et al. (1986) reported that in a Nebraska study conducted in 1980 through 1983, corn yield goals were overestimated by an average of 2.0 Mg ha^{-1} (32 bu/acre). This resulted in an average overapplication of N fertilizer of 52, 14, 30, and 43 kg ha^{-1}, respectively, for the 4 years. Padgitt (1985) reported that some producers in the Iowa Big Spring watershed were applying N in excess of rates required for the grain yields being achieved. For both states, it is doubtful that N application was the factor preventing the farmers from achieving their yield goals.

Vitosh and Jacobs (1990) stated that critical factors in selecting yield goals were a knowledge of soil productivity and realistic understanding of the crop management systems in use. They recommended that if the yield goal was not achieved at least 2 out of 5 years, or nearly 50% of the time, the entire soil and crop management system should be reassessed to identify the factors other than soil fertility that were limiting crop yields.

The establishment of overly optimistic yield goals is one factor for which the U.S. agricultural chemical industry is often criticized. Karlen and Zublena (1990) cited this as one mistake made by the industry in the maximum yield research (MYR) program initiated in 1980 by the Potash and Phosphate Institute (PPI) and the Foundation for Agronomic Research (FAR). Although reference was generally being made to small experimental areas or test fields, Wallingford (1989) suggested that yield goals should be set as much as 30% above previous high yield levels. For most farmers, this was probably unrealistic and would tend to encourage excessive fertilizer and other external inputs. The rationale for suggesting that higher yield goals were needed was that they would encourage farmers to give ample attention to all production practices. Wallingford (1989) also stated that in less variable fields, maximum economic yield (MEY) was often very close to previous maximum yields.

Soil survey maps published by the USDA-Soil Conservation Service and their cooperating state institutions provide a good basis for establishing realistic crop yield goals (Miller, 1987; Hudson, 1990). Use of soil map units as management units takes into consideration the nonanthropogenic factors that influence crop production and soil fertility status and then allows anthropogenic input to use differential fertilization and selection of optimum inputs according to the potential productivity of the soil (Luellen, 1985; Buchholz and Wollenhaupt, 1989; Karlen et al., 1990).

Full-Service Soil Fertility and Plant Nutrient Consulting

Site-Specific Nutrient Management Strategies

Full-service soil and plant nutrient consulting programs is a sustainable soil fertility management strategy that we anticipate will emerge during the 1990s because of the need and ability to use differential fertilization and crop management techniques based on potential soil productivity (Luellen, 1985; Buchholz and Wollenhaupt, 1989; Karlen et al., 1990). Developing technologies such as geographical information systems (GIS), which can transfer information collected for individual soil map units across spatial or temporal scales, is being developed and will facilitate implementation and transfer of these strategies. Bliss and Reybold (1989) have already used GIS techniques to study natural resource problems. Similarly, by using currently available data bases such as the Soil Survey Geographic Data Base (SSURGO), the State Soil Geographic Data Base (STATSGO), or the National Soil Geographic Data Base (NATSGO), interpretive maps can be made by overlaying soil data with other spatial resource data (Reybold and TeSelle, 1989). These services will be useful for both conventional and alternative farmers. For example, Karlen and Colvin (1991) found significant corn yield and profile N concentration differences among soil map units on adjacent fields managed for more than 20 years, using either conventional or alternative management practices. They suggested that use of differential fertilization may be environmentally and economically advantageous for both farming systems.

Contribution of Agricultural Industry

The U.S. fertilizer and agricultural chemical industry can assume a major role in developing sustainable soil fertility services. Key leadership roles which industry can provide include (1) identifying integrated management practices that can be optimized for both anthropogenic and nonanthropogenic factors at site-specific locations; (2) facilitating improved communication between the urban and agriculture communities; and (3) providing management services for nontraditional nutrient sources including municipal sewage sludge, food-processing, and biotechnology by-products.

Integrated Management Practices. A review of the MYR/MEY research/implementation programs by Karlen and Zublena (1990) showed that the real impact of those programs was improved site-specific recommendations and production guides for several crops.

The MYR program, which was defined as "the study of variables and their interactions in a multidisciplinary system that strives for the highest yield possible for the soil and climate at the research site" (Wagner, 1985), was conducted between 1980 and 1985 by more than 35 U.S. scientists at state and federal research locations. The participants were small groups of agricultural scientists who were interested in and capable of organizing interdisciplinary research teams so that all aspects of the crop production system could be evaluated. Researchers participating in MYR programs were asked to provide all management and production inputs at nonlimiting levels for the soil and environmental conditions at the study site.

Karlen and Zublena (1990) found that some researchers focused on one or two critical inputs, but most studies examined interactions between two or more production factors. These included plant populations, fertilizer rates and/or soil-test levels, cultivar selection, and pesticides. The multidisciplinary approach used in MYR studies showed that fully integrated "packages" of management practices were essential to raise current yield plateaus (Gravelle et al., 1988). This program provided a strong basis for full-service soil fertility and plant nutrient consulting because no single combination of factors could guarantee increased yields owing to soil-to-soil and environmental differences encountered from year to year and location to location.

After the MYR program had established potential yields for several crops and environments, adopting optimum practices for each location was encouraged through the MEY concept. As a result of MYR/MEY efforts, more than 50 technical publications and numerous extension fact sheets and guidelines have been developed. The MYR results demonstrated that it was important to consider not only the primary crop but all crops that may be grown as part of the rotation. A recent extension publication from Virginia (Alley et al., 1990) is an example of how MYR results have been used to develop on-farm implementation plans. Their guidelines for N fertilization are based on crop growth stage and N accumulation patterns. The authors chose this format based on their MYR results, recognizing that efficient N fertilization is crucial for both economic wheat production and protection of ground and surface water resources.

Integrator for Urban and Agricultural Communities. Full-service soil and plant nutrient consulting services will also improve communication between urban and agricultural communities concerning environmental quality. To facilitate this communication, scientists and full-service consultants must develop better educational programs using scientific facts. For example, use of animal manures rather than commercial fertilizer is often suggested as being more environmentally sound, but the amount of nutrients available from manure largely depends on how it is stored, handled, and how much is applied. Good manure management is essential for successful use of that material as a nutrient source (NRC, 1989). Furthermore, intensive animal feeding and handling operations that provide enough manure to replace substantial amounts of chemical fertilizers have also been identified as local point sources for nitrate contamination of ground water (Hallberg 1989a; Wallingford et al., 1975).

Outlet for Nontraditional Nutrient Sources. A full-service soil and plant nutrient management service could help address the manure management issue by providing a redistribution service for organic wastes, providing analytical services, and using the organic nutrient sources as an integral part of the overall fertilizer program. In addition to assisting with manure management, these services could also provide outlet for sewage sludge from community wastewater treatment facilities or biotechnology and food-processing by-products such as whey or molasses fermentation solubles (Watson et al., 1977; Henning and Krogmeier, 1990).

By developing full-service soil and plant nutrient management services such as these, sustainable soil fertility programs that are environmentally and economically achievable can be achieved in the U.S. and around the world.

MODEL EVALUATION OF SUSTAINABLE SOIL FERTILITY SYSTEMS

Obtaining reliable information on the long-term effects of agricultural production on sustainable soil fertility is a lengthy process which can be expensive, labor intensive, and site specific. Thus, there is an immediate need for development and application of models simulating the effect of agricultural production on the dynamics of processes controlling soil fertility. The use of computer simulation models in evaluating nutrient mass balances for sustainable soil fertility systems requires a comprehensive description of nutrient inputs, cycling, and losses.

Nutrient Inputs

The Classen and Barber (1976) model simulating soil nutrient flux to plant roots has been developed to assess the effect of fertilizer rate and placement on nutrient uptake by several crops (Anghioni and Barber, 1980; Kovar and Barber, 1987). A user-oriented model, "Decide", was developed by Bennett and Ozanne (1972) and Helyar and Godden (1976) to simulate residual fertilizer P efficiency and used to give objective advice on fertilizer use to farmers. The model constructs a fertilizer response curve for each farming situation by combining research information with the farmer's knowledge of his own farming conditions, in terms of past soil fertility and future yield goals. The complex factors controlling the input of nutrients through residue incorporation and organic matter decomposition and mineralization have been simulated by Jenkinson and Rayner (1977) and Van Veen and Paul (1981).

Nutrient Cycling

Much effort has been spent on simulating the complex processes involved in nutrient cycling and influence of soil, climate and plant variables. For example, models have been developed to simulate specific soil processes such as denitrification (Rolston et al., 1980; Smith, 1981), ammonia N volatilization (Parton et al., 1983), N cycling through soil microbial biomass (Jenkinson and Parry, 1989), N leaching (Addiscott, 1981), inorganic (Mansell et al., 1977) and organic P transformations (Mishra et al., 1979), P availability (Barrow and Carter, 1978; Cox et al., 1981; Russell, 1977), ion uptake (Bouldin, 1989), and liming and soil nutrient availability (Kirk and Nye, 1986; Nye and Ameloko, 1987).

More comprehensive models have been developed for N (Frissel and Van Veen, 1981; Van Keulen and Seligman, 1987; Wolf et al., 1989) and P cycling (Cole et al., 1977; Harrison, 1978). The N component of the CERES models, which simulate growth, phenology, and yield of several crops, can predict the effects of varying N fertilizer rate, timing, placement depth, and source on fertilizer use efficiency within a growing season (Godwin and Jones, 1991; Jones and Ritchie, 1991). However, few attempts have been made to formulate research- or management-oriented holistic models that can be used to evaluate the sustainability of agricultural production systems.

Research Oriented

Many of the concepts used in earlier P (Cole et al., 1977) and soil organic matter models (Parton et al., 1983), were incorporated into the "Century" model (Parton et al., 1987, 1988). The model simulates the dynamics of C, N, P, and S cycling in soil-plant systems using a monthly time step and incorporates the effects of moisture, temperature, soil properties, plant phenology, and organic matter decomposition on nutrient flows. Century has been used to simulate the impact of cultivation (100 years) on soil organic matter dynamics, nutrient mineralization, and plant production (Parton et al., 1988). Although its use is limited by availability of detailed soil and plant information, Century has revealed gaps in our knowledge of processes such as the contribution of organic matter cycling to sustainable soil fertility as a function of management (Parton et al., 1988). Thus, the model may provide valuable direction for future research.

Management Oriented

The simulation of soil C, N, and P cycling in the erosion-productivity impact calculator (EPIC) is described as an example of a management-oriented model (Williams et al., 1984; Sharpley and Williams, 1990). EPIC is composed of physically based components for simulating erosion, plant growth and related processes, and economic components for such assessments as the cost of erosion and determining optimum management strategies.

Soil organic matter and N transformations closely follow Seligman and Van Keulen's (1981) PAPRAN model. They are divided into fresh pools, consisting of decomposing crop residue and microbial biomass N, and stable pools, consisting of organic matter humus and humus N. For P, the model simulates uptake and transformations between several inorganic and organic pools in up to 10 soil layers of variable thickness (Jones et al., 1984). Fertilizer P is added to the labile inorganic P pool (i.e., plant-available P), which rapidly achieves equilibrium with active inorganic P. Movement of inorganic P from the active to the stable pools simulates slow adsorption of inorganic P. Crop P uptake from a soil layer is sensitive to crop P demand and the amounts of labile P, soil water, and roots in the layer. Stover and root P are added to the fresh organic P pool upon their death and/or incorporation into the soil. Fresh and stable organic matter decomposition may result in net labile P immobilization or net organic P mineralization.

Regression equations were developed to estimate labile P from soil test P, organic P from total N or organic C, and fertilizer availability index from soil chemical and taxonomic characteristics (Sharpley et al., 1984). Thus, except for mineral N and labile P, the minimum data set required to run the N and P models can be obtained from soil survey information. Processes described are sensitive to soil chemical and physical properties, crop nutrient requirements, tillage practice, fertilizer rate, soil temperature, and soil water content,

Some of the most extensive field data on the effects of long-term cultivation on soil organic C, N, and P cycling were obtained by Haas et al. (1957, 1961) for several Great Plains locations. These data were used to test EPIC predictions of climatic, soil, and crop rotation effects on nutrient cycling (Table 7). The decline in soil organic C, N, and P content was accurately predicted, with no significant difference (at 0.1 probability level) between measured and predicted values after the period of cultivation. The slight overestimation of organic C, N, and P may have resulted in part from the fact that soil erosion by both water and wind was kept minimal during the simulation. The slightly greater variability in labile P prediction may be a result of the dynamic nature of this pool, as all inorganic-organic P transformations and P inputs, pass through the labile P pool.

Accurately predicting the effect of long-term cultivation on soil nutrient cycling (Table 6) demonstrates one potential use of EPIC for evaluating relative effects of alternative management strategies on sustainability of soil fertility. For example, Meisinger et al. (1991) showed that EPIC adequately predicted the effect of cover crops on N availability and leaching in a range of soils. In addition, EPIC has been used to evaluate alternative soil conservation and crop tillage practices (Benson et al., 1988), furrow dike management and use (Kirshna et al., 1987; Williams et al., 1988), and alternative dryland crop rotations (Lacewell et al., 1988).

Nutrient Losses

Many comprehensive models have been developed to simulate the loss of nutrients in surface and ground water, with the purpose of aiding

Table 7. Measured and Simulated Changes in Surface Soil Organic P, Total N, Labile P, and Organic C in the Great Plains

Location	Duration of study (years)	Rotation	Organic C Virgin	Organic C Cultivated Meas.	Organic C Cultivated Sim.	Total N Virgin	Total N Cultivated Meas.	Total N Cultivated Sim.	Organic P Virgin	Organic P Cultivated Meas.	Organic P Cultivated Sim.	Labile P Virgin	Labile P Cultivated Meas.	Labile P Cultivated Sim.
Havre, MT	31	SWF[b]	17.5	8.3	12.4	1510	900	1135	157	102	108	11	13	15
Moccasin, MT	39	SWF	32.4	21.9	19.6	3000	2050	1787	308	183	169	14	14	20
Dickinson, ND	41	SWF	36.3	15.1	21.5	2930	1490	1957	292	148	174	10	12	14
Mandan, ND	31	SWF	24.2	17.0	12.8	1600	1160	1172	139	132	97	9	12	7
Sheridan, WY	30	SWF	16.6	11.9	12.6	1590	1210	1149	120	93	86	12	14	9
Laramie, WY	34	SWF	13.3	7.8	9.9	1220	820	900	142	91	96	13	24	9
Akron, CO	39	SWF	14.2	7.7	10.0	1340	800	911	115	82	81	26	45	19
Colby, KS	31	WWF[b]	18.3	10.1	10.4	1650	1050	952	158	61	92	34	30	27
Hays, KS	30	W. wheat	24.7	12.1	14.9	2200	1220	1360	174	97	108	11	40	8
Lawton, OK	28	W. wheat	17.3	8.0	9.9	1540	740	904	128	71	73	8	9	8
Dalhart, TX	29	Maize	7.2	4.4	4.9	670	420	444	84	39	53	17	13	9
Big Springs, TX	41	W. wheat	6.7	4.0	3.6	600	410	328	55	30	29	12	12	6
Mean	34		19.1	10.7	11.9	1654	1023	1083	156	94	97	15	20	13

[a]Labile P represents plant-available P.
[b]SWF and WWF represent spring wheat-fallow and winter wheat-fallow, respectively.
Measured data from Haas et al. (1957 and 1961).

selection of management practices capable of minimizing associated water quality problems.

Surface Runoff Models

Many agricultural nonpoint source models have been developed that simulate sediment and nutrient loss in surface runoff. For more detailed information on these models and approaches used, reviews by Leavesley et al. (1990) and Rose et al. (1990) are available. In general, these models describe the loss of N and P in soluble and particulate or sediment-bound forms. Soluble P is predicted by equilibrium extraction coefficients or kinetic approaches (Sharpley and Smith, 1989). The amount of nitrate N in surface soil and runoff are not closely related because of nitrate N mobility in the soil profile with infiltrating water (Sharpley et al., 1985); therefore accurate prediction of its loss in runoff is precluded. Particulate N and P loss is described by enrichment ratio approaches (Sharpley, 1985c). Accurate predictions of N and P loss in runoff as a function of agricultural management have been obtained (Sharpley et al., 1991b; Smith et al., 1991).

Ground Water Models

Modeling ground water and associated N movement lags behind that for runoff because of the complexity and heterogeneity of soil and geologic layers through which water passes enroute to an aquifer or water body. Detailed information on specific models and approaches used to simulate solute leaching is available in the reviews of Duffy et al. (1990), Vachaud et al. (1990), and Wagenet et al. (1990).

Accurately simulating N loss from agricultural production systems will remain a formidable task, because N ranges in valence from +5 to -3, is present in both inorganic and organic forms, and exists in solid, liquid, and gaseous phases. Furthermore, N loss is strongly influenced by temperature, water, microbial, and plant processes. Many of the above models have received limited testing because of these complexities and a lack of adequately detailed field data. However, for those models to be useful for developing sustainable soil fertility management practices, increased effort must be made to validate the predictions using field data from several different physiographic regions.

Model Limitations

Clearly, there are many models available for simulating input, cycling, and loss of plant nutrients. Many could be applied for planning alternative nutrient management strategies or identifying basic research needs.

However, in choosing a model suited to a certain problem or need, a user is often faced with the problem of selecting the most appropriate model to obtain the level of detail required. Once an appropriate model is chosen, the major limitation is then obtaining the necessary data on the various parameters to run the model. Lack of input data most frequently limits model use, because many models require detailed information on soil physical, chemical, and biological properties as well as crop and tillage operations. A user must recognize that model output or predictions will only be as reliable as the data input. Consequently, use of these models to provide quantitative estimates under specific conditions is limited. Therefore, models are recommended only for comparing relative effects of management strategies on sustainable soil fertility.

Some model use and application limitations may be overcome by developing and generating adaptable data bases to fill essential information gaps. Weather generators and soil data bases are being developed to provide needed inputs so that all a user needs to know is the soil name, location, and management practice to be imposed. Linkage of models simulating nutrient inputs, cycling, and losses may be necessary to evaluate the effect of agricultural management strategies on nutrient mass balances. This approach is being used to combine the soil productivity EPIC and water quality GLEAMS (Groundwater Loading Effects of Agricultural Management Systems) models.

As stated at the beginning of this section, models evaluating the effect of management strategies on sustainable soil fertility will require a comprehensive description and linkage of nutrient inputs, cycling, and losses for each system. However, in being comprehensive it is difficult to visualize these models being used or applied without understanding how they work.

THE RESEARCH CHALLENGES OF SUSTAINABLE SOIL FERTILITY

To develop sustainable soil fertility, a holistic and interdisciplinary research approach, considering both environmental and economic benefits and limitations, is needed to optimize nutrient transformations and transfer in soil-plant-water systems. The major challenges of this research will be to (1) increase inherent soil nutrient availability and cycling with the aim of replacing only the nutrients exported from the site; (2) increase fertilizer use efficiency for novel rotations and cropping systems; (3) minimize nutrient losses; (4) evaluate production or environmental compromises that may be necessary to increase sustainability.

Specific challenges or needs are to:

1. Identify soils, management systems, nutrient sources, and crop types most suitable for a given situation.
2. Develop soil specific management systems where use of fertilizer mixtures and other amendments may enhance initial and residual availability while reducing losses from systems.
3. Determine the feasibility of enhancing microfauna and microflora activity to increase soil nutrient availability for different management systems.
4. Develop management techniques to better synchronize nutrient applications and crop needs.
5. Develop viable rotation practices which efficiently utilize inherent and accumulated soil nutrients.
6. Implement cover crop management strategies to retain nutrients in desired locations within the agricultural system.
7. Utilize waste materials as alternative nutrient sources, including the possible use of cost-sharing programs.
8. Develop and apply soil test procedures to better account for plant nutrients supplied by organic and residual pools.
9. Improve fertilizer requirement prediction technology that it is more accessible to general agricultural systems and "farmer-friendly".
10. Link of soil fertility and water quality models to anticipate the long-term sustainability of alternate management strategies on water resources and soil fertility status.

REFERENCES

Abbott, J. L. 1978. Importance of the organic phosphorus fraction in extracts of calcareous soils. *Soil Sci. Soc. Am. J.* 42:81–85.

Addiscott, T. M. 1981. Leaching of nitrate in structured soils. In M. J. Frissel and J. A. Van Veen (Eds.), *Simulation of Nitrogen Behavior of Soil-Plant Systems*. PUDOC, Wageningen, Netherlands.

Adepetu, J. A. and R. B. Corey. 1976. Organic phosphorus as a predictor of plant available phosphorus in soils of Southern Nigeria. *Soil Sci.* 122:159–164.

Adepoju, A. Y., P. F. Pratt, and S. V. Mattigod. 1982. Availability and extractability of phosphorus from soils having high residual phosphorus. *Soil Sci. Soc. Am. J.* 46:583–588.

Agboola, A. A. and B. Oko. 1976. An attempt to evaluate plant available P in Western Nigerian soils under shifting cultivation. *Agron. J.* 68:798–801.

Alley, M. M., D. E. Brann, W. E. Baethgen, P. Scharf, G. W. Hawkins, and S. J. Donohue. 1990. Efficient N fertilization of winter wheat: principles and recommendations. Bull. No. 424–026. Virginia Cooperative Extension Service, Blacksburg, VA.

Alston, A. M. 1980. Response of wheat to deep placement of nitrogen and phosphorus fertilizers on a soil high in phosphorus in the surface layer. *Aust. J. Agric. Res.* 31:13–24.

Amundson, R. L. and F. E. Koehler. 1987. Utilization of DRIS for diagnosis of nutrient deficiencies in winter wheat. *Agron. J.* 79:472–476.

Anghinoni, I. and S. A. Barber. 1980. Predicting the most efficient phosphorus placement for corn. *Soil Sci. Soc. Am. J.* 44:1016–1020.

Angle, J. S., G. McClung, M. C. Mcintosh, P. M. Thomas, and D. C. Wolf. 1984. Nutrient losses in runoff from conventional and no-till corn watersheds. *J. Environ. Qual.* 13:431–435.

Arnold, R. W. 1983. *Pedogenesis and Soil Taxonomy*, Vol. 1, *Concepts and Interactions.* L. P. Wilding, N. E. Smeck, and G. F. Hall (Eds.). Elsevier Publishing Co., Amsterdam.

Bailey, L. D. and C. A. Grant. 1989. Fertilizer phosphorus placement studies on calcareous and noncalcareous chernozemic soils: growth, P-uptake and yield of flax. *Commun. Soil Sci. Plant Anal.* 20:635–654.

Barisas, S. G., J. L. Baker, H. P. Johnson, and J. M. Laflen. 1978. Effect of tillage systems on runoff losses of nutrients, a rainfall simulation study. *Trans. Am. Soc. Agric. Eng.* 21:893–897.

Barrow, N. J. and E. D. Carter. 1978. A modified model for evaluating residual phosphate in soil. *Aust. J. Agric. Res.* 29:1011–1021.

Beaufils, E. R. 1961. Les desequilibres chimiques chez l'*Hevea brasiliensis*. La methode dite du diagnostic physiologique. Ph. D. dissertation. Sorbonne, Paris.

Bennett, D. and P. G. Ozanne. 1972. Annual Report. CSIRO Division of Plant Industry, Australia. 45 pp.

Benson, V. W., H. C. Bogusch, Jr., and J. R. Williams. 1988. Evaluating alternative soil conservation and crop tillage practices with EPIC. In P. W. Unger, W. R. Jordan, T. V. Sneed, and R. W. Jensen (Eds.), *Challenges in Dryland Agriculture—A Global Perspective.* Texas Agricultural Experiment Station, College Station, TX. pp. 91–94.

Beverly, R. B., M. E. Sumner, W. S. Letzsch, and C. O. Planck. 1986. Foliar diagnosis of soybeans by DRIS. *Commun. Soil Sci. Plant Anal.* 17:237–256.

Binford, G. D., A. M. Blackmer, and N. M. El-Hout. 1990. Tissue test for excess nitrogen during corn production. *Agron. J.* 82:124–129.

Bjork, S. 1972. Swedish lake restoration program gets results. *Ambio.* 1:153–165.

Blackmer, A. M., D. Pottker, M. E. Cerrato, and J. Webb. 1989. Correlations between soil nitrate concentrations in late spring and corn yields in Iowa. *J. Prod. Agric.* 2:103–109.

Bliss, N. B. and W. U. Reybold. 1989. Small-scale digital soil maps for interpreting natural resources. *J. Soil Water Conserv.* 44:30–34.

Bouldin, D. R. 1989. A multiple ion uptake model. *J. Soil Sci.* 40:309–319.

Bowman, R. A. and C. V. Cole. 1978a. Transformations of organic phosphorus substrates in soils as evaluated by $NaHCO_3$ extraction. *Soil Sci.* 125:49–53.

Bowman, R. A. and C. V. Cole. 1978b. An exploratory method for fractionation of organic phosphorus from grassland soils. *Soil Sci.* 125:95–101.

Brams, E. 1973. Residual soil phosphorus under sustained cropping in the humid tropics. *Soil Sci. Soc. Am. Proc.* 37:579–583.

Bray, R. H. and L. T. Kurtz. 1945. Determination of total, organic, and available forms of phosphorus in soils. *Soil Sci.* 59:39–45.

Brinsfield, R., K. Staver, and W. Magette. 1988. The role of cover crops in reducing nitrate leaching to groundwater. In *Agricultural Impacts on Groundwater*. National Well Water Association, Dublin, OH.

Broadbent, F. E. and A. B. Carlton. 1978. Field trials with isotopically labeled nitrogen fertilizer. In D. R. Nielsen and J. G. MacDonald (Eds.) *Nitrogen in the Environment*, Vol. 1, *Nitrogen Behavior in Field Soil*. Academic Press, New York. pp. 1–41.

Brookes, P. C., D. S. Powlson, and D. S. Jenkinson. 1982. Measurement of microbial biomass phosphorus in soil. *Soil Biol. Biochem.* 14:319–329.

Brookes, P. C., D. S. Powlson, and D. S. Jenkinson. 1984. Phosphorus in the soil microbial biomass. *Soil Biol. Biochem.* 16:169–175.

Brown, J. R. (Ed.). 1987. Soil testing: *Sampling, Correlation, Calibration, and Interpretation*. Soil Science Society of America, Madison, WI.

Buchholz, D. D. and N. C. Wollenhaupt. 1989. Fertilizing soils, not fields in sub-humid regions. *Agron. Abstr.*, p. 314.

Bundy, L. G., K. A. Kelling, and L. Ward Good. 1990. Using legumes as a nitrogen source. *Univ. Wis. Coop. Extn. Serv. Bull. No. A3517*.

Campbell, R. B., D. C. Reicosky, and C. W. Doty. 1974. Physical properties and tillage of Paleudults in the southeastern Coastal Plains. *J. Soil Water Conserv.* 29:220–224.

Campbell, R. B., D. L. Karlen, and R. E. Sojka. 1984. Conservation tillage for maize production in the U.S. southeastern Coastal Plain. *Soil Tillage Res.* 4:511–529.

Carignan, R. and J. Kalff. 1980. Phosphorus sources for aquatic weeds: water or sediments? *Science* 207:987–989.

Carter, D. L. and R. D. Berg. 1990. Growing sweet corn after alfalfa to reduce nitrate leaching on furrow irrigated land. *Agron. Abstr.* p. 311.

Carter, D. L. and R. D. Berg. 1991. Crop sequences and conservation tillage to control irrigation furrow erosion and increase farmer income. *J. Soil Water Conserv.* 46:139–142.

Carter, D. L., R. D. Berg, and B. J. Sanders. 1991. Producing no-till cereal or corn following alfalfa on furrow irrigated land. *J. Prod. Agric.* 4:174–179.

Cassel, D. K. and D. R. Nielsen. 1986. Field capacity and available water capacity. In A Klute (Ed.), *Methods of Soil Analysis*. Part 1. 2nd ed. Agron. Monogr. 9 ASA and SSSA, Madison, WI. pp. 901–926.

Cassel, D. K. and E. H. Vasey. 1974. How fertilizer moves in soil. *N. Dak. State Univ. Agric. Ext. Bull. 21*, Fargo, ND.

Chapman, H. D. (Ed.). 1966. *Diagnostic Criteria for Plants and Soils*. Division of Agriculture, University of California, Berkeley.

Claassen, N. and S. A. Barber. 1976. Simulation model for nutrient uptake from soil by a growing plant root system. *Agron. J.* 68:961–964.

Cole, C. V., G. S. Innis, and J. W. B. Stewart. 1977. Simulation of phosphorus cycling in semiarid grasslands. *Ecology.* 58:1–15.

Cooke, G. W. 1984. The present use and efficiency of fertilisers and their future potential in agricultural production systems. In F. P. W. Winteringham (Ed.) *Environment and Chemicals in Agriculture*. Elsevier Applied Science Publishers, New York.

Cooperband, L. R. and T. J. Logan. 1991. Measuring in situ changes in labile soil P in a humid tropical silvopastoral system with anion exchange membrances. *Agron. Abstr.*, p. 241.

Cox, F. R., E. J. Kamprath, and R. E. McCollum. 1981. A descriptive model of soil test nutrient levels following fertilization. *Soil Sci. Soc. Am. J.* 45:529–532.

Cruse, R. M., G. A. Yakle, T. C. Colvin, D. R. Timmons, and A. C. Mussleman. 1983. Tillage effects on corn and soybean production in farmer-managed, university monitored field plots. *J. Soil Water Conserv.* 38:512–514.

Dahnke, W. C. and E. H. Vasey. 1973. Testing soils for nitrogen. In L. M. Walsh and J. D. Beaton (Eds.) *Soil Testing and Plant Analysis*. Soil Science Society of America, Madison, WI.

Daughtrey, Z. W., J. W. Gilliam, and E. J. Kamprath. 1973. Soil test parameters for assessing plant-available P of acid organic soils. *Soil Sci.* 115:438–446.

Denmead, O. T., J. R. Freney, and J. R. Simpson. 1976. A closed ammonia cycle within a plant canopy. *Soil Biol. Biochem.* 8:161–164.

Doran, J. W., D. L. Karlen, and R. L. Thompson. 1989. Seasonal variations in soil microbial biomass and available N with varying tillage and cover crop management. *Agron. Abstr.* p. 213.

Doran, J. W., L. N. Mielke, and J. F. Power. 1990. Microbial activity as regulated by soil water-filled pore space. Symp. Ecology of Soil Microorganisms in the Microhabitat Environment. In Trans. 14th Int. Congr. Soil Sci. III:94–99. Kyoto, Japan.

Duffy, C. J., C. Kincaid, and P. Huyakorn. 1990. A review of groundwater models for assessment and prediction of nonpoint-source pollution. In D. G. DeCoursey (Ed.), Proc. Int. Symp. Water Quality Modeling Agricultural Nonpoint Sources, Part I. USDA-ARS 81. U.S. Government Printing Office, Washington, DC. pp. 253–278.

Eghball, B., D. H. Sander, and J. Skopp. 1990. Diffusion, adsorption, and predicted longevity of banded phosphorus fertilizer in three soils. *Soil Sci. Soc. Am. J.* 54:1161–1165.

Ellis, B. G. and R. J. Olson. 1986. Economic, agronomic and environmental implications of fertilizer recommendations. North Central Regional Res. Pub. 310. Agricultural Experiment Station, Michigan State University, East Lansing, MI.

Escano, C. R., C. A. Jones, and G. Uehara. 1981. Nutrient diagnosis in corn grown on Hydric Dystrandepts. II. Comparison of two systems of tissue diagnosis. *Soil Sci. Soc. Am. J.* 45:1140–1144.

Fixen, P. E. and J. H. Grove. 1990. Testing soils for phosphorus. In R. L. Westerman (Ed.), *Soil Testing and Plant Analysis*, 3rd ed. SSSA Book Series, No. 3. Soil Science Society of America, Madison, WI. pp. 141–180.

Fixen, P. E. and A. E. Ludwick. 1982. Residual available phosphorus in near-neutral and alkaline soils. II. Persistence and quantitative estimation. *Soil Sci. Soc. Am. J.* 46:335–338.

Fox, R. H. and W. P. Piekielek. 1978. Field testing of several nitrogen availability indexes. *Soil Sci. Soc. Am. J.* 42:747–750.

Freney, J. R., J. R. Simpson, and O. T. Denmead. 1983. Volatilization of ammonia. In J. R. Freney and J. R. Simpson (Eds.), *Gaseous Loss of Nitrogen from Plant-Soil Systems.* Martinus Nijhoff/Dr. W. Junk Publishers. The Hague. pp. 1–32.

Freney, J. R. and A. S. Black. 1988. Importance of ammonia volatilization as a loss process. pp. 156–173. In J. R. Wilson (Ed.), *Advances in Nitrogen Cycling in Agricultural Ecosystems.* CAB International. Wallingford, U.K.

Fried, M., K. K. Tanji, and R. M. Van De Pol. 1976. Simplified long term concept for evaluating leaching of nitrogen from agricultural land. *J. Environ. Qual.* 5:197–200.

Frissel, M. J. and J. A. Van Veen (Eds.). 1981. Simulation of nitrogen behavior of soil-plant systems. Centre for Agricultural Publication and Documentation. Wageningen, Netherlands.

Frye, W. W., R. L. Blevins, M. S. Smith, S. J. Corak, and J. J. Varco. 1988. Role of annual legume cover crops in efficient use of water and nitrogen. In W. L. Hargrove (Ed.) *Cropping Strategies for Efficient Use of Water and Nitrogen.* ASA Spec. Publ. 51. ASA, CSSA, and SSSA, Madison, WI.

Frye, W. W. and R. L. Blevins. 1989. Economically sustainable crop production with legume cover crops and conservation tillage. *J. Soil Water Conserv.* 44:57–60.

Gilliam, J. W., T. J. Logan, and F. E. Broadbent. 1985. Fertilizer use in relation to the environment. In O. P. Englestad (Ed.), *Fertilizer Technology and Use,* 3rd ed. Soil Science Society of America, Madison, WI. pp. 561–588.

Godwin, D. C. and C. A. Jones. 1991. Nitrogen dynamics in soil-plant systems. In J. Hanks and J. T. Ritchie (Eds.), *Modeling Plant and Soil Systems.* Agron. Monogr. 31. American Society of Agronomy, Madison, WI.

Gravelle, W. D., M. M. Alley, D. E. Brann, and K. D. S. M. Joseph. 1988. Split spring nitrogen application effects on yield, lodging, and nutrient uptake of soft red winter wheat. *J. Prod. Agric.* 1:249–256.

Grecu, S. J., M. B. Kirkham, E. T. Kanemasu, D. W. Sweeney, L. R. Stone, and G. A. Milliken. 1988. Root growth in a claypan with a perennial-annual rotation. *Soil Sci. Soc. Am. J.* 52:488–494.

Haas, H. J., C. E. Evans, and E. F. Miller. 1957. Nitrogen and carbon changes in Great Plains soils as influenced by cropping and soil treatments. *U.S. Department of Agriculture Tech. Bull.,* 1164.

Haas, H. J., D. L. Grunes, and G. A. Reichman. 1961. Phosphorus changes in Great Plains soils as influenced by cropping and manure applications. *Soil Sci. Soc. Am. Proc.* 24:214–218.

Hallberg, G. R. 1989a. Nitrate in ground water in the United States. In R. F. Follett (Ed.) *Nitrogen Management and Groundwater Protection.* Elsevier Science Publications, Amsterdam.

Hallberg, G. R. 1989b. Pesticide pollution of groundwater in the humid United States. *Agriculture, Ecosystems and Environment.* 26:299–367.

Halvorson, A. D. and A. L. Black. 1985. Long-term dryland crop responses to residual phosphorus fertilizer. *Soil Sci. Soc. Am. J.* 49:928–933.

Halvorson, A. D., B. R. Bock, and J. J. Meisinger. 1989. Advances in N use efficiency. In G. H. Heichel, R. F. Follett, J. F. Power, and D. R. Keister (Eds.), Current Advances and Future Priorities in Nitrogen Research. USDA-ARS, Tech. Rep. Government Printing Office, Washington, DC.

Hanway, J. J. 1962. Corn growth and composition in relation to soil fertility. III. Percentages of N, P, and K in different plant parts in relation to stage of growth. *Agron. J.* 54:222–229.

Hanway, J. J. 1964. North Central Regional potassium studies. III. Field studies with corn. *Iowa State Univ., Res. Bull. 503.* Agric. and Home Ec. Exp. Stn., Ames, IA.

Harrison, A. F. 1978. Phosphorus cycles of forest and upland grassland systems and some effects of land management practices. In *Phosphorus in the Environment: Its Chemistry and Biochemistry.* CIBA Foundation Symp. 57, Amsterdam, Elsevier/North-Holland. pp. 175–195.

Harrison, A. F. 1987. *Soil Organic Phosphorus. A Review of World Literature.* CAB International. Wallingford, U.K.

Hauck, R. D., and J. M. Bremner. 1976. Use of tracer for soil and fertilizer nitrogen research. *Adv. Agron.* 28:219–266.

Havlin, J. L. 1991. Sampling and evaluating residual value of band applied phosphorus. pp. 1–9. In F. J. Sikora (Ed.), Future directions for agricultural phosphorus research. National Fertilization and Environmental Research Center, TVA, Muscle Shoals, AL.

Havlin, J. L., D. E. Kissel, L. D. Maddux, M. M. Claassen, and J. H. Long. 1990. Crop rotation and tillage effects on soil organic carbon and nitrogen. *Soil Sci. Soc. Am. J.* 54:448–452.

Hedley, M. J. and J. W. B. Stewart. 1982. Method to measure microbial phosphate in soils. *Soil Biol. Biochem.* 14:377–385.

Hedley, M. J., J. W. B. Stewart, and B. S. Chauhan. 1982. Changes in inorganic and organic soil phosphorus fraction induced by cultivation practices and by laboratory incubations. *Soil Sci. Soc. Am. J.* 46:970–976.

Helyar, K. R. and D. P. Godden. 1976. The phosphorus cycle—What are the sensitive areas? In G. J. Blair (Ed.), Prospects for Improving Efficiency of Phosphorus Utilization. Proc. of Symp. at Univ. of New England, Armidale, N.S.W., Australia *Reviews in Rural Sci.* p. 23–30.

Henning, S. J. and M. J. Krogmeier. 1990. Condensed molasses fermentation solubles as a nitrogen source for grain crops. *Agron. Abstr.*, p. 270.

Hofstetter, B. 1988. Cover crop guide. *New Farm.* 10:17–30.

Holford, I. C. R. 1989. Efficacy of different phosphate application methods in relation to phosphate sorptivity in soils. *Aust. J. Soil Res.* 27:123–133.

Hooker, M. L. 1976. Soil Sampling Intensities Required to Estimate Available N and P in Five Nebraska Soil Types. M.S. thesis, University of Nebraska, Lincoln (Cat. No. LD 3656 H665X 1976).

Hoyt, P. B. and R. M. Leitch. 1983. Effect of forage legume species on soil moisture, nitrogen and yield of succeeding barley crops. *Can. J. Soil Sci.* 62:125–136.

Hudson, B. D. 1990. Concepts of soil mapping and interpretation. *Soil Survey Horizons.* 31:61–82.

Hunt, P. G., T. A. Matheny, and A. G. Wollum II. 1985. *Rhizobium japonicum* nodular occupancy, nitrogen accumulation, and yield for determinate soybean under conservation and conventional tillage. *Agron. J.* 77:579–584.

Iversen, K. V., R. H. Fox, and W. P. Piekielek. 1985. The relationships of nitrate concentrations in young corn stalks to soil nitrogen availability and grain yields. *Agron. J.* 77:927–932.

Jackson, G., D. Keeney, D. Curwen, B. Webendorfer. 1987. Agricultural management practices to minimize groundwater contamination. Cooperative Extension Service, University of Wisconsin, Madison, 115 pp.

Jackobs, J. A. and D. M. Gossett. 1956. Seeding alfalfa over corn with a cultipacker type seeder. *Agron. J.* 48:194–195.

Jenkinson, D. S. 1968. Chemical tests for potentially available nitrogen in soil. *J. Sci. Food Agric.* 19:160–168.

Jenkinson, D. S. and J. H. Rayner. 1977. The turnover soil organic matter in some of the Rothamsted classical experiments. *Soil Sci.* 123:298–305.

Jenkinson, D. S. and L. C. Parry. 1989. The nitrogen cycle is the Broadbalk wheat experiment: a model for the turnover of nitrogen through the soil microbial biomass. *Soil Biol. Biochem.* 21:535–541.

Jenny, H. 1941. *Factors of Soil Formation, a System of Quantitative Pedology.* McGraw-Hill, New York.

Johnson, H. P., J. L. Baker, W. D. Shrader, and J. M. Laflen. 1979. Tillage system effects on sediment and nutrients in runoff from small watersheds. *Trans. Am. Soc. Agric. Eng.* 22:1110–1114.

Johnson, J. W., L. R. Welch, and L. T. Kurtz. 1975. Environmental implications of N fixation by soybeans. *J. Environ. Qual.* 4:303–306.

Jones, C. A., C. V. Cole, A. N. Sharpley, and J. R. Williams. 1984. A simplified soil and plant phosphorus model. I. Documentation. *Soil Sci. Soc. Am. J.* 48:800–805.

Jones, J. W. and J. T. Ritchie. 1991. Crop growth models. In G. J. Hoffman, T. A. Howell, and K. H. Solomon (Eds.), *Management of Farm Irrigation Systems.* American Society for Agricultural Engineering, St. Joseph, MI. pp. 63–89.

Karlen, D. L. 1990. Conservation tillage research needs. *J. Soil Water Conserv.* 45:365–369.

Karlen, D. L. and T. S. Colvin. 1992. Alternate soil and crop management effects on profile nitrogen concentrations on two Iowa farms. *Soil Sci. Soc. Am. J.* 56:1249–1256.

Karlen, D. L. and J. W. Doran. 1991. Cover crop management effects on soybean and corn growth and nitrogen dynamics in an on-farm study. *Am. J. Altern. Agric.* 6:71–82.

Karlen, D. L. and E. J. Sadler. 1989. Aerial dry matter accumulation rates for soybean at different yield levels. In R. D. Munson (Ed.) *Proc. Physiology, Biochemistry, Nutrition, and Bioengineering of Soybeans: Implications for Future Management.* Foundation for Agronomy Research and Potash and Phosphate Institute, Atlanta, GA.

Karlen, D. L. and E. J. Sadler. 1990. Nutrient accumulation rates for wheat in the southeastern Coastal Plains. *Commun. Soil Sci. Plant Anal.* 21:1329-1352.

Karlen, D. L. and J. P. Zublena. 1990. Maximum yield research (MYR) in the USA. Benefits and detriments. In Proc. Symp. Maximum Yield Research. 14th Intl. Congr. Soil Sci., August 12 to 18., Kytot, Japan. pp. 110-119.

Karlen, D. L., P. G. Hunt, and R. B. Campbell. 1984. Crop residue removal effects on corn yield and fertility of a Norfolk sandy loam. *Soil Sci. Soc. Am. J.* 48:868-872.

Karlen, D. L., M. J. Kasperbauer, and J. P. Zublena. 1987a. Row-spacing effects on corn in the southeastern U.S. *Appl. Agric. Res.* 2:65-73.

Karlen, D. L., R. L. Flannery, and E. J. Sadler. 1987b. Nutrient and dry matter accumulation rates for high yielding maize. *J. Plant Nutr.* 10:1409-1417.

Karlen, D. L., E. J. Sadler, and C. R. Camp. 1987c. Dry matter, nitrogen, phosphorus, and potassium accumulation rates by corn on Norfolk loamy sand. *Agron. J.* 79:649-656.

Karlen, D. L., R. L. Flannery, and E. J. Sadler. 1988. Aerial accumulation and partitioning of nutrients by corn. *Agron. J.* 80:232-242.

Karlen, D. L., W. R. Berti, P. G. Hunt, and T. A. Matheny. 1989. Soil-test values after eight years of tillage research on a Norfolk loamy sand. *Commun. Soil Sci. Plant Anal.* 20:1413-1426.

Karlen, D. L., E. J. Sadler, and W. J. Busscher. 1990. Crop yield variation associated with Coastal Plain soil map units. *Soil Sci. Soc. Am. J.* 54:859-865.

Karlen, D. L., E. C. Berry, T. S. Colvin, and R. S. Kanwar. 1991. Twelve-year tillage and crop rotation effects on yields and soil chemical properties in northeast Iowa. *Commun. Soil Sci. Plant Anal.* 22:1985-2003.

Keeney, D. R. 1982. Nitrogen-availability indices. In A. L. Page (Ed.) Methods of Soil Analysis, Part 2, 2nd ed. *Agronomy* 9:649-658.

Keeney, D. R. 1989. Sources of nitrate to groundwater. In R. F. Follett (Ed.) *Nitrogen Management and Groundwater Protection*. Elsevier Science Publishers, Amsterdam.

Keeney, D. R. and J. M. Bremner. 1966. Comparison and evaluation of laboratory methods of obtaining an index of soil nitrogen availability. *Agron. J.* 58:498-503.

Killorn, R. 1988. Interpretation of soil test results. Coop. Ext. Bull. Ser. Pm-1310. Iowa State University, Ames, IA. 14 pp.

King, L. D. 1990. Sustainable soil fertility practices. In C. A. Frances, C. B. Flora, and L. D. King (Eds.) *Sustainable Agriculture in Temperate Zones*. John Wiley & Sons, New York. pp. 144-177.

King, L. D., J. C. Burns, and P. W. Westerman. 1990. Long-term swine lagoon effluent applications on "Coastal" Bermudagrass. II. Effects on nutrient accumulation in soil. *J. Environ. Qual.* 19:756-760.

Kirk, G. J. D. and P. H. Nye. 1986. A simple model for predicting the rates of dissolution of sparingly soluble calcium phosphates in soil. I. The basic model. *J. Soil Sci.* 37:529-540.

Kirshna, H. H., G. F. Arkin, J. R. Williams, and J. R. Mulkey. 1987. Simulating furrowdike impacts on runoff and sorghum yields. *Trans. Am. Soc. Agric. Eng.* 20:1013–1017.

Kissel, D. E., O. W. Bidwell, and J. F. Kientz. 1982. Leaching classes of Kansas soils. Bull. 641. Kansas State University, Agricultural Experiment Station, Manhattan, KS.

Kitchen, N. R., J. L. Havlin, and D. G. Westfall. 1990. Soil sampling under no-till banded phosphorus. *Soil Sci. Soc. Am. J.* 54:1661–1665.

Klausner, S. D., P. J. Zwerman, and D. S. Ellis. 1974. Surface runoff losses of soluble nitrogen and phosphorus under two systems of soil management. *J. Environ. Qual.* 3:42–46.

Kovar, J. L. and S. A. Barber. 1987. Placing phosphorus and potassium for greatest recovery. *J. Fert. Issues* 4:1–6.

Lacewell, R. D., J. G. Lee, C. W. Wendt, R. J. Lascamo, and J. W. Keeling. 1988. Implications of alternative dryland crop rotation: Texas High Plains. In P. W. Unger, W. R. Jordan, T. V. Sneed, and R. W. Jensen (Eds.), *Challenges in Dryland Agriculture—A Global Perspective*. Texas Agricultural Experiment Station, College Station, TX. pp. 635–638.

Laflen, J. M., and M. A. Tabatabai. 1984. Nitrogen and phosphorus losses from corn-soybean rotations as affected by tillage practices. *Trans. Am. Soc. Agric. Eng.* 27:58–63.

Lammond, R. E. 1987. Comparison of fertilizer solution placement methods for grain sorghum under two tillage systems. *J. Fert. Issues* 4:43–47.

Langdale, G. W., R. A. Leonard, and A. W. Thomas. 1985. Conservation practice effects on phosphorus losses from Southern Piedmont watersheds. *J. Soil Water Conserv.* 40:157–160.

Leavesley, G. H., D. B. Beasely, H. B. Pionke, and R. A. Leonard. 1990. Modeling of agricultural nonpoint-source surface runoff and sediment yield—a review from the modeler's perspective. pp. 171–196. In D. G. DeCoursey (Ed.), Proc Int. Symp. Water Quality Modeling of Agricultural Non-point Sources, Part 1. USDA-ARS 81. U.S. Government Printing Office, Washington, DC.

Linn, D. M. and J. W. Doran. 1984a. Aerobic and anaerobic microbial populations in no-till and plowed soils. *Soil Sci. Soc. Am. J.* 48:794–799.

Linn, D. M. and J. W. Doran. 1984b. Effect of water-filled pore space on carbon dioxide and nitrous oxide production in tilled and nontilled soils. *Soil Sci. Soc. Am. J.* 48:1267–1272.

Luellen, W. R. 1985. Fine-tuned fertility: tomorrow's technology here today. *Crops Soils*. 37(3):18–22.

MacKay, D. C., J. M. Carefoot, and T. Entz. 1987. Evaluation of the DRIS for assessing the nutritional status of potato (*Solanum tuberosum* L.). *Commun. Soil Sci. Plant Anal.* 18:1331–1353.

Magdoff, F. R., D. Ross, and J. Amadon. 1984. A soil test for nitrogen availability to corn. *Soil Sci. Soc. Am. J.* 48:1301–1304.

Mansell, R. S., H. M. Selim, and J. G. A. Fiskell. 1977. Simulated transformations and transport of phosphorus in soil. *Soil Sci.* 124:102–109.

Martinez, J. and G. Guiraud. 1990. A lysimeter study of the effects of a ryegrass catch crop, during a winter wheat/maize rotation, on nitrate leaching and on the following crop. *J. Soil Sci.* 41:5–16.

Mathers, A. C., B. A. Stewart, and B. Blair. 1975. Nitrate removal from soil profiles by alfalfa. *J. Environ. Qual.* 4:403–405.

McClenahan, E. J. and R. Killorn. 1988. Relationship between basal corn stem nitrate N content at V6 growth stage and grain yield. *J. Prod. Agric.* 1:322–326.

McCracken, D. V., S. J. Corak, M. S. Smith, W. W. Frye, and R. L. Blevins. 1989. Residual effects of nitrogen fertilization and winter cover cropping on nitrogen availability. *Soil Sci. Soc. Am. J.* 53:1459–1464.

McDowell, L. L. and K. C. McGregor. 1984. Plant nutrient losses in runoff from conservation tillage corn. *Soil Tillage Res.* 4:79–91.

McLaughlin, M. J. and A. M. Alston. 1986. The relative contribution of plant residues and fertilizer to the phosphorus nutrition of wheat in a pasture/cereal rotation. *Aust. J. Soil Res.* 24:517–526.

McLaughlin, M. J., A. M. Alston, and J. K. Martin. 1988a. Phosphorus cycling in wheat-pasture rotations. II. The role of the microbial biomass in phosphorus cycling. *Aust. J. Soil Res.* 26:333–342.

McLaughlin, M. J., A. M. Alston, and J. K. Martin. 1988b. Phosphorus cycling in wheat-pasture rotations. III. Organic phosphorus turnover and phosphorus cycling. *Aust. J. Soil Res.* 26:343–353.

Mehlich, A. 1984. Mehlich 3 soil test extractant: a modification of Mehlich 2 extractant. *Commun. Soil Sci. Plant Anal.* 15:1409–1416.

Meisinger, J. J. 1984. Evaluating plant-available nitrogen in soil-crop systems. In R. D. Hauck (Ed.) *Nitrogen in Crop Production.* ASA, CSSA, and SSSA, Madison, WI. pp. 391–416.

Meisinger, J. J., W. L. Hargrove, R. Mikkelsen, J. R. Williams, V. W. Benson. 1991. Effect of cover crops on groundwater quality. In W. L. Hargrove (Ed.) *Cover Crops for Clean Water.* Soil and Water Conservation Society, Ankeny, IA. pp. 57–68.

Melsted, S. W. 1954. New concepts of management of Corn Belt soils. *Adv. Agron.* 6:121–142.

Melsted, S. W. and T. R. Peck. 1973. The principles of soil testing. In L. M. Walsh and J. D. Beaton (Eds.) *Soil Testing and Plant Analysis.* Soil Science Society of America, Madison, WI.

Menon, R. G., S. H. Chien, and L. L. Hammond. 1989a. Comparison of Bray I and P_i tests for evaluating plant-available phosphorus from soils treated with different partially acidulated phosphate rocks. *Plant Soil.* 114:211–216.

Menon, R. G., L. L. Hammond, and H. A. Sissingh. 1989b. Determination of plant-available phosphorus by the iron hydroxide-impregnated filter paper (P_i) soil test. *Soil Sci. Soc. Am. J.* 52:110–115.

Menon, R. G., S. H. Chien, L. L. Hammond, and J. Henoa. 1989c. Modified techniques for preparing paper strips for the new P_i soil test for phosphorus. *Fert. Res.* 19:85–91.

Menon, R. G., S. H. Chien, L. L. Hammond, and B. R. Arora. 1990. Sorption of phosphorus by the iron oxide-impregnated filter paper (P_i soil test) embedded in soils. *Plant Soil.* 126:287–294.

Miller, G. A. 1987. Establishing realistic yield goals. Coop. Ext. Ser. Bull. Pm-1268. Iowa State University, Ames, IA.

Mishra, B., P. K. Khanna, and B. Ulrich. 1979. A simulation model for organic phosphorus transformation in a forest soil ecosystem. *Ecol. Model.* 6:31-46.

Moody, J. E. 1961. Growing corn without tillage. *Soil Sci. Soc. Am. Proc.* 25:516-517.

Muir, J., J. S. Boyce, E. C. Seim, P. N. Mosher, E. J. Deibert, and R. A. Olson. 1976. Influence of crop management practices on nutrient movement below the root zone in Nebraska soils. *J. Environ. Qual.* 5:255-259.

Muir, J., E. C. Seim, and R. A. Olson. 1973. Principles and practices in plant analysis. In R. L. Westerman (Ed.) *Soil Testing and Plant Analysis.* 3rd edition. Soil Science Society of America, Madison, WI.

Munson, R. D. and W. L. Nelson. 1990. Principles and practices in plant analysis. In R. L. Westerman (Ed.) *Soil Testing and Plant Analysis.* 3rd edition. Soil Science Society of America, Madison, WI.

Musgrave, G. W. 1955. How much of the rain enters the soil? In Water—the Yearbook of Agriculture. U.S. Government Printing Office, Washington, DC. pp. 151-159.

National Research Council. 1989. *Alternative Agriculture.* John Pesek (Chm.) National Academy Press, Washington, DC.

Novias, R. and E. J. Kamprath. 1978. Phosphorus supplying capacities of previously heavily fertilized soils. *Soil Sci. Soc. Am. J.* 42:931-935.

Nye, P. H. and A. T. Ameloko. 1987. Predicting the rate of dissolution of lime in soil. *J. Soil Sci.* 38:641-649.

O'Halloran, I. P., R. G. Kachanoski, and J. W. B. Stewart. 1985. Spatial variability of soil phosphorus as influenced by soil texture and management. *Can. J. Soil Sci.* 65:475-487.

Olsen, S. R., C. V. Cole, F. S. Watanabe, and L. A. Dean. 1954. Estimation of available phosphorus in soils by extraction with sodium bicarbonate. USDA Circ. 939. U.S. Government Printing Office, Washington, DC.

Olson, R. J., R. F. Hensler, O. J. Attoe, S. A. Witzel, and L. A. Peterson. 1970. Fertilizer nitrogen and crop rotation in relation to movement of nitrate nitrogen through soil profiles. *Soil Sci. Soc. Am. Proc.* 34:448-452.

Padgitt, S. 1985. Farming operations and practices in Big Spring Basin. Iowa Coop. Ext. Ser. Rpt. CRD 229. Iowa State University, Ames, IA.

Parton, W. J., D. W. Anderson, C. V. Cole, and J. W. B. Stewart. 1983. Simulation of soil organic matter formations and mineralization in semiarid agroecosystems. In R. R. Lowrance, R. L. Todd, L. E. Asmussen, and R. A. Leonard (Eds.) *Nutrient Cycling in Agricultural Ecosystems.* Spec. Publ. 23, College of Agriculture Exp. Stn., University of Georgia, Athens, GA. pp. 533-550.

Parton, W. J., D. S. Schimel, C. V. Cole, and D. S. Ojima. 1987. Analysis of factors controlling soil organic matter levels in Great Plains grasslands. *Soil Sci. Soc. Am. J.* 51:1173-1179.

Parton, W. J., J. W. B. Stewart, and C. V. Cole. 1988. Dynamics of C, N, P and S in grassland soils: a model. *Biogeochemistry.* 5:109-131.

Peters, R. H. 1981. Phosphorus availability in Lake Memphremagog and its tributaries. *Limnol. Oceanogr.* 26:1150-1161.

Peterson, A. E. 1956. Get good seedings in corn without special equipment. *Crops Soils.* 9:16–18.

Peterson, G. A. and W. W. Frye. 1989. Fertilizer nitrogen management. In R. F. Follett (Ed.) *Nitrogen Management and Ground Water Protection. Developments in Agricultural and Managed-Forest Ecology Vol. 21.* Elsevier Press, Amsterdam. pp. 183–219.

Pierzynski, G. M., T. J. Logan, and S. J. Traina. 1990a. Phosphorus chemistry and mineralogy in excessively fertilized soils: solubility equilibria. *Soil Sci. Soc. Am. J.* 54:1589–1595.

Pierzynski, G. M., T. J. Logan, S. J. Traina, and J. M. Bingham. 1990b. Phosphorus chemistry and mineralogy in excessively fertilized soils: quantitative analysis of phosphorus-rich particles. *Soil Sci. Soc. Am. J.* 54:1576–1583.

Pierzynski, G. M., T. J. Logan, S. J. Traina, and J. M. Bingham. 1990c. Phosphorus chemistry and mineralogy in excessively fertilized soils: descriptions of phosphorus-rich particles. *Soil Sci. Soc. Am. J.* 54:1583–1589.

Pieters, A. J. and R. McKee. 1938. The use of cover and green-manure crops. pp. 431–444. In Soils and Men: the Yearbook of Agriculture. U.S. Government Printing Office, Washington, D.C.

Potash and Phosphate Institute. 1990. Plant Food Uptake. Folder. Atlanta, GA.

Power, J. F. 1987. *The Role of Legumes in Conservation Tillage Systems.* Soil Conservation Society of America, Ankeny, IA.

Power, J. F. and F. E. Broadbent. 1989. Proper accounting for N in cropping systems. In R. F. Follet (Ed.) *Nitrogen Management and Groundwater Protection. Developments in Agriculture and Managed-Forest Ecology* Vol. 21. Elsevier Press. Amsterdam. pp. 160–181.

Powers, W. L., G. W. Wallingford, L. S. Murphy, D. A. Whitney, H. L. Manges, and H. E. Jones. 1974. Guidelines for applying beef feedlot manure to fields. Kansas Coop. Ext. Serv. Cir. C-502, Kansas State University, Manhattan, KS.

Ram, D. N. and P. J. Zwerman. 1960. Influence of management systems and cover crops on soil physical conditions. *Agron. J.* 52:473–476.

Ram, D. N., M. T. Vittum, and P. J. Zwerman. 1960. An evaluation of certain winter cover crops for the control of splash erosion. *Agron. J.* 52:479–482.

Randall, G. W., K. L. Wells, and J. J. Hanway. 1985. Modern techniques in fertilizer application. In O. P. Engelstad (Ed.) *Fertilizer Technology and Use.* Soil Science Society of America, Madison, WI. pp. 521–560.

Rauschkolb, R. S., A. L. Brown, J. Quick, J. D. Prato, R. E. Pelton, and F. R. Kegel. 1974. Rapid tissue testing for evaluating nitrogen nutritional status of (1) corn and (2) sorghum. *Calif. Agric.* 6:10–14.

Read, D. W. L., E. D. Spratt, L. D. Bailey, F. G. Warder, and W. S. Ferguson. 1973. Residual value of phosphatic fertilizer on chernozemic soils. *Can. J. Soil Sci.* 53:389–398.

Rehm, G. 1991. Management of phosphate fertilizers in conservation tillage production systems: the Midwest. pp. 1–8. In F. J. Sikora (Ed.) Future directions for agricultural phosphorus research. National Fertilization and Environmental Research Center, TVA, Muscle Shoals, AL.

Reybold, W. U. and G. W. TeSelle. 1989. Soil geographic data bases. *J. Soil Water Conserv.* 44:28–29.

Rodale Research Institute. 1990. *The Thompson Farm On-Farm Research*. Rodale Institute, Emmaus, PA.

Rogers, T. H. and J. E. Giddens. 1957. Green manure and cover crops. In Soil, the 1957 Yearbook of Agriculture. U.S. Government Printing Office, Washington, DC. pp. 252–257.

Rolston, D. E., F. E. Broadbent, and D. A. Goldhamer. 1979. Field measurement of denitrification: II. Mass balance and sampling uncertainty. *Soil Sci. Soc. Am. J.* 43:703–708.

Rolston, D. E., A. N. Sharpley, D. W. Toy, D. L. Hoffman, and F. E. Broadbent. 1980. Denitrification as affected by irrigation frequency of a field soil. EPA-600/2-80-066, U.S. Environmental Protection Agency, Ada, OK.

Rolston, D. E., A. N. Sharpley, D. W. Toy, and F. E. Broadbent. 1982. Field measurement of denitrification. III. Rates during irrigation cycles. *Soil Sci. Soc. Am. J.* 46:289–296.

Rose, C. W., W. T. Dickenson, H. Ghadiri, and S. E. Jorgensen. 1990. Agricultural nonpoint-source runoff and sediment yield water quality (NPSWQ) models: modeler's perspective. In D. G. DeCoursey (Ed.), Proc. Int. Symp. Water Quality Modeling of Agricultural Non-point Sources, Part 1. USDA-ARS 91. U.S. Government Printing Office, Washington, DC. pp. 145–170.

Russell, E. W. 1973. *Soil Conditions and Plant Growth*, 10th ed. Longman, New York.

Russell, J. S. 1977. Evaluation of residual nutrient effects in soils. *Aust. J. Agric. Res.* 28:461–475.

Ryden, J. C., J. K. Syers, and R. F. Harris. 1973. Phosphorus in runoff and streams. *Adv. Agron.* 25:1–45.

Ryden, J. C., L. J. Lund, J. Letey, and D. D. Focht. 1979. Direct measurement of denitrification loss from soils: II. Development and application of field methods. *Soil Sci. Soc. Am. J.* 43:110–117.

Sander, D. H., E. S. Penas, and B. Eghball. 1990. Residual effects of various phosphorus application methods on winter wheat and grain sorghum. *Soil Sci. Soc. Am. J.* 54:1473–1478.

Sawyer, C. N. 1947. Fertilization of lakes by agricultural and urban drainage. *J. New Engl. Waterworks Assoc.* 61:109–127.

Schepers, J. S. and R. H. Fox. 1989. Estimation of N budgets for crops. pp. 221–246. In R. F. Follett (Ed.) *Nitrogen Management and Ground Water Protection. Developments in Agricultural and Managed-Forest Ecology.* Vol. 21. Elsevier Press, Amsterdam.

Schepers, J. S., K. D. Frank, and C. Bourg. 1986. Effect of yield goal and residual soil nitrogen considerations on nitrogen fertilizer recommendations for irrigated maize in Nebraska. *J. Fert. Issues.* 3:133–139.

Schepers, J. S., D. D. Francis, and M. T. Thompson. 1989. Simultaneous determination of total C, total N, and ^{15}N on soil and plant material. *Commun. Soil Sci. Plant Anal.* 20:949–959.

Schepers, J. S., D. D. Francis, Ferguson, R. B., and R. D. Lohry. 1990a. Comparison of early season stem nitrate and leaf total N concentrations across corn hybrids. *Commun. Soil Sci. Plant Anal.* 21:1381–1390.

Schepers, J. S., D. D. Francis, and C. Clausen. 1990b. Techniques to evaluate N status of corn. *Agron. Abstr.*, p. 280.

Schindler, D. W. 1977. Evolution of phosphorus limitation in lakes. *Science.* 195:260–262.

Schuman, G. E., R. G. Spomer, and R. F. Piest. 1973a. Phosphorus losses from four agricultural watersheds on Missouri Valley loess. *Soil Sci. Soc. Am. Proc.* 37:424–427.

Schuman, G. E., R. E. Burwell, R. F. Piest, and R. G. Spomer. 1973b. Nitrogen losses in surface runoff from agricultural watersheds on Missouri Valley loess. *J. Environ. Qual.* 2:299–302.

Seligman, N. G. and H. Van Keulen. 1981. PAPRAN: A simulation model of annual pasture production limited by rainfall and nitrogen. In *Simulation of Nitrogen Behavior of Soil-Plant Systems.* PUDOC, Wageningen, The Netherlands. pp. 192–221.

Shapiro, C. A. 1988. Soil sampling fields with a history of fertilizer bands. In *Soil Science News-Nebraska Cooperative Extension Service.* Vol. 10, No. 5.

Sharpley, A. N. 1985a. Phosphorus cycling in unfertilized and fertilized agricultural soils. *Soil Sci. Soc. Am. J.* 49:905–911.

Sharpley, A. N. 1985b. Depth of surface soil-runoff interactions as affected by rainfall, soil slope and management. *Soil Sci. Soc. Am. J.* 49:1010–1015.

Sharpley, A. N. 1985c. The selective erosion of plant nutrients in runoff. *Soil Sci. Soc. Am. J.* 49:1527–1534.

Sharpley, A. N. 1986. Disposition of fertilizer phosphorus applied to winter wheat. *Soil Sci. Soc. Am. J.* 50:953–958.

Sharpley, A. N. 1991. Soil phosphorus extracted by iron-aluminum oxide-impregnated filter paper. *Soil Sci. Soc. Am. J.* 55:1038–1041.

Sharpley, A. N. and R. G. Menzel. 1987. The impact of soil and fertilizer phosphorus on the environment. *Adv. Agron.* 41:297–324.

Sharpley, A. N. and S. J. Smith. 1983. Distribution of phosphorus forms in virgin and cultivated soil and potential erosion losses. *Soil Sci. Soc. Am. J.* 47:581–586.

Sharpley, A. N. and S. J. Smith. 1989. Prediction of soluble phosphorus transport in agricultural runoff. *J. Environ. Qual.* 18:313–316.

Sharpley, A. N. and S. J. Smith. 1991. Effects of cover crops on surface water quality. In W. L. Hargrove (Ed.) *Cover Crops for Clean Water.* Soil Water Conservation Society, Ankeny, IA. pp. 41–49.

Sharpley, A. N. and J. R. Williams (Eds.). 1990. EPIC-Erosion/Productivity Impact Calculator. I. Model documentation. U.S. Department of Agriculture Tech. Bull. 1768.

Sharpley, A. N., C. A. Jones, C. Gray, and C. V. Cole. 1984. A simplified soil and plant phosphorus model. II. Prediction of labile, organic, and sorbed phosphorus. *Soil Sci. Soc. Am. J.* 48:805–809.

Sharpley, A. N., S. J. Smith, W. A. Berg, and J. R. Williams. 1985. Nutrient runoff losses as predicted by annual and monthly soil sampling. *J. Environ. Qual.* 14:354–360.

Sharpley, A. N., S. J. Smith, and J. W. Naney. 1987. The environmental impact of agricultural nitrogen and phosphorus use. *J. Agric. Food Chem.* 36:812–817.

Sharpley, A. N., W. W. Troeger, and S. J. Smith. 1991a. The measurement of bioavailable phosphorus in agricultural runoff. *J. Environ. Qual.* 20:235-238.

Sharpley, A. N., J. J. Meisinger, J. Power, and D. L. Suarez. 1992. Root extraction of nutrients associated with long-term soil management. In J. L. Hatfield and B. A. Stewart (Eds.) *Advances in Soil Science.* Vol. 19. Springer-Verlag, New York. pp. 151-217.

Sharpley, A. N., S. J. Smith, J. R. Williams, O. R. Jones, and G. A. Coleman. 1991b. Water quality impacts associated with sorghum culture in the Southern Plains. *J. Environ. Qual.* 20:239-244.

Sibbesen, E. 1978. An investigation of the anion-exchange resin method for soil phosphate extraction. *Plant Soil* 50:305-321.

Sibbesen, E. 1983. Phosphate soil tests and their suitability to assess the phosphate status of soil. *J. Sci. Food Agric.* 34:1368-1374.

Smith, K. A. 1981. A model of denitrification in aggregated soils. In *Simulation of Nitrogen Behavior of Soil-Plant Systems.* Centre for Agricultural Publication and Documentation. Wageningen, The Netherlands, pp. 259-266.

Smith, M. S., W. W. Frye, and J. J. Varco. 1987. Legume winter cover crops. *Adv. Soil Sci.* 7:95-139.

Smith, S. J. and L. B. Young. 1975. Distribution of nitrogen forms in virgin and cultivated soils. *Soil Sci.* 120:354-360.

Smith, S. J. and A. N. Sharpley. 1990. Soil nitrogen mineralization in the presence of surface and incorporated crop residues. *Agron. J.* 82:112-116.

Smith, S. J. and D. K. Cassel. 1991. Estimating nitrate leaching in soil materials. In R. F. Follett, D. R. Keeney, and R. Cruse (Eds.), Managing nitrogen for ground water quality and farm profitability. *Soil Sci. Soc. Am. Spec. Pub.,* Madison, WI. pp. 165-168.

Smith, S. J., J. S. Schepers, and L. K. Porter. 1990. Assessing and managing agricultural nitrogen losses to the environment. *Adv. Soil Sci.* 14:1-43.

Smith, S. J., A. N. Sharpley, J. W. Naney, W. A. Berg, and O. R. Jones. 1991. Water quality impacts associated with wheat culture in the Southern Plains. *J. Environ. Qual.* 20:244-249.

Soil Science Society of America (SSSA). 1987. *Glossary of Soil Science Terms.* Soil Science Society of America. Madison, WI.

Spratt, E. D., F. G. Warder, L. D. Bailey, and D. W. L. Read. 1980. Measurement of fertilizer phosphorus residues and its utilization. *Soil Sci. Soc. Am. J.* 44:1200-1204.

Stanford, G. 1982. Assessment of soil nitrogen availability. In F. J. Stevenson (Ed.) Nitrogen in Agricultural Soils. *Agronomy.* 22:651-688.

Staver, K. W. and R. B. Brinsfield. 1990. Patterns of soil nitrate availability in corn production systems: Implications for reducing groundwater contamination. *J. Soil Water Conserv.* 45:318-323.

Stewart, B. A., F. G. Viets, and G. L. Hutchinson. 1968. Agriculture's effect on nitrate pollution of groundwater. *J. Soil Water Conserv.* 23:13-15.

Stewart, J. W. B. and A. N. Sharpley. 1987. Controls on dynamics of soil and fertilizer phosphorus and sulfur. In R. F. Follett, J. W. B. Stewart, and C. V. Cole (Eds.) Soil fertility and organic matter as critical components of production systems. *Soil Sci. Soc. Am. Spec. Publ. 19,* Madison, WI. pp. 101-121.

Stewart, J. W. B. and H. Tiessen. 1987. Dynamics of soil organic phosphorus. *Biogeochemistry* 4:41–60.

Sumner, M. E. 1977. Application of Beaufils diagnostic indices to maize data published in the literature irrespective of age and conditions. *Plant Soil.* 46:359–369.

Terman, G. L. 1979. Volatilization of losses of nitrogen as ammonia from surface-applied fertilizers, organic amendments, and crop residues. *Adv. Agron.* 31:189–223.

Tiessen, H., J. W. B. Stewart, and J. O. Muir. 1983. Changes in organic and inorganic P composition of two grassland soils and their particle size fractions during 60–90 years of cultivation. *J. Soil Sci.* 34:815–823.

Triplett, G. B., Jr. 1966. Herbicide systems for no-tillage corn (*Zea mays* L.) following sod. *Agron. J.* 58:157–159.

Triplett, G. B., Jr. and D. M. Van Doren. 1969. Nitrogen phosphorus and potassium fertilization of non-tilled maize. *Agron. J.* 61:637–639.

U.S. Environmental Protection Agency. 1984. Report to Congress: Nonpoint Source Pollution in the U.S. U.S. Government Printing Office, Washington, DC.

Vachaud, G., M. Vauclin, and T. M. Addiscott. 1990. Solute transport in the vadose zone: A review of models. In D. G. DeCoursey (Ed.), Proc. Int. Symp. Water Quality Modeling of Agricultural Non-point Sources, Part 1. USDA-ARS 81. U.S. Government Printing Office, Washington, DC. pp. 81–104.

Van Veen, J. A. and E. A. Paul. 1981. Organic carbon dynamics in grassland soil. I. Background information and computer simulation. *Can. J. Soil Sci.* 61:185–201.

Van Keulen, H. and N. G. Seligman. 1987. *Simulation of water use, nitrogen nutrition and growth of a spring wheat crop.* 310 pp. Simulation Monographs, PUDOC, Wageningen, Netherlands.

Vitosh, M. L. and L. W. Jacobs. 1990. Nutrient management to protect water quality. Ext. Bull. WQ25. Cooperative Extension Service, Michigan State University, E. Lansing, MI.

Vollenweider, R. A. and J. Kerekes. 1980. The loading concept as a basis for controlling eutrophication: Philosophy and preliminary results of the OECD program on eutrophication. *Progr. Water Technol.* 12:5–38.

Wagenet, R. J., M. J. Shaffer, and R. E. Green. 1990. Predictive approaches for leaching in the unsaturated zone. In D. G. DeCoursey (Ed.), Proc Int. Symp. Water Quality Modeling of Agricultural Non-pont Sources, Part 1. USDA-ARS 81. U.S. Government Printing Office, Washington, DC. pp. 63–80.

Wagger, M. G. and D. B. Mengel. 1988. The role of nonleguminous cover crops in the efficient use of water and nitrogen. In Am. Soc. Agron. Spec. Publ. 51, Madison, WI. pp. 115–127.

Wagner, R. E. 1985. Maximum yield research — New horizons in agronomy. *Better Crops.* Vol. 69 (winter) Potash and Phosphate Institute, Atlanta, GA. pp. 3–5.

Wallingford, G. W. 1989. Setting realistic and challenging yield goals. *Better Crops.* Vol. 73 (spring) Potash and Phosphate Institute, Atlanta, GA. pp. 3–5.

Wallingford, G. W., L. S. Murphy, W. L. Powers, and H. L. Manges. 1975. Disposal of beef-feedlot manure: effects of residual and yearly applications on corn and soil chemical properties. *J. Environ. Qual.* 4:526–531.

Walton, C. P. and G. F. Lee. 1972. A biological evaluation of the molybdenum blue method for orthophosphate analysis. *Tech. Int. Ver. Limnol.* 18:676–684.

Walworth, J. L., M. E. Sumner, R. A. Isaac, and C. O. Plank. 1986. Preliminary DRIS norms for alfalfa in the southeastern United States and a comparison with midwestern norms. *Agron. J.* 78:1046–1052.

Ward, R. and D. F. Leikam. 1986. Soil sampling techniques for reduced tillage and band fertilizer application. Great Plains Soil Fertility Workshop. March 4 to 5, 1986, Denver, CO.

Watson, R. A., A. E. Peterson, and R. D. Powell. 1977. Benefits of spreading whey on agricultural land. *J. Water Pollut. Control Fed.* 49:24–35.

Welch, L. F., D. L. Mulvaney, L. V. Boone, G. E. McKibben, and J. W. Pendleton. 1966. Relative efficiency of broadcast versus banded phosphorus for corn. *Agron. J.* 58:283–287.

Westermann, D. T. 1992. Lime effects on phosphorus availability in a calcareous soil. *Soil Sci. Soc. Am. J.* 56:489–494.

Wikner, I. 1990. Crop management research and groundwater quality. In Proc. Best Management Practice to Maintain Groundwater Quality. Coop. Ext. Serv., Iowa State University, Ames, IA and Pioneer Hi-Bred Intl., Inc., Johnston, IA.

Williams, J. R., C. A. Jones, and P. T. Dyke. 1984. A modeling approach to determining the relationship between erosion and soil productivity. *Trans. Amer. Soc. Agric. Eng.* 27:129–144.

Williams, J. R., G. L. Wistrand, V. W. Benson, and J. H. Kirshna. 1988. A model for simulating furrow dike management and use. In P. W. Unger, W. R. Jordan, T. V. Sneed, and R. W. Jensen (Eds.) *Challenges in Dryland Agriculture—A Global Perspective.* Texas Agricultural Experiment Station, College Station, TX. pp. 255–257.

Wolf, J., C. T. De Wit, and H. Van Keulen. 1989. Modeling long-term crop response to fertilizer and soil nitrogen. I. Model description and application. *Plant Soil.* 120:11–22.

Yang, J. E., E. O. Skogley, S. J. Georgitis, B. E. Schaff, and A. H. Ferguson, 1991. Phytoavailability soil test: development and verification of theory. *Soil Sci. Soc. Am. J.* 55:1358–1365.

Yerokum, O. A. and D. R. Christenson. 1990. Relating high soil test phosphorus concentrations to plant phosphorus uptake. *Soil Sci. Soc. Am. J.* 54:796–799.

Yoo, K. H., J. T. Touchton, and R. H. Walker. 1988. Runoff, sediment and nutrient losses from various tillage systems of cotton. *Soil Tillage Res.* 12:13–24.

Yost, R. S., E. J. Kamprath, G. C. Naderman, and E. Lobato. 1981. Residual effects of phosphorus applications on a high phosphorus absorbing Oxisol of Central Brazil. *Soil Sci. Soc. Am. J.* 45:540–543.

Zhu, J. C., C. J. Gantzer, S. H. Anderson, E. E. Alberts, and P. R. Beuselinck. 1989. Runoff, soil, and dissolved nutrient losses from no-till soybean with winter cover crops. *Soil Sci. Am. J.* 53:1210–1214.

4

Soil Management

C. A. Robinson, R. M. Cruse, and K. A. Kohler

TABLE OF CONTENTS

INTRODUCTION

In the formation of a new conventional agriculture, it will be necessary to scrutinize current management systems to determine those practices that sustain the soil resource and those that degrade it. The soil resource will need to be managed and maintained for the future of agriculture. Good stewardship of natural resources will ensure future generations the ability to provide for their needs and those of their progeny.

Stewardship must be an ongoing practice. The concept of individual ownership with no outside accountability is a somewhat new philosophical stance. Several ancient cultures had ideas about land "ownership" and use that differed from those widely held today. The ancient Hebrews believed that Jehovah had made and therefore "owned" everything, according to what might be called "Creator's rights." The Hebrews were to manage the land as stewards, caring for it and protecting it for future generations. Although the Mosaic law mandated a 1-year rest from crop production for the land out of every 7, faith waned, greed did not, and the system broke down. Some North American Indians were guided by the concept that they were one with the land. No one would willingly harm or exploit something of which they were a part. Imperialism was the unfortunate end of this widespread philosophy. The philosophical battle today includes the search for an agricultural system that maintains, rather than exploits, the land and that is economically viable.

Soil management involves the manipulation of the soil resource to obtain the optimal output per given level of input. Philosophical issues often have been thrust aside in the search for the greatest short-term economic reward. Agriculturalists have an opportunity to adopt sustainable practices. If they do not, the public, through governmental regulations, will eventually force the adoption of such practices, because the concern for water quality, energy conservation, and other aspects of resource management continues to rise.

Though its role in soil tilth is only somewhat understood, soil organic matter (SOM) may contribute more to soil productivity than any single soil component. Biological, physical, and chemical properties are affected by the stability of SOM. This chapter will address historical SOM levels and practices necessary for SOM level maintenance.

The soil furnishes skeletal support for plants. There is growing concern regarding the management of this "skeleton" for agricultural sustainability. Soil structure is an important skeletal component because it affects availability and transport of water, air, and nutrients and because it influences soil temperature and strength. Soil structure is economically important because it affects erosion and energy consumption during tillage operations. Therefore, this chapter also will address characterization, development, natural amelioration, and stability of soil structure.

Soil management encompasses the diverse areas of cropping and tillage systems, fertilization, conservation, and pest management, and how these practices influence the soil and its properties. Soil organic matter and soil structure conditions will be used to evaluate the sustainability of selected management systems, particularly those involving tillage.

SOIL ORGANIC MATTER CHARACTERIZATION

Many reactions occurring in the soil are affected by diverse SOM components. Kononova (1966) suggested that the responsible SOM substances be divided into two major groups. Primary products of normal plant decomposition, such as lignin and cellulose, along with animal residues composed of common organic substrates, comprise the first of these groups. Included in this broad category are resynthesized components of proteins, carbohydrates, organic acids, fats, resins, and other nitrogenous and nonnitrogenous substances. Schnitzer and Khan (1978) suggested that carbohydrates comprise 50 to 70% of most plant tissue dry weight and hence are the most abundant plant residue materials added to the soil. Carbohydrates become constituents of soil microbes, performing many roles in their development. Because these rather simple, low molecular weight compounds are used primarily as substrates for soil microbes, their presence is transient and depends greatly upon infusion of fresh organic sources (Vaughan and Malcolm, 1985).

The greater portion of SOM, perhaps 85 to 90% of it, is composed of the humic substances (Kononova, 1966). This somewhat generic term refers to a host of microbial and chemical degradation products. Vaughan and Malcolm (1985) described problems identifying specific components of the humic substances as a result of variations in extraction techniques. The common terms "humic acid," "fulvic acid," and "humin" suggest sequential extractions with acids and alkalis. A well-outlined evolution of extraction history and the resulting constituent nomenclature can be found in the literature (Kononova, 1966; Vaughan and Malcolm, 1985). Research continues on the precise synthesis of these substances. Evidently, these components of SOM are structurally similar long-chain polymers with differences in oxygen, carbon, and nitrogen members of the functional groups. Humic substances are stable, displaying great resistance to further breakdown relative to nonhumic substances.

Quantity Changes

Concern over the maintenance of SOM has been cyclical. In early agriculture, SOM management may have been incidental, for fertility (and subsequent productivity) directly responded to organic matter additions. Recent use of high energy inputs, i.e., synthetic fertilizers, contributed to good yields, thus disguising the potential long-term importance of the SOM fraction. At present, SOM maintenance or increase is emphasized as a sustainable practice favoring fertility, erosion control, and soil tilth.

Man's influence on SOM levels undoubtedly began as agriculture flourished. Virgin state SOM levels differed as a result of local factors such as indigenous plant species, climate, animal influence, soil parent material, and other site-specific considerations. The development and widespread implementation of modern day agriculture has resulted in destruction of the SOM virgin state quantities, with soil levels decreasing at various rates toward new equilibriums. Production system aspects affecting SOM dynamics include cropping system, tillage, residue management, and fertilization practices. Unger (1968) concurred, noting that SOM declines could be caused by factors such as crop species, crop rotations, disposal of residues, climate, soil characteristics, fertilizer applications, and tillage practices. The effect on SOM of altering these factors is the topic of much research.

To monitor the status of SOM levels, some authors suggest formation of national data bases allowing evaluation of changes caused by various management schemes. Follett et al. (1987) suggest that because SOM and fertility status are routinely assessed during soil surveys, a national effort to synthesize organic matter and fertility status by soil type, soil association, and major land resource area would allow evaluation of subtle changes in these parameters. This approach may help soil scientists develop economic assessments of SOM and quantify its contributions to crop productivity (Stewart et al., 1987).

SOIL STRUCTURE

Soil structure has been defined in terms of the solid matrix or the pore geometry. In terms of the solid matrix, structure is the arrangement of primary soil particles into aggregates (Donahue et al., 1983). Small aggregates may come together to form larger ones. Pore geometry, which is influenced by such structural units, has been defined in terms of the pore size distribution (Hillel, 1982) and the pore continuity (Gupta et al., 1989).

Two very different approaches have been used to characterize soil structure. Subjective field observations, based on appearance and feel of the soil, were probably the first characterizations of structure. Morphologists have refined this approach, and their classifications appear in soil series descriptions. Objective systems have also been used extensively to describe soil structure. Engineering approaches use strength as an indication of structure: tensile strength, shear strength, penetration resistance, and mechanical stability (Spangler and Handy, 1982). Strength is related to mechanical resistance, which is the ability of the soil body to resist deformation and failure when subjected to an external force. Agronomic approaches tend to emphasize bulk density, porosity, aggregate and pore size distributions, aggregate stability, and soil water retention characteristics (Hillel, 1982). Soil structure strongly influences tilth. Most definitions and discussions of tilth have been directly or indirectly related to soil structure (Hillel, 1982; Karlen et al., 1990). Karlen et al. (1990) propose a broader definition of soil tilth, but in such terms that certain aspects can be quantified. Tilth is a multifaceted feature of soil that will continue to make quantitative determinations difficult. Some workers have proposed the use of a few quick, simple field determinations to describe tilth or structure (Colvin et al., 1984; Koppi and Douglas, 1991) whereas others have advocated some form of plant response as a measure of tilth (Ketcheson, 1982). Additional work is needed before adequate understanding of the relation of plant growth to soil structural conditions is attained. Current knowledge is insufficient to adequately model plant growth as a function of soil structure, except under certain well-defined conditions (Russell, 1977).

Soil Structure Formation

To understand soil structure, an appreciation for the processes of soil structural formation is required. Several factors may be involved in the formation of bonds between soil particles leading to the formation of structural units or aggregates. Over very small distances, van der Waal forces bond particles together, especially organic polymers and clay-sized particles (Gast, 1977; Hillel, 1982). Thickness of the diffuse or electrical double layer (EDL) influences efficacy of these forces. Thickness of the EDL is determined by the cation exchange capacity (CEC) of the soil, and by the intensity and nature of the ionic species present in the soil solution and at the exchange sites (Gast, 1977; Hillel, 1982). The EDL also affects bonding of soil minerals. Cations may bridge clay particles and thus form clay domains (Russell, 1973). Thixotropy, or age hardening of soils, also affects both structure and strength of soil. After a soil is disturbed, new bonds are formed, and/or existing bonds are strengthened (Dexter et al., 1988).

Several different bonding agents are active in the soil. Phyllosilicates are the most active among clay minerals, and the activity of the various clay minerals depends upon their crystalline structure and upon the extent of isomorphic substitution. The mineralogy affects the CEC and the specific surface area of the clays, both of which influence their activity in bond formation (Gast, 1977; Russell, 1973).

Alumino-iron compounds are the dominant mineral bonding agents in some soils, whereas carbonates or silicates dominate in others (Birkeland, 1984; Black, 1968). The strengthening of existing bonds in thixotropy is due mostly to mineral cementation as minerals are precipitated at contact points (Kemper and Rosenau, 1986). Once formed, these bonds tend to persist.

Soil organic matter may complex with soil minerals to form aggregates. This activity results from charges on functional groups associated with SOM. These bonds have been classified as persistent, temporal, and transient by Tisdall and Oades (1982). Humic substances are associated with persistent bonds. This persistence is, in part, a result of the physical structure of humic substances that resists microbial degradation. The presence of these substances in complexes with soil particles and clay domains decreases their availability to microbes as well. Fungal hyphae and plant roots form temporal bonds in many aggregates. Polysaccharides and other plant and microbial exudates often form transient bonds. These bonds are not persistent, because of their solubility in water and their susceptibility to microbial degradation. These substances may contribute to the formation of new bonds as thixotropy occurs, but the presence of SOM and the subsequent formation of new bonds may mask or prohibit the strengthening mechanism (Dexter et al., 1988).

There are no easy ways to predict which bonding agent or mechanism will dominate in a given soil. This characteristic will be a function of the frequency of disturbance (or tillage) and the time allowed for bonding processes. Particle rearrangement may begin when tillage is completed and may be completed in 1 week in soils at intermediate water contents (Dexter et al., 1988). Polysaccharides, fungal hyphae, and exudates may contribute to somewhat rapid bond formation. These bonds may be weak if insufficient time is allowed for cementation. The formation of complexes may primarily be a function of position, occurring when clay particles, cations, and organic polymers are proximate. A special case of soil structure formation occurs in structurally damaged soils.

Natural Amelioration

Structural regeneration occurs as a result of several natural processes: wetting/drying cycles, freezing/thawing cycles, root growth, faunal activity, and time (Dexter, 1991). Depending upon the severity of the damage, compaction may leave a platy or massive structure and affect primarily interaggregate porosity (Dexter, 1988). With many soils, drying results in crack formation, which Dexter (1991) suggests is one of the most effective forms of natural amelioration. Wetting results in closing of cracks and often causes mechanical failure of the soil. This results in slaking and in formation of microaggregates (< 250 μm) or at least microcracks if the soil is initially drier than -1 MPa water potential. Time is important because soil is a thixotropic material. It loses strength after tillage because bonds have been disrupted. Over time, the soil gains strength as previously discussed. In practical terms, it is suggested that a 5 to 7 day rest be allowed between a tillage operation and any other management practice (Dexter, 1991). Bullock et al. (1985) found that an Albaquic Paleudalf had regained initial porosity in the top 5 cm under a bare surface 18 months after compaction. More regeneration occurred at the surface than at depth because at the surface there was greater water content variation, caused by more wetting/drying and freezing/thawing cycles. Pagliai (1987) found regeneration of similar morphology in a clay loam soil under a bare surface in 1 year. There were, however, fewer large pores (greater than 1000 μm diameter) than were present before compaction.

Root activity also effectively ameliorates compacted soils. Roots perform several processes in soil: dehydration, enmeshment, SOM enrichment, and biopore creation (Goss, 1987). They enhance wetting/drying cycles and thus encourage crack formation. Roots bind large aggregates and, through associated fungal hyphae, contribute to bonding of smaller aggregates. Microaggregates may be stabilized by root exudates. Grass roots are especially effective in contributing to aggregate formation. Upon death, their decomposition, often occurring within 1 year for most crop and grass species, contributes to maintenance of, or increases in, SOM. Biopores remain and contribute to increased macroporosity (Goss, 1987). Biopores are more abundant at the surface but have greater longevity at depth (Dexter, 1991). Dexter (1991) also suggests the use of "biological tillage" to disrupt hardpans. Elkins (1985) used grass roots to penetrate a hardpan and to establish biopores as root channels for subsequent crops.

Finally, soil faunal activity, especially that of earthworms, contributes to amelioration of compacted soils. Earthworms can tunnel through compacted soil that, in its previous state, would have greatly limited root extension (Dexter, 1991). The soil excreted in the tunnels of earthworms has a bulk density of about 1.15 Mg m^{-3} (Dexter, 1991). Thus, these tunnels also provide preferential growth zones for roots in compacted soils.

Natural amelioration occurs as a result of the combined effects of all these factors. Dexter (1991) suggests that managers use or enhance such natural processes to hasten amelioration. Research in this area is definitely needed.

Stability

Stability is the ability of a soil to resist disruption of its aggregates. The percentage of stable aggregates is typically determined through a wet or dry sieving process (Kemper and Rosenau, 1986). Soil stability, a dynamic aspect of cropped soils, is affected by seasonal cycles as well as by cropping and tillage systems. Wetting/drying cycles and freezing/thawing cycles tend to disrupt existing aggregates and to form new aggregates (Terpstra, 1989). The relative stability of structural aggregates is thus influenced by the time of the season and by the point in the cycle. Stability is also affected by soil faunal activity, plant root extension (Payne, 1988), and aggregate size. Smaller aggregates are more stable than larger aggregates (Braunack and Dexter, 1989). These concepts are important in the understanding of water erosion because unstable aggregates are more easily destroyed by high energy raindrops than are stable aggregates (Davies and Payne, 1988). In contrast to cropped soils, undisturbed soils have stable structures and are less affected by seasonal variations (Reinert et al., 1991).

For many soils, structure and tilth play major roles in productivity (Langdale and Shrader, 1982). The most favorable soil structural condition likely to occur for most soils is that of a virgin (unfarmed) condition. Well-structured conditions promote relatively rapid water infiltration and drainage, favorable aeration, adequate water storage, improved soil warming, and low root-growth resistance (Karlen et al., 1990). For optimal soil structure conditions, the soil would be continuously covered with living or dead plant material. Maintaining these conditions is a goal for sustainable soil management.

CROP MANAGEMENT PRACTICES, SOIL ORGANIC MATTER, AND STRUCTURE

Management practices affect SOM content and soil structure in many ways. Cropping systems and rotations may be managed to improve soil conditions. Proper crop residue and fertility management schemes also may be beneficial.

Cropping Practices

Commercial agriculture has measurably altered SOM makeup and quantity. Haas et al. (1957) suggests that organic carbon (directly related to SOM) levels in the upper 30 cm of medium-textured fields in the Great Plains ranged from 20 to 36% of that found in adjacent virgin grassland. These losses seemed the direct result of cropping practices using rigorous tillage and subsequent fallow. SOM decomposition, as a result of these practices, enhances erosion losses.

Soil management practices also affect SOM maintenance. Ridley and Hedlin (1968) explored SOM level differences attributed to fallowing. After continuous systems had been in place for 37 years, soils cropped every year had higher SOM levels than did those fallowed every other year. The amount of SOM remaining was 7.2 and 3.7%, respectively (in a soil assumed to have an initial SOM content somewhat less than 10%). Soils fallowed after every two or three crops had intermediate SOM contents. The more rapid decline in the frequently fallowed plots was probably due to the greater number of tillage operations, which had caused greater organic carbon oxidation. These authors attributed the slower rate of SOM decline on the intensively cropped plots on greater residue additions as roots. The above-ground portion of the residue had no effect because it was removed from the plot. Similarly, Unger (1968) found that when tillage treatments were kept constant, continuously cropped plots had significantly higher SOM contents than did plots in a crop-fallow system. Fallowing is typically practiced for water management purposes; it thus seems an antagonism exists between water and SOM management ideologies.

Davidson et al. (1967) studied crop species effect on SOM levels. The cropping scheme consisted of continuous cotton and continuous lespedeza on adjacent plots for 24 years. Soil organic matter measurements with depth after the 24 years of continuous cropping showed significant cropping effects, with the surface SOM levels for cotton 55.8% of that for lespedeza. However, at greater depths, similar SOM levels were found. The large differences in the upper profile suggest that crop species may have an effect on SOM level.

Different rotations maintain various equilibrium SOM contents. Virgin soils and rotations including legumes have relatively high levels of SOM, which contribute to favorable structural conditions (Becher and Martin, 1988; Karlen et al., 1990). Rotations including legumes and/or grasses are beneficial to aggregate stability. Increases in aggregate stability are coincident with decreases in wind and water erosion (Karlen et al., 1990). Many studies report stability increases, or higher stabilities due to the use of grasses or legumes in a rotation (Angers and Avon, 1991; Becher and Martin, 1988; Hussain et al., 1988; Stone, 1991). These improved conditions contribute to higher saturated hydraulic conductivities and available water contents than are found in conventional, continuous cropping systems (Becher and Martin, 1988; Hussain et al., 1988).

Residue, Manure, and Fertilizer Additions

The quantity of crop residue allowed to remain within the cropping system has a major affect on SOM level. Black (1973) added various quantities of wheat straw to a series of plots. At the initiation of the study, SOM values did not differ significantly among residue plots. After four crop-fallow cycles, however, SOM increased significantly and proportionately to the quantity of straw incorporated at all three soil-depth increments sampled. The data indicated that SOM equilibrium levels will be raised by addition of relatively large quantities of carbonaceous materials. Increased residue levels contributed to increased SOM levels, decreased soil bulk densities, and increased grain yields. Other classic long-term studies showing linear increases of soil organic carbon with crop residue additions may be reviewed in Rasmussen et al. (1980) and Larson et al. (1971).

Addition to the soil system of materials other than those from indigenous species or crops also may affect SOM levels. Manure additions, as studied by Sommerfeldt et al. (1988), can increase SOM levels significantly. Large SOM accumulations as induced by long-term, repeated manure additions may have extensive beneficial effects, producing soils with high overall fertility and nutrient reserves. This situation deserves monitoring, however, to alleviate justifiable concerns that crop nutrient utilization may be less than the natural release level. Such applications also may benefit soil structural conditions, contributing to improved water retention and saturated hydraulic conductivities (Anderson et al., 1990).

Applications of inorganic fertilizer also may affect SOM levels, although the association may be indirect. Larson et al. (1971) proposed that adequate amounts of fertilizer allowing high yields in a continuous corn system may produce enough residues to prevent the loss of SOM. Rasmussen and Rohde (1988) point out, however, that in nearly all studies involving soil carbon retention, crop residue effects have been positive whereas fertilizer application effects have been inconsistent. In instances when beneficial effects have been associated with fertilizer applications, it is often unclear whether these effects are independent of or associated with increases in residue production.

TILLAGE PRACTICES, SOIL ORGANIC MATTER, AND STRUCTURE

Tillage is as old as agriculture and has symbolized agricultural production for centuries. Although technology has changed dramatically through the centuries, the reasons for tillage have remained quite consistent. Cook (1962) identified weed control, seed bed preparation, and coverage of existing vegetation, manures, and crop residues as primary reasons for conducting preplant tillage. These activities contribute to favorable growth of the planted crop and ultimately to favorable crop yield. With the somewhat recent invention of mechanical planters, incorporation of surface plant materials and manures became increasingly important to enhance planter performance.

Inorganic fertilizers and pesticides have played an increasing role in agricultural production in recent years. Pesticide expenditures in the U.S. have increased from $3,538.6 million in 1980 to $5,721.5 million in 1989 (TVA, 1991). Fertilizer, particularly nitrogen (N), use has increased dramatically (see Figure 1). Use of these materials has introduced another reason for tillage—incorporation for more efficient material performance. This is particularly true for selected soil-applied pesticides and nitrogen fertilizers (Timmons and Cruse, 1990).

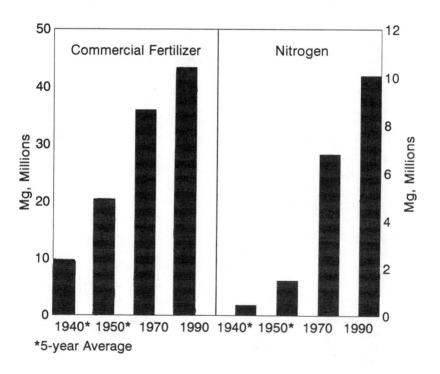

FIGURE 1. United States fertilizer usage (TVA, 1991).

Soil Environment

Tillage changes the packing arrangement of soil particles and the distribution of plant residues and other materials applied near the surface. The short-term effects of tillage are derived from the redistribution of soil, plant residues, and other materials. The long-term effects of tillage are due to the influence that this redistribution has on processes such as heat and air exchange between the soil and the atmosphere, evaporation, and movement of water and heat in the soil. By influencing soil particle arrangement, surface exchange, and subsurface transport properties, tillage greatly affects the soil environment. Because plant growth and crop yield are directly related to this environment, tillage directly affects them.

Conventional clean tillage, commonly conducted with a combination of the moldboard plow, disk, and/or harrow, has been popular throughout many regions of the U.S. for several decades. Moldboard plowing buries plant residues and loosens the soil surface. This enhances the surface energy and water exchange, permitting relatively rapid drying and warming in the spring. This advantage can be critical for early spring planting, which is an important factor influencing crop yields (Cruse, 1985; Logan et al., 1991). Loosening of the soil surface layer reduces compaction and improves soil aeration due to the initial loss of water by evaporation. Selected fertilizers, liming materials, and pesticides may be surface applied and incorporated or mixed to desired depths. Incorporating fertilizers enlarges the soil volume for plant root-fertilizer contact. With these favorable characteristics associated with "clean tillage agriculture", one might ask why a better system is sought. In short, the long-term impact of tillage reaches far beyond the immediate reasons for it. The unintended consequences of tillage on the environment and on natural resources prompt immense concern about the sustainability of conventional agriculture.

Effects on Soil Organic Matter

Soil organic matter level differences as a result of tillage practices have been documented extensively. Unger (1968) studied the effect of various tillage practices on SOM in 24-year-old wheat plots. Several tillage treatments were imposed on these established plots. In a wheat-fallow cropping system, the tillage treatments resulted in significant SOM differences. The number of tillage operations performed annually ranged from 4 to 10, depending upon the treatment. Bauer and Black (1981) also reported differences in organic carbon (directly related to SOM) concentrations between conventional tillage management and a less intensive system. For some soils, up to 44% more organic carbon was found in the less tilled plots. These authors hypothesized that organic carbon concentration differences between the two tillage systems were compounded by differences in erosion, amount of crop residues returned (related to yield), and crop rotation.

Relations between other soil parameters and SOM may be important in a final evaluation of tillage effects. Although percentage organic carbon is consistently smaller in surface soils of perennially tilled plots, sometimes total mass of organic carbon in the top 30-cm profile (computed from percentage organic carbon and bulk density) may be equal to or even greater in tilled than in undisturbed soils. This is a result of consistently high bulk densities in compacted tilled soils (Coote and Ramsey, 1983). Mann (1986) also cautions that the initial, or virgin, organic carbon storing capabilities of a soil may largely affect interpretations of organic carbon loss patterns. Farming practices may allow "exploitation" of soils initially high in organic carbon, but management practices of soils low in initial organic carbon must enhance productivity and therefore conserve or increase SOM. The cultivation of such soils could result in an increase in SOM if factors such as tillage, manure, water, cultivation, and crop selection are managed properly.

Soil organic matter loss from virgin soils is widely chronicled. The results of previously cited authors and of many others suggest that cultivation of virgin prairie soils has cost approximately 30 to 60% of the initial SOM (Campbell and Souster, 1982; Dormaar, 1979; Haas et al., 1957; Martel and Paul, 1974; Rasmussen et al., 1980; Smith and Young, 1975; Voroney et al., 1981). Therefore, improved management practices are worthy of attention to sustain a soil fraction so important in nutrient storage, CEC, carbon dioxide contribution, and structure stabilization.

Although an increase in tillage intensity (directly related to the total energy imparted to the soil in the tillage operation[s]) results in a decrease in SOM (Tisdall and Oades, 1982), a decrease in intensity allows SOM to rebound and to attain a new equilibrium consistent with the new tillage intensity. For these reasons, reduced and no-till systems show promise for SOM management. These practices alter and somewhat control the direction of the SOM level changes in cultivated fields. Wood et al. (1991) suggest that the introduction of no-till slowed SOM losses and at times replenished soils previously managed under a tilled system. The greatest changes understandably occur in the surface horizons because burial of plant residues is decreased with reduced tillage systems. Doran (1980) similarly concluded that although conventional tillage greatly accelerates the loss of carbon from the surface of grasslands, previously tilled cropland adapted to no-till may result in eventual carbon increases. Evidently, these levels will fall between those of virgin grassland and cultivated soils.

Effects on Soil Structure

It has been stated that well-structured soils facilitate relatively rapid water infiltration and drainage, favorable aeration, adequate water storage, improved soil warming, and low root growth resistance. Cropping and tillage systems strongly influence these aspects of structure, as well as aggregate stability and soil strength. The following discussion considers tillage system effects on soil structure.

Most virgin soils were initially tilled for the purposes of residue burial, seedbed preparation, and weed control. Blanket or uniform tillage across the entire field was used, which disturbed all soil found near the surface (as opposed to disturbing only soil in the immediate zone of crop seed placement). This distinction becomes important because soil disturbance by tillage, particularly when conducted at inappropriate soil water contents, has negative effects on soil structure. One tillage event can destroy or completely modify structural conditions that required years to develop (Reinert et al., 1991).

As soil structure deteriorates, tillage becomes increasingly relied upon to create ephemerally favorable soil tilth. Tillage loosens the soil, reducing compaction in the tilled zone and creating favorable soil aeration, seedbed, and water relations. Although these tillage-created conditions favor the planted crop, the created conditions are transitory. That is, the low soil structural stability resulting from previous tillage causes the soil to reconsolidate to its original condition within weeks (and sometime days) after the operation. The more intensive the tillage practices used, the more reliant the soil becomes on tillage practices to maintain physical conditions favorable to crop production. In essence, soil becomes "addicted" to tillage.

The primary aspects of tillage that influence soil structure are timing and intensity. Timing of tillage is important in relation to soil water content (Karlen et al., 1990). If tillage is performed at water contents above the lower plastic limit, or the lowest water content at which the soil will form a ribbon, compression of the soil may occur (Davies and Payne, 1988). This compression results in the formation of clods with great strength. The preparation of a favorable seedbed in such soils is very difficult. The water content of some soils at field capacity is above the lower plastic limit, so tilling these soils should be delayed to avoid formation of poor structure (Davies and Payne, 1988). Tillage in soils that are too dry increases energy consumption because these soils usually have high strengths (strength is inversely proportional to water content). In some areas, tillage of dry soils can contribute to wind erosion due to excessive fracturing and fragmentation of aggregates (Davies and Payne, 1988).

Tillage intensity affects many aspects of the soil condition, including residue cover and erosion. An increase in tillage intensity results in decreased surface residue coverage. Residue acts as a shock absorber during rainfall events, decreasing the kinetic energy of raindrops before their contact with the soil surface. Thus, aggregate disruption is decreased and crust formation is prevented or delayed, allowing relatively high infiltration rates to be maintained for extended periods (Radcliffe et al., 1988). Residue also reduces runoff velocity and thus decreases the amount of detached soil that can be carried in the flowing water. Residue alters the surface configuration and changes the surface wind flow patterns, thereby protecting the soil. A bare surface is most susceptible to both wind and water erosion. The amount of residue required to control erosion changes with location, soil type, and crop. Within limits, however, increasing the quantity of residue left on the surface increases the effectiveness of erosion control. Clean tillage systems are therefore the most detrimental tillage systems to the soil because they leave the soil surface completely exposed to both wind and rain, enhancing erosion.

If tillage intensity is altered, stability changes will reflect those of SOM, as a new equilibrium consistent with the new intensity is attained (Cochrane and Aylmore, 1991; Reinert et al., 1991; Wood et al., 1991). Increases in SOM are usually associated with increases in aggregate stability (Black, 1968). Aggregate stability decreases are often detectable in aggregate size reductions (Reinert et al., 1991). Tillage also increases the seasonal variability of aggregate stability (Reinert et al., 1991). Figure 2 is a conceptualization of the effects of tillage on soil stability over time. The most stable soils have never been tilled and show the least seasonal variation.

Soil Strength and Compaction

Soil strength gradually increases with time after a tillage event (Carter, 1988). This increase in strength is associated primarily with increased bulk density. Tillage "fluffs" the soil and creates a loose soil condition characterized by high potential (or free) energy. The soil will attempt to regain the lowest potential energy state it can through three mechanisms: redistribution of soil water, particle rearrangements and flocculation, and cementation (Dexter, 1988, 1991). Decreased volume and increased bulk density result. Aggregate stability and soil strength levels are at their lowest in the early spring and at their highest in the late summer, but these generalizations ignore weather patterns. Bradford and Grossman (1982) reported that, due to differences in compaction and wetting, soil strength was greater during wet years than during dry years.

Compacted zones may be created as a result of either wheel traffic or tillage implements. Uncontrolled wheel traffic results in somewhat uniform compaction over the whole field, characterized by high bulk

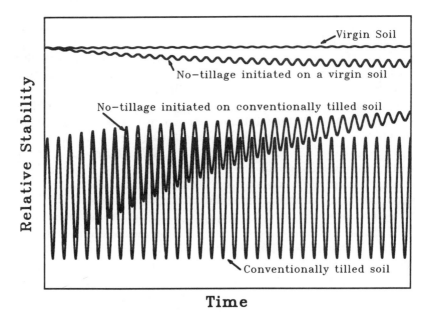

FIGURE 2.   A conceptualization of tillage system effects on the relative stability of soils. Adapted from Pierce, F. J. and Lal, R., 1991. With permission.

densities and soil strength. These bulk density increases come almost entirely at the expense of interaggregate macroporosity (Dexter, 1988). Decreased macroporosity results in decreased water infiltration and increased runoff. The continued use of some tillage implements (moldboard plows, cultivators, and rotary hoes) at the same depth may result in the formation of a plowpan (Donahue et al., 1983), a soil layer of increased bulk density and strength just below the depth of tillage. Areas of great soil strength limit root growth (Carter, 1988) and affect soil hydraulic properties.

Soil tillage addiction is exemplified by deep tillage to alleviate compaction problems. An exception is the use of deep tillage to ameliorate dense genetic soil horizons. Soils that have undergone compaction and remedial tillage may be susceptible to recompaction with lesser loads than those that caused the compacted condition (Cooper, 1971).

Using light equipment and limiting field operations to soils at or below their lower plastic limit water content are the preferred means of reducing and/or avoiding soil compaction problems. If light equipment is not an option, controlling wheel traffic to permanent traffic lanes such that no row receives traffic on both sides will minimize or eliminate most yield-related compaction problems.

Hill et al. (1985) studied the effects of three tillage systems on water retention and pore size distribution of two Mollisols. Their observations were that the conventionally tilled soil had more large pores (>15 μm) but was more subject to densification, whereas the reduced and no-till treatments had larger proportions of pores in the 15 to 0.1 μm range. These observations were made during late spring or early summer, and densification over the remainder of the growing season probably altered the proportions. The researchers concluded that the conventionally tilled soil should have higher saturated hydraulic conductivities and should therefore drain better. The conservation tilled soils would provide more available water for plant growth and have higher unsaturated conductivities. Benjamin (1991) found few differences in hydraulic conductivity, pore size distribution, or moisture retention characteristics of eight soils under three tillage systems. Tillage had significant effects only on two soils with low organic matter contents. García-Préchac (1991) conducted tillage studies in wet and drought years. He determined that of three tillage systems, conventional, reduced, and no-till, the reduced tillage system had the most desirable water status for plant growth in both moisture extremes. Van Es et al. (1991) used infiltration characteristics to find that controlled wheel traffic in a ridge system resulted in compaction in the interrow but in good structure in the ridge.

Impact on Sustainability

Tillage intensity influences many factors, including soil organic matter, soil erosion, water quality, energy consumption, weed control, fertilizer incorporation, planter performance, and crop yield. Technological innovations over the last three decades, however, have changed the relation between tillage intensity and these factors, challenging the classical reasons for preplant tillage operations (particularly as they relate to crop yield). Some quite general relations between tillage intensity and various crop production factors are given in Figure 3. A comparison of the period prior to the development of "high performance" planters and cultivators, fertilizer and manure injectors, and herbicides (prior to 1960), with the present period suggests several interesting relations. Increasing tillage generally decreases erosion control and increases energy consumption, despite available technology (Figure 3a). In contrast, Figure 3b suggests that other factors are altered by technological innovation. Weed control is less reliant on intensive tillage operations today than it was before and may be accomplished more effectively with improved cultivators and/or herbicides. Fertilizer and/or manure incorporation is less reliant on separate tillage operations now than in the past. Planter performance is much improved and no longer requires a bare soil

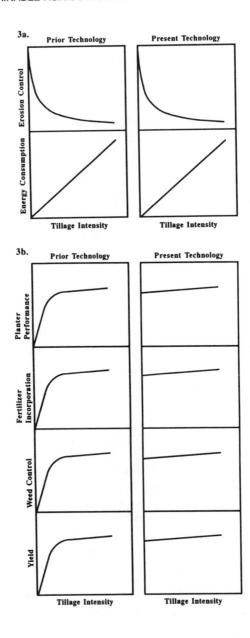

FIGURE 3. Conceptual relations between tillage intensity and various factors affecting crop production for the period before modern technological innovations became widespread (approximately pre-1960) and for the period since (post-1960).

surface, reducing the need for preplant tillage and/or residue incorpora- tion. As a result of these technologies, yield is less responsive to tillage today than in past years.

New tillage-related technologies have important implications. Envi- ronmental factors of primary concern in sustainable agriculture include erosion control, energy conservation, and water quality maintenance (Logan et al., 1991; Pierce and Lal, 1991). Less preplant tillage seems required now than in the past for optimal crop production, and therefore tillage is having less environmental impact.

The role of conservation tillage in crop production exemplifies this change. Surface residues associated with conservation tillage can dramat- ically reduce both wind and water erosion (Unger and McCalla, 1980). This is seemingly favorable, as is the concurrent increased water infiltra- tion, which is normally observed over the long term with conservation tillage systems (Baker, 1987). Yet, the ground water threat of soil-applied pesticides and supplemental plant nutrients (particularly nitrogen) may increase with infiltration rates (Knox and Moody, 1991). Thus, the im- portance of appropriate technology use, balanced with reduced tillage, becomes evident if soil and water are to be equally protected. There seems little to gain if one resource is saved while the other is damaged or destroyed.

The energy demands of tillage are significant, although considerably smaller than those associated with nitrogen fertilization. (Typically, N energy costs in diesel fuel equivalents [DFE] for production, transporta- tion, and application to corn range from 1.5 to 2.1 l/kg of N applied.) Energy demands for selected field operations, including tillage, are given in Table 1. The fossil fuel dependence of tillage is of great concern. The potential to reduce energy use by increasing tillage efficiency or by using reduced tillage practices is important. Future energy costs and availabil- ity, however, may have greater effects on tillage system selection than will the noble desire to reduce fuel consumption to conserve energy resources. Practices requiring minimal amounts of energy, fortunately, tend to leave plant residues on or near the surface (favoring soil conser- vation) and concurrently reduce fuel consumption (reducing costs). Till- age systems less energy intensive than the traditional moldboard plow system can be as productive as, or more productive than, that system.

Tillage System Selection

A multitude of tillage systems, each incorporating a unique combina- tion of tillage operations, is available to farmers. Farmer managerial ability will determine to a great extent the minimal tillage necessary to consistently produce optimal yields. Plant residues should remain on or near the soil surface. Ideally, tillage should occur only to the depth and in

Table 1. Diesel Fuel Requirements for Selected Field Operations

Field Operation	Soil Draft Rating[a] (1/ha DFE[b])		
	Low	Mod.	High
Shred cornstalks	7.0	7.0	7.0
Subsoil (33 cm)	10.3	15.9	27.1
Moldboard plow (20 cm)	13.1	17.3	23.4
Chisel plow (20 cm)	9.4	11.2	15.4
Offset disk	7.5	10.3	12.6
Field cultivate (plowed ground)	7.5	8.4	9.4
Field cultivate (disked ground)	5.1	5.6	6.1
Tandem disk (not tilled)	6.1	7.0	7.9
Tandem disk (plowed ground)	5.6	6.1	6.5
Tandem disk (second time)	5.1	5.6	6.1
Form ridges (fall)	3.7	4.2	4.7
Harrow	3.7	3.7	3.7
Apply NH_3 (not tilled)	6.1	8.4	10.3
Apply NH_3 (tilled)	5.6	6.5	7.5
Field cultivate + plant	8.9	9.8	10.8
Till-plant	4.2	5.1	6.1
Strip rotary till + plant	7.9	8.9	9.8
Plant (tilled)	4.2	5.1	6.1
Plant (not tilled)	3.7	4.7	5.6
Cultivate (sweeps)	4.2	4.7	5.1
Cultivate (disk hillers)	4.7	5.1	5.6
Cultivate (rolling tines)	4.2	4.7	5.1
Rotary hoe	2.3	2.3	2.3
Apply fertilizer (on surface)	1.9	1.9	1.9
Apply pesticides	1.9	1.9	1.9

[a]Soil types associated with the draft ratings include: low = sands and sandy loams; moderate = loams and silt loams; high = clay loams and clays. Actual fuel requirements for a particular field operation on a particular soil type may vary as much as 20 to 50% from the values shown here.
[b]DFE—diesel fuel equivalents. To convert diesel to gasoline equivalent, multiply by 1.4.
Adapted from Siemens, J. C., D. R. Griffith, and S. D. Parsons, 1986.

the positions needed to accomplish operation objectives. One hectare of soil uniformly tilled for seedbed preparation to a 15 cm depth would require tilling 1500 m³ of soil. But if tillage were limited to the seedbed (approximately 4 cm² by 5 to 7 cm deep), the volume of soil disturbed is only 1% of that for blanket tillage. Tilling an entire plow layer for seedbed preparation thus seems excessive. Although the technology required for such precision tillage is not available (except as arguably exists with no-till planters), the inefficient resource use found in many fields is exemplified. Strip tillage systems and no-till systems seem to address this problem more directly than do either broadcast or blanket tillage systems.

SUMMARY

Characterizing historical soil organic matter levels in agricultural systems required much insight and research. Predictions of future SOM levels will require even more. Although the importance of the SOM resource is largely unquestioned, its effect on the productivity of a soil is so subtle and diverse that correcting destructive systems (those involving exploitation) has been underemphasized. The task of the concerned agriculturalist is one of balancing the need for a conservation ethic appropriate for this inherent resource with that for managing crop production at a desirable economic level. This will require greater understanding of long-term management effects on organic matter reserves, overall functions of SOM, and economic assessments to characterize the *value* of SOM. Such insight should provide impetus to consider conservation strategies designed to sustain SOM.

Soil structure is a dynamic property affected by SOM levels, cropping and tillage management systems, climatic conditions, and time. Several means of soil structure characterization exist, but none has gained widespread acceptance as the most useful in soil structure description. Structure formation is affected by cropping and tillage practices, weather patterns, plant root growth, and soil faunal activity. Natural amelioration is a special case of soil structure reformation in compacted soils. Well-structured soils are more resistant to erosion and otherwise more favorable to plant growth than are poorly structured soils. Tilth is an often used, vaguely defined term. Many attempts are in progress to define tilth to facilitate quantification. Management systems contribute to degradation of, or improvement of, soil structure. Though stability changes with time, management affects both mean and range of the variation. Tillage of virgin soils has resulted in structural degradation. Once degraded, soil structure is susceptible to further, or repeated, damage by similar practices. This degradation increases the need for tillage to create favorable conditions for seedling establishment, but concurrently decreases the longevity of the created conditions. Thus, many agricultural soils suffer from a dependency that could be called "tillage addiction."

Sustainable Management Practices

A sustainable soil management system would incorporate the following concepts: controlling wheel traffic; timing tillage operations according to soil water contents; limiting depth, width, and number of tillage operations; maintaining surface residue cover; and using crop rotations that include legumes. Controlling wheel traffic limits the area affected by compaction, resulting in more field area with favorable structural properties. Appropriate timing of tillage operations decreases the risks of compaction and wind erosion. Limiting the tillage zone decreases the volume of soil disturbed, maintaining SOM levels and existent favorable structural properties in the untilled zones. Maintaining surface residue cover assists erosion control and alters the surface energy budget and the soil temperature. Rotations including legumes increase SOM, improve aggregate stability, and contribute nitrogen to the system.

REFERENCES

Anderson, S. H., C. J. Gantzer, and J. R. Brown. 1990. Soil physical properties after 100 years of continuous cultivation. *J. Soil Water Conserv.* 45:117–121.

Angers, D. A. and D. Avon. 1991. Influence of crops on aggregation of Quebec soils. In J. A. Stone, B. D. Kay and D. A. Angers (Eds.) *Soil Structure Research in Eastern Canada*. Proc. E. Canada Soil Structure Workshop, Guelph, Ont. Sept. 10 to 11, 1990. Herald Press Limited, Windsor, Ont. pp. 105–110.

Baker, J. L. 1987. Hydrologic effects of conservation tillage and their importance relative to water quality. In T. J. Logan, J. M. Davidson, J. L. Baker, and M. Overcash (Eds.) *Effects of Conservation Tillage on Groundwater Quality*. Lewis Publishers, Chelsea, MI. pp. 113–124.

Bauer, A. and A. L. Black. 1981. Soil carbon, nitrogen, and bulk density comparisons in two cropland tillage systems after 25 years and in virgin grassland. *Soil Sci. Soc. Am. J.* 45:1166–1170.

Becher, H. H. and W. Martin. 1988. Selected physical properties of three soil types as affected by land use. In J. Drescher, R. Horn, and M. deBoodt, (Eds.) *Impact of Water and External Forces on Soil Structure*. Catena Suppl. 11. Catena Verlag, Cremlingen-Destedt, West Germany. pp. 93–100.

Benjamin, J. G. 1991. Tillage effects on soil hydraulic properties. *Agron. Abstr.* ASA, Madison, WI. p. 325.

Birkeland, P. W. 1984. *Soils and Geomorphology*. Oxford University Press, New York.

Black, A. L. 1973. Soil property changes associated with crop residue management in a wheat-fallow rotation. *Soil Sci. Soc. Am. Proc.* 37:943–946.

Black, C. A. 1968. *Soil-Plant Relationships*. John Wiley & Sons, New York.

Bradford, J. M. and R. B. Grossman. 1982. In-situ measurement of near-surface soil strength by the fall cone device. *Soil Sci. Soc. Am. J.* 46:685–688.

Braunack, M. V. and A. R. Dexter. 1989. Soil aggregation in the seedbed: a review. I. Properties of aggregates and beds of aggregates. *Soil Tillage Res.* 14:259–279.

Bullock, P., A. C. D. Newman, and A. J. Thomasson. 1985. Porosity aspects of regeneration of soil structure after compaction. *Soil Tillage Res.* 5:325–341.

Campbell, C. A. and W. Souster. 1982. Loss of organic matter and potentially mineralizable nitrogen from Saskatchewan soils due to cropping. *Can. J. Soil Sci.* 62:651–656.

Carter, M. R. 1988. Penetration resistance to characterize the depth and persistence of soil loosening in tillage studies. *Can. J. Soil Sci.* 68:657–668.

Cochrane, H. R. and L. A. G. Aylmore. 1991. Assessing management-induced changes in the structural stability of hardsetting soils. *Soil Tillage Res.* 20:123–132.

Colvin, T. S., D. C. Erbach, W. F. Buchele, and R. M. Cruse. 1984. Tillage index based on created soil conditions. *Trans. ASAE* 27:370–371.

Cook, R. L. 1962. *Soil Management for Conservation and Production.* John Wiley & Sons, New York.

Cooper, A. W. 1971. Effects of tillage on soil compaction. In K. K. Barnes et al. (Eds.) *Compaction of Agricultural Soils.* ASAE Monograph. American Society of Agricultural Engineers, St. Joseph, MI. pp. 315–364.

Coote, D. R. and J. F. Ramsey. 1983. Quantification of the effects of over 35 years of intensive cultivation on four soils. *Can. J. Soil Sci.* 63:1–14.

Cruse, R. M. 1985. Watch soil temperatures: managing crop residues brings several benefits. *Crops Soils.* 38(3):17–18.

Davidson, J. M., Fenton Gray, and D. I. Pinson. 1967. Changes in organic matter and bulk density with depth under two cropping systems. *Agron. J.* 59:375–378.

Davies, D. B. and D. Payne. 1988. Management of soil physical properties. In A. Wild (Ed.) *Russell's Soil Conditions and Plant Growth.* Longman Scientific and Technical, London, pp. 412–448.

Dexter, A. R. 1988. Strength of soil aggregates and aggregate beds. In J. Drescher, R. Horn, and M. deBoodt, (Eds.) *Impact of Water and External Forces on Soil Structure.* Catena Suppl. 11. Catena Verlag, Cremlingen-Destedt, West Germany. pp. 35–52.

Dexter, A. R. 1991. Amelioration of soil by natural processes. *Soil Tillage Res.* 20:87–100.

Dexter, A. R., R. Horn, and W. D. Kemper. 1988. Two mechanisms for age-hardening of soil. *J. Soil Sci.* 39:163–175.

Donahue, R. L., R. W. Miller, and J. C. Schickluna. 1983. *Soils: An Introduction to Soils and Plant Growth,* 5th ed. Prentice-Hall, Englewood Cliffs, NJ.

Doran, J. W. 1980. Soil microbial and biochemical changes associated with reduced tillage. *Soil Sci. Soc. Am. J.* 44:765–771.

Dormaar, J. F. 1979. Organic matter characteristics of undisturbed and cultivated Chernozemic and Solonetzic A horizons. *Can J. Soil Sci.* 59:349–356.

Elkins, C. B. 1985. Plant roots as tillage tools. pp. 519–523. In *Proc. Int. Conf. on Soil Dynamics.* Vol. 3. Auburn University, Office of Continuing Education, Auburn University, Auburn, AL.

Follett, R. F., S. C. Gupta, and P. G. Hunt. 1987. Conservation Practices: Relation to the Management of Plant Nutrients for Crop Production. In R. F. Follett et al. (Eds.) *Soil Fertility and Organic Matter as Critical Components of Production Systems.* Soil Sci. Soc. Am. Spec. Publ. No. 19.

García-Préchac, F. 1991. Strip Position, Tillage, and Water Regime Effects on a Strip Intercropping Rotation. Ph.D. dissertation. Iowa State University, Ames. (Diss. Abstr. Intl. DA9202354).

Gast, R. G. 1977. Surface and colloid chemistry. In J. B. Dixon and S. B. Weed, (Eds.) *Minerals in Soil Environments.* Soil Science Society of America, Madison, WI. pp. 27–74.

Goss, M. J. 1987. The specific effects of roots on the regeneration of soil structure. In G. Monnier and M. J. Goss (Eds.) *Soil Compaction and Regeneration: Proc. of the Workshop on Soil Compaction: Consequences and Structural Regeneration Processes.* Published for the Commission of European Communities by A. A. Balkema, Boston. pp. 145–155.

Gupta, S. C., P. P. Sharma, and S. A. DeFranchi. 1989. Compaction effects on soil structure. *Adv. Agron.* 42:311–338.

Haas, H. J., C. E. Evans, and E. F. Miles. 1957. Nitrogen and carbon changes in Great Plains soils as influenced by cropping and soil treatments. *U.S. Dep. Agric. Tech. Bull. No. 1164.*

Hill, R. L., R. Horton, and R. M. Cruse. 1985. Tillage effects on soil water retention and pore size distribution of two Mollisols. *Soil Sci. Soc. Am. J.* 49:1264–1270.

Hillel, D. 1982. *Introduction to Soil Physics.* Academic Press, San Diego, CA.

Hussain, S. K., L. L. Mielke, and J. Skopp. 1988. Detachment of soil as affected by fertility management and crop rotations. *Soil Sci. Soc. Am. J.* 52:1463–1468.

Karlen, D. L., D. C. Erbach, T. C. Kaspar, T. S. Colvin, E. C. Berry, and D. R. Timmons. 1990. Soil tilth: a review of past perceptions and future needs. *Soil Sci. Soc. Am. J.* 53:153–161.

Kemper, W. D. and R. C. Rosenau. 1986. Aggregate stability and size distribution. In A. Klute (Ed.) *Methods of Soil Analysis.* Part 1. *Physical and Mineralogical Methods,* 2nd Ed. Agron. Monogr. No. 9. ASA-SSSA. Madison, WI. pp. 425–442.

Ketcheson, J. W. 1982. Toward a practical assessment of soil tilth. *Soil Tillage Res.* 2:1–2.

Knox, E. and D. W. Moody. 1991. Influence of hydrology, soil properties, and agricultural land use on nitrogen in groundwater. In R. F. Follett et al. (Eds.) *Managing Nitrogen for Groundwater Quality and Farm Profitability.* Soil Science Society of America, Madison, WI. pp. 19–57.

Kononova, M. M. 1966. *Soil Organic Matter.* Pergamon Press. New York.

Koppi. A. J. and J. T. Douglas. 1991. A rapid, inexpensive and quantitative procedure for assessing soil structure with respect to cropping. *Soil Use Manage.* 7:52–56.

Langdale, G. W. and W. D. Shrader. 1982. Soil erosion effects on soil productivity of cultivated cropland. In B. L. Schmidt et al. (Eds.) *Determinants of Soil Loss Tolerance.* ASA Spec. Publ. No. 45. American Agronomy Society, Madison, WI. pp. 41–51.

Larson, W. E., C. E. Clapp, W. H. Pierre, and Y. B. Morachan. 1971. Effects of increasing amounts of organic residues on continuous corn. II. Organic carbon, nitrogen, phosphorus, and sulfur. *Agron. J.* 64:204–208.

Logan, T. J., R. Lal, and W. A. Dick. 1991. Tillage systems and soil properties in North America. *Soil Tillage Res.* 20:241–270.

Mann, L. K. 1986. Changes in soil carbon storage after cultivation. *Soil Sci.* 142:279–288.

Martel, Y. A. and E. A. Paul. 1974. Effects of cultivation on the organic matter of grassland soils as determined by fractionation and radiocarbon dating. *Can. J. Soil Sci.* 54:419–426.

Pagliai, M. 1987. Micromorphometric and micromorphological investigations on the effect of compaction by pressures and deformations resulting from tillage and wheel traffic. In G. Monnier and M. J. Goss (Eds.) *Soil Compaction and Regeneration: Proc. of the Workshop on Soil Compaction: Consequences and Structural Regeneration Processes.* Published for the Commission of European Communities by A. A. Balkema, Boston. pp. 31–38.

Payne, D. 1988. Soil structure, tilth and mechanical behavior. In A. Wild (Ed.) *Russell's Soil Conditions and Plant Growth.* Longman Scientific and Technical, London. pp. 378–411.

Pierce, F. J. and R. Lal. 1991. Soil management in the 21st century. In R. Lal and J. J. Pierce (Eds.) *Soil Management for Sustainability.* Soil and Water Conservation Society, Ankeny, IA. pp. 175–179.

Radcliffe, D. E., E. W. Tollner, W. L. Hargrove, R. L. Clark, and M. H. Golabi. 1988. Effect of tillage practices on infiltration and soil strength of a Typic Hapludult soil after ten years. *Soil Sci. Soc. Am. J.* 52:798–804.

Rasmussen, P. E., R. R. Allmaras, C. R. Rohde, and N. C. Roager, Jr. 1980. Crop residue influences on soil carbon and nitrogen in a wheat-fallow system. *Soil Sci. Soc. Am. J.* 44:596–600.

Rasmussen, P. E. and C. R. Rohde. 1988. Long-term tillage and nitrogen fertilization effects on organic nitrogen and carbon in a semiarid soil. *Soil Sci. Soc. Am. J.* 52:1114–1117.

Reinert, D. J., F. J. Pierce and M.-C. Fortin. 1991. Temporal variation in structural stability induced by tillage. In J. A. Stone, B. D. Kay, and D. A. Angers (Eds.) *Soil Structure Research in Eastern Canada.* Proc. E. Canada Soil Structure Workshop, Guelph, Ont. Sept. 10 to 11, 1990. Herald Press Limited, Windsor, Ont. pp. 63–72.

Ridley, A. O. and R. A. Hedlin. 1968. Soil organic matter and crop yields as influenced by frequency of summer fallowing. *Can. J. Soil Sci.* 48:315–322.

Russell, E. W. 1973. *Soil Conditions and Plant Growth,* 10th ed., Longman, New York.

Russell, R. S. 1977. *Plant Root Systems: Their Function and Interaction with the Soil.* McGraw-Hill, London.

Schnitzer, M. and S. U. Khan. 1978. *Soil Organic Matter.* Elsevier North-Holland, New York.

Siemens, J. C., D. R. Griffith, and S. D. Parsons. 1986. Energy requirements for corn tillage-planting systems. NCH-24. Iowa State University Cooperative Extension Service, Ames.

Smith, S. J. and L. B. Young. 1975. Distribution of nitrogen forms in virgin and cultivated soils. *Soil Sci.* 120:354–360.

Sommerfeldt, T. E., C. Chang, and T. Entz. 1988. Long-term annual manure applications increase soil organic matter and nitrogen, and decrease carbon to nitrogen ratio. *Soil Sci. Soc. Am. J.* 52:1668–1672.

Spangler, M. G. and R. L. Handy. 1982. *Soil Engineering*, 4th ed. Harper and Row, New York.

Stewart, J. W. B., R. F. Follett, and C. V. Cole. 1987. Integration of organic matter and soil fertility concepts into management decisions. In R. F. Follett et al. (Eds.) *Soil Fertility and Organic Matter as Critical Components of Production Systems*. Soil Sci. Soc. Am. Spec. Publ. No. 19.

Stone, J. A. 1991. Several forages and the rate of change in soil structure following corn on a clay loam soil. In J. A. Stone, B. D. Kay, and D. A. Angers (Eds.) *Soil Structure Research in Eastern Canada*. Proc. E. Canada Soil Structure Workshop, Guelph, Ont. Sept. 10 to 11, 1990. Herald Press Limited, Windsor, Ont. pp. 143–152.

Tennessee Valley Authority (TVA), National Fertilizer and Environmental Research Center (NFERC). 1991. Fertilizer Summary Data: 1990. TVA/NFERC-91/6. Bull. Y-19. Muscle Shoals, AL.

Terpstra, R. 1989. Formation of new aggregates under laboratory-simulated field conditions. *Soil Tillage Res.* 13:13–21.

Timmons, D. R. and R. M. Cruse. 1990. Effect of fertilization method and tillage on nitrogen-15 recovery by corn. *Agron. J.* 82:777–784.

Tisdall, J. M. and J. M. Oades. 1982. Organic matter and water-stable aggregates in soils. *J. Soil Sci.* 33:141–163.

Unger, P. W. 1968. Soil organic matter and nitrogen changes during 24 years of dryland wheat tillage and cropping practices. *Soil Sci. Soc. Am. Proc.* 32:426–429.

Unger, P. W. and T. M. McCalla. 1980. Conservation Tillage Systems. *Adv. Agron.* 33:1–56.

Van Es, H. M., S. M. L. Verheijden, and J. Mt. Pleasant. 1991. Temporal changes in soil hydraulic properties as affected by soil management. In J. A. Stone, B. D. Kay, and D. A. Angers (Eds.) *Soil Structure Research in Eastern Canada*. Proc. E. Canada Soil Structure Workshop, Guelph, Ont. Sept. 10 to 11, 1990. Herald Press Limited, Windsor, Ont. pp. 197–210.

Vaughan, D. and R. E. Malcolm. 1985. Influence of humic substances on growth and physiological processes. In D. Vaughan and R. E. Malcolm (Eds.) *Soil Organic Matter and Biological Activity*. Kluwer Academic Publishers, Hingham, MA.

Voroney, R. P., J. A. Van Veen, and E. A. Paul. 1981. Organic C dynamics in grassland soils. II. Model validation and simulation of long-term effects of cultivation and rainfall erosion. *Can. J. Soil Sci.* 61:211–224.

Wood, C. W., D. G. Westfall, and G. A. Peterson. 1991. Soil carbon and nitrogen changes on initiation of no-till cropping systems. *Soil Sci. Soc. Am. J.* 55:470–476.

5

Crop Management

Max D. Clegg and Charles A. Francis

TABLE OF CONTENTS

INTRODUCTION

Crop management decisions address a number of biological and economic challenges, regardless of whether a farm consists of a few or hundreds of hectares. Although scale will dictate some choice of practices or timing, there is a common goal of efficient use of both renewable and nonrenewable resources. Both large and small farmers are seeking sustainable cropping systems that will provide consistent returns for their efforts and investment. The producers' sometimes elusive objectives are to achieve efficient and profitable crop production with systems that have a minimal impact on the broader environment.

In crop production enterprises, there are variables over which a producer has complete control: crop species, variety or hybrid, plant density, row spacing, fertilizer, and most insects. Over others there is moderate control: moisture stored in seedbed, soil structure, diseases, time of sowing, length of sowing period, and date of harvesting. Over many variables there is little or no control: frost, moisture at harvest, moisture during growth (except irrigation), wind, flood, large-scale plagues of insects and diseases (Makeham and Malcom, 1981). Zandstra (1977) describes crop production mathematically, where yield ($\bar{y}$) is a function of management ($\bar{M}$) and environment ($\bar{E}$):

$$\bar{y} = f\,(\bar{M}, \bar{E})$$

$\bar{M}$ includes the choice of variety, plant establishment, arrangement in time and space, fertilization, pest control, and harvesting. $\bar{E}$ relates to soil, climatic variables, and economic resources.

Practically all of these variables are addressed in specific terms in other chapters. This chapter considers the direct interaction of crops with the environment and cultural practices. The challenge is to determine how cropping and cultural strategies can be chosen for the most efficient use of resources, both renewable and nonrenewable. We recognize that crop management systems in the future will move toward greater reliance on efficient use of renewable resources, and their substitution wherever possible for scarce fossil-fuel based, nonrenewable resources.

CLIMATE

The climate of a region (length of season, rain, temperature) is the first determinant for the type of farming that is chosen. Altitude, availability of natural resources, and location of major topographical features are local factors that can modify the temperature and moisture availability at any location and directly influence the choice of crops and management methods. Humans must be included as biotic factors because of their manipulation of cultural practices and plants, often including new crops that were not originally adapted to each zone. Rubel (1935) refers to these adjustments or substitutions as factor replaceability.

Temperature

Biological activities of crops are limited to a specific temperature range (0 to 50°C). The lower end of the range is mean freezing point of water and the upper end of the range is the temperature for protein denaturation (Leopold and Kriedmann, 1975). Plants vary in their upper and lower tolerances; thus, choice of crop is important for a given site. For example, in cool root temperature zones (15 to 20°C) wheat (*Triticum aestivum*), barley (*Hordeum vulgare*), oats (*Avena sativa*), peas (*Pisum sativum*) or crambe (*Crambe abyssinica*) may be the crops of choice. At higher root temperature zones (25 to 30°C) maize (*Zea mays* L.), sorghum (*Sorghum bicolor*), or cotton (*Gossypium hirsutum*) may be the crops of choice (Nielsen, 1971; Martin et al. 1976). To pursue a more active direction toward greater diversity and sustainability, other species chosen for a particular cropping sequence or spatial strategy will need to be chosen for adaptation within the local temperature constraints.

Light

Light energy is intercepted by a crop and through photosynthesis is converted to carbohydrate which is then used by the plant for growth and storage. The amount of light available for a crop varies seasonally and daily depending on latitude, slope, and cloud cover. These factors determine the total radiation received by crops during the growing season (Clegg, 1972).

Light interception studies with pearl millet (*Pennisetum americanum*) and groundnut (*Arachis hypogaea*) showed that increased dry weights were linearly related to increased interception of photosynthetic active radiation (PAR) (Marshall and Willey, 1983). Board et al. (1990) reported that for the normal planting date of soybean (*Glycine max*), seed yield was strongly correlated with light interception duration.

Although the amount of light utilized by a plant is very small in relation to the amount of total light received (1.0 to 1.5%), level of light significantly influences yield, especially when environmental conditions are near optimal (Chang, 1968). Choosing optimum planting dates, cultivars, and other cropping strategies can maximize light utilization and lead to high and sustained yields.

Water

Plants are composed of 80 to 90% water. Whether from rain or irrigation, water is one of the most important factors in crop production. If conditions exist that cause a reduction in the water status of crop plants (stress), then increased leaf water potential, increased canopy temperatures, and reduction of photosynthesis occur, and these reduce plant growth and ultimately productivity.

Leaf area per unit ground area, or leaf area index (LAI), is directly related to transpiration and water use (Chang, 1968; Brun et al., 1972). LAI of sorghum and soybean contributed approximately 50% of the total evapotranspiration at a LAI of 2 and as much as 95% at a LAI of 4 (Brun et al., 1972). At Temple, Texas, soil evaporation was estimated as 20% of total evaporation; for sorghum and soybean, the transpiration fraction was estimated at 75% (Ritchie and Burnet, 1971).

The moisture status of the soil can influence crop choice because of the amount of water each crop and variety uses to successfully reach maturity. Sunflower (*Helianthus annuus* L.) used 6.2 mm d^{-1} of water as compared to a group of five other species that used an average of 5.1 mm d^{-1} (Hattendorf et al., 1988). The water content in the upper 30 cm was generally higher after continuous sorghum and least with continuous soybean in Nebraska (Roder et al., 1989). Early maturity corn generally yielded more grain in the northern plains because moisture was more favorable during grain fill (Alessi and Power, 1974).

In summary, crop plants require continuity of moisture availability and water flow through the plant in order to survive and function efficiently. Water availability can be controlled through irrigation and also through a range of other management techniques to be discussed. Many of these variables have been discussed in the recent books by Edwards et al. (1990) and Francis et al. (1990) in relation to the sustainability of agriculture. There is also a wide variation both among and within species in their tolerance to suboptimum moisture conditions. The choice of crops, varieties, and hybrids interacts with the prevailing climate in each location.

CROP DECISIONS

To increase sustainability of systems, choice of a specific crop to fit a niche within specific cropping systems depends on many factors. Most crops grown within a region are well adapted there after a long process of introduction of new species, trial and error by farmers, and economic sorting out of options. Both biological and economic changes occur over time due to the development and availability of new varieties and hybrids and changes in markets. There may also be changes in the most prevalent cropping conditions or most economically viable systems. Crop breeders play a major role in this process.

Choice of Crops

A major factor in crop choice is the range of climatic conditions and constraints already described. New crops are introduced and tested by researchers and farmers, but very few are successful. Soybeans were introduced from China into the U.S. and other parts of the temperate zone over 70 years ago and were initially grown with maize for silage. Now they have become a staple feed and oil crop in the U.S. and the temperate countries where they were once unknown. Grain sorghum hybrids replaced open pollinated varieties and made possible their economical culture in areas once planted in maize, for example, in the southeastern part of Nebraska. Most new crops that are tested are not successful, due to lack of adaptation or complications of marketing. Yet a strategy that includes periodic testing of newly available cultivars and genetic combinations is a valuable part of any search for optimum crop species (Fehr, 1987).

Variety or Hybrid Choice

A healthy debate continues on the issue of developing varieties or hybrids that are specifically designed for sustainable cropping systems. Potentials of adaptation to nonoptimum growing conditions have been reviewed by Christiansen and Lewis (1982), Blum (1985), and Francis (1990). The answer to this question revolves around the magnitude of the genotype by cropping system or genotype by cultural condition interaction. If all varieties or hybrids respond in the same way to a change in tillage, water availability, mineral stress, or extremes in temperature, there is no need to consider different cultivars for systems that may include these changes in crop growing conditions. If there is a strong interaction or difference in response among cultivars, then there is reason to expect that specific crop selections for the new conditions will enhance productivity, stability of yield, and sustainability of the system (Francis, 1990).

In many states and national research organizations, a uniform testing program provides nonbiased evaluation of available varieties and hybrids (e.g., Nelson et al., 1991a,b,c). This can be an easily accessible and current source of objective information for the producer. Depending on how well testing sites are characterized, uniform testing may not answer the questions about adaptation of genetic materials to a specific system or location. Tests may not include what farmers consider sustainable systems.

Whether farmers should develop their own varieties and produce seed is a large question in many developing countries, though less important in a more developed agriculture. When seed cost may represent 100% of the purchased inputs to the Colombian maize farmer with one hectare, this is far different from the Nebraska maize farmer whose seed cost is only 5% of total production costs over 1000 acres. Through experience, a farmer can produce seed of high quality that will perform well and may represent a variety closely adapted to the specific climate and cultural practices on that farm. However, use of that variety for many years will preclude the opportunity to use new cultivars that are emerging from both public and private breeding programs, and will end up costing the producer more in the long term. High quality seed is one of the most cost effective purchased inputs according to most farmers.

Specificity of Adaptation to Stress Conditions

Whether a new or different variety is needed to make a system more sustainable, perhaps with reduced use of nonrenewable inputs, depends on the different stress conditions in the new system. Genetic differences exist in tolerance to a range in stress conditions — drought, high and low temperature, soil conditions, excessive water, and biotic effects of insects, pathogens, or weeds. The critical question is whether these conditions will change as a result of modification in the cropping system (Francis, 1990). For example, a change in system is unlikely to make a difference in the need for adaptation to a new range of temperature conditions, unless the system also dictates a change in planting dates. In a change to a more intensive intercrop system, there may be a need for tolerance in a lower crop to shading by an upper story species or to the increased water use by two or more crops. If a change is made to a reduced fertilizer or limited irrigation system, it is important to know if other cultivars are more efficient in using the newly scarce resource. These are all susceptible to testing by planting a range of cultivars in both the present and the new system, and testing for a genotype by system interaction.

Differences in genetic reaction to drought among genotypes were reviewed by Quisenberry (1982). Breeding for tolerance to drought is complicated by the facts that timing and intensity of drought can vary widely and that genetic control of drought tolerance at different stages of the life cycle of a crop may be different. Blum (1985) suggests that selection for specific physiological measurements may be more effective than direct selection for drought tolerance per se. Greater drought resistance could be important to the sustainability of yields under reduced irrigation conditions or the effects of global warming.

Crop management in sustainable systems may include genetic adaptation to mineral stress conditions in new areas or systems. Devine (1982) suggests three strategies for success in this situation: (1) changing the soil environment by adding lime or fertilizers, (2) selecting for tolerance traits and yield in wild relatives, and (3) selecting desirable individuals from existing high yielding cultivars that include tolerance to the stress conditions. The latter two approaches would appear best suited to most sustainable agricultural strategies. The last method is the most rapid and cost efficient, if there is variability in genetic reaction to the specific stress condition. Reduced application of fertilizer may require selection for greater efficiency of uptake or use in crops. Zweifel et al. (1987) measured both significant grain sorghum hybrid by N level interaction and difference in N use efficiency among hybrids tested. Some of these differences may result from differential partitioning of nutrients between vegetative and reproductive organs (Maranville et al., 1980). There are also observed differences in P uptake, accumulation, and distribution in grain sorghum (Clark et al., 1978). Such differences suggest that breeding for adaptation to distinct soil nutrient conditions is possible, and thus breeding for sustainable systems is one available strategy.

Crop tolerance to insects and plant pathogens has been a high priority for breeders over the past several decades. Current commercial cultivars have acceptable tolerance to a range of potential pests. For reviews of breeding progress, three chapters in Christiansen and Lewis (1982) describe tolerance or resistance to insects (Jenkins, 1982), to pathogens (Bell, 1982), and to nematodes (Sasser, 1982). Basic tolerance mechanisms are found in these chapters as well as in a more recent review by Bird et al. (1990).

CULTURAL PRACTICES

The energy budget equation for a plant can be written as

$$Rn - H - \lambda E - P - R = 0$$

where Rn is net radiation, H is sensible heat, λE is latent heat flux density, P is photosynthesis, and R is respiration (Campbell, 1977). Although P and R are important for plant productivity, they are negligibly small components in the energy budget and are usually omitted to simplify the equation.

Efficient agricultural cropping systems are designed to create a near-optimum environment for the desired crop or crops to maximize productivity, or to derive greatest possible benefit from available resources. Various cultural practices designed to enhance water availability and reduce plant stress are used to obtain these objectives. Cultural practices, for the most part, are designed to optimize growth conditions so that in the energy budget equation, P and R become significant.

Changes in cultural practices for sustainable production systems will reflect changes in farmers' objectives. Rather than maximum productivity, the goal may be maximum economic yield, or maximum return per unit of invested capital, or maximum biological yield per unit of scarce rainfall. More likely, the goal will be to optimize return to a number of these scarce inputs, and to introduce greater stability of yield and economic return into the system. Only by making goals explicit and carefully evaluating a range of potential strategies to achieve those goals can the farmer make rational decisions about inputs and the cultural practices that are more appropriate for a given field or form.

Planting Date

Soil temperature, water availability, soil fertility, and the specific crop need to be considered when selecting an optimum planting date. It is also important to evaluate where each crop fits into a rotation or complex intercrop system.

There is a tendency to plant as early as possible with long-season hybrids or varieties in the temperate zone, or to use the entire rainy season in the tropics. Late maturing sorghum hybrids yielded more than early hybrids (Dalton, 1967). The relationship was linear with an increase in 227 $\pm$ 70 kg ha^{-1} d^{-1} of delayed maturity. Late maturity maize hybrids yielded more than medium or early maturing hybrids when planted at the optimum time, April 26 to May 7, in southeastern North Dakota (Albus et al., 1988). However, with later planting dates, intermediate and earlier maturing hybrids reduced the risk of frost damage to immature kernels and resultant low test weight. Delay in planting date after May 7 decreased corn yields 250 kg ha^{-1} if delayed 1 week, 878 kg ha^{-1} if delayed 2 weeks, and 2006 kg ha^{-1} if delayed 3 weeks (Albus et al., 1988). Delay in planting date of barley (*Hordeum vulgare* L.) from April 20 to May 19 at Powell, Wyoming, resulted in a decrease in yield and seed plumpness (Lauer and Partridge, 1990). Water is available for planting in early April in southern Georgia and other areas of the southern U.S. However, highest yields were obtained with May to early June plantings (Parker et al., 1981). Maturity types are not available for the earlier planting; however, they speculated that cultivars could be developed for earlier planting. Some varieties in their tests were longer maturing but grew too tall and lodged.

There are certain instances when planting is delayed by weather, or when one crop follows another in a double-cropping pattern. Maranville et al. (1971) showed early maturing sorghum planted July 10 after wheat in eastern Nebraska with the application of 187 l ha^{-1} nitrogen-phosphorus (10–34–0) starter solution resulted in yields of 4.7 Mg ha^{-1} as compared to 1.9 Mg ha^{-1} without the starter. The phosphorus contributed to earlier maturing of the sorghum and this resulted in significantly higher yield. Villar et al. (1989) obtained equal or higher yields by planting early maturing sorghum in narrow rows (0.355 m) as compared to longer season hybrids in the more traditional 0.75-m rows. In a study of postoptimum planting dates for soybean, Boquet (1990) found that the earlier date (June 20) resulted in greater productivity than later planting (July 3).

In most cases, the earliest planting date possible allows for use of a large fraction or the entire season (e.g., increase light interception and water use). If planting date is delayed, then crop type and/or production management can be changed to insure that the crop matures and high yields can be attained.

Row Spacing/Plant Density

Both plant density and row spacing directly influence the growth and development of crops and how these species interact with the environment.

Light interception was increased as the row spacing became narrower (1.02, 0.76, and 0.51 m) in grain sorghum (Clegg et al., 1974). Light transmission characteristics were also different for clear and cloudy days.

The rate at which the canopy covers the ground is also influenced by row spacing. Ten days elapsed between the time narrow row sorghum (30-cm rows) intercepted 50% of available light, as compared to the widest row spacing (120-cm rows) in one trial. At maturity, the 30- and 60-cm row spacings intercepted 95% of the light compared to 75% and 68% for 90- and 120-cm rows, respectively (Clegg, 1972). Grain yield of sorghum was highly positively correlated with light interception.

Seeding soybean in narrow (0.48 m) rows resulted in a soybean canopy that covered the soil sooner than in wider (0.81 m) rows (Bitzer et al., 1983b). Also, the soybean in the narrower rows yielded more than in wider rows. For determinate soybean, highest yield occurred in narrow 0.5-m rows as compared to 1.0-m rows (Boquet, 1990). Similar results were obtained by Board et al. (1990). In their study, the yield increase in narrower rows was related to increased light interception.

Changes in plant population will affect plant type and yield components more than changes in row spacing. Schulze (1971) studied the morphological characters of three hybrids and two varieties of grain sorghum. He observed that tillering, stem diameter, leaf area per plant, and leaf angle increased with decreasing plant population, in the range from 54.9 to 2.3 plants per square meter. The yield components of grain sorghum (seed weight, seeds/plant, seeds/head) increased and grain yield decreased as plant population decreased. Among the phenological traits, leaf area per plant increased and stover yield and LAI decreased as plant population decreased. Days to 50% flowering, grain/stover ratio and plant height were affected differently with change in plant population, depending on the variety or hybrid.

Similarly, plant density (in a range from 0.8 to 15.4 plants m^{-2}) exerted a strong influence on growth and yield of maize (Tetio-Kagho and Gardner, 1988). As plant population density pressure on prolific maize hybrids was reduced, yield was adjusted first by higher kernel number per ear and kernel number per ear row. This was followed by higher ear number per plant, kernel number per ear, and finally kernel weight.

In Kentucky, the highest grain yields for dryland maize were obtained at 49,400 plants per hectare (Bitzer et al., 1983a). With irrigation, maize plant population needed to be increased to 69,000 to 74,000 plants per hectare for higher yields. Also, there was no grain yield advantage of maize planted in a 0.76-m row compared to a 0.91-m row spacing unless the crop was irrigated.

In southeastern Queensland, Australia, sorghum severely stressed during boot-anthesis yielded more in a double (1.67 + 0.33 m) row as compared to a narrow (0.33 m) row (Fukai and Foale, 1988). However, there was no significant effect of row spacing for an early flowering cultivar that headed before severe stress developed. Population density trials were conducted with sorghum from 1980 to 1984 when severe water stress occurred (Rees, 1986). When conditions of favorable water availability occurred, increasing plant density resulted in better plant growth and grain yield. The opposite occurred when conditions were poor. Optimum densities for sorghum grain production were 10,000 plants ha^{-1} when it was very dry to over 120,000 plants ha^{-1} in moist conditions. Medium sorghum plant populations were recommended to reduce the risk of complete crop failure.

As previously mentioned, early maturing crops are equated with low yields. However, plants of early maturing crops are usually smaller and often planted in the same row spacings and at the same plant populations as the larger, later maturing cultivars. Most evidence indicates that as row spacing decreases, light interception increases. Thus, for proper comparison productivity should be related to equal light interception, e.g., adjust conditions (row spacing and plant density) so the LAI is the same for all maturities. Comparisons based on these premises were conducted by Villar et al. (1989). A medium maturity hybrid (0.76-m row spacing and 162,600 plants ha^{-1}) was compared to an early maturity hybrid (0.355-m row spacing and 370,500 plants ha^{-1}) at three planting dates. Results showed the short season hybrid yielded equally well or higher than the more conventional hybrid. This was especially evident at the later plantings. In North Dakota, results of population and row spacing studies with early grain sorghum hybrids showed altering plant population had only minor effects on agronomic traits (Schatz et al., 1990). However, sorghum grain yields were highest when the crop was grown in 0.38-m rows with a plant population of 171,900 to 222,300 plants ha⁻¹.

If growing conditions are optimal, a plant density and row spacing for maximum light interception will result in the highest yields. However, if conditions are not optimum (limited water), the canopy will absorb energy regardless of water status and result in plant stress (Chang, 1968). The amount of yield reduction for the crop will depend on the species and the growth stage or stages at which the crop experiences stress. Maximum benefits can be achieved by using a row spacing and plant population best suited to the most likely environmental conditions for each location and season.

Residue/Tillage

Tillage has been considered necessary to form a seedbed for good seed-soil contact, uniform emergence, and stand establishment. Adopting a system that leaves residue on the surface for erosion control results in seedbed characteristics which may not favor plant establishment.

Increasing mulch rates (0 to 12 Mg ha⁻¹) delayed the time at which soil reached a favorable temperature for sorghum germination (Unger, 1978). This was of little consequence at planting, since germination temperatures were adequate for all residue treatments, but growth was slowed early in the season with the higher levels of residue. When water became limiting in the control plots, the sorghum showed stress, and this did not occur in the high mulch plots. Sorghum stands were reduced (by 20%) with no-tillage treatments in a semiarid, subtropical environment (Thomas et al., 1990). Early crop growth was less vigorous in no-tillage plots than with tilled plots, but the growth differences became less as the season progressed. When comparable stands were obtained, yields were virtually the same.

In the northern United States where the season is shorter and low temperatures are encountered, Imholte and Carter (1987) studied the influence of residues on maize growth and production. However, soil temperatures occurred under no tillage and caused reduced plant emergence, delayed emergence and silking, and increased harvest moisture as compared to conventional tillage. However, they concluded no-tillage could be applicable for use if the proper maturity hybrid were chosen and seeding rates were increased to overcome the reduced emergence of maize seedlings.

Munawar et al. (1990) studied the effects of tillage systems, N fertilizer rates and cover crop management on soil temperature, soil moisture, and corn yield. Highest corn yields were obtained using the no-tillage system (4.41 Mg ha^{-1}) as compared to conventional tillage (2.25 Mg ha^{-1}). Cover crop comparisons showed yields were higher if the rye cover crop were killed 3 weeks before planting, as compared to the cover crop that was killed at planting. Early killing of the cover crop resulted in the field retaining more moisture for the crop.

In many instances crop growth is delayed with increased amounts of residue on the surface. Fortin and Pierce (1990) showed plant development was significantly delayed with straw mulch as compared to a bare soil control. However, when comparisons were based on developmental stages, the rate of development of maize was the same for mulch-treated and bare soil.

In a study comparing clean tillage to reduced tillage or no-tillage systems for soybean, Deibert (1989) indicated little difference could be observed for seed yield, seed weight, seed moisture, seed oil concentration, and seed oil yield due to tillage. More water seemed to be available for soybean grown in a no-till system (wheat straw on the surface) since plants showed no stress during a 14-day drought period as compared to soybean in a clean tillage system (Harper et al., 1989).

Experiments by Elmore (1990) compared the influence of tillage, planting date, and variety on soybean growth and yield. Soil temperatures were slightly reduced early in the season for no-till as compared to tilled soybean plots. However, the temperature differences were reduced as the season progressed. Soybean yields were not greatly different between tillage practices. There were no-tillage × variety or tillage × planting date × variety interactions. The data also indicated that the best-yielding varieties in tilled performance tests would most likely be the same varieties in a no-till performance test.

Soil disturbance-residue management studies on winter wheat were conducted by Wilhelm et al. (1989). The amount of residue applied was 0, 0.5, and 1.0 of the amount produced by the wheat crop. Reduced soil disturbance and increased residue decreased maximum and increased minimum soil temperatures. Soil water content was increased with increased residue application. Plant growth was delayed by residues and absence of tillage because of the cooler soil temperatures.

In Oklahoma, reduced tillage delayed maturity of spring wheat (Dao and Nguyen, 1989). Wheat yields were similar for all tillage treatments, but no-tillage gave higher yields in years with cold autumns that had erosive rain or years with dry spring weather.

To improve seed-soil placement, numerous ridge tillage systems have been developed. Ridge tillage is a system that maintains the residue on the surface and then prior to planting the top of the row (or ridge) is removed. This has the advantage of early soil warm up and maintaining erosion control (Griffith et al., 1990). Since the same rows are maintained, traffic can be confined to certain rows and thus localize or control soil compaction. However, the greatest difficulty with ridge tillage is maintenance of the ridge, and erosion may occur on "droughty soils" that may not have adequate residue.

Reduced tillage systems which leave residue on the surface for erosion control will decrease the soil temperature, reduce plant emergence, and delay plant development. Moisture is also conserved with reduced tillage. If adequate stands are obtained with increased planting rates or an improved planting system, yield is not reduced; crop yields are generally improved if conditions occur that will cause drought stress.

It is generally assumed that reduced tillage will lead to greater crop residue on the soil surface, less erosion, and higher yields of crops, especially if moisture is limiting. This would appear to be desirable to increase the stability or sustainability of crop production. As shown in these examples, the situation is much more complicated, and design of an optimum system requires study of a number of factors for each specific site and cropping sequence.

Rotations/Intercropping/Crop Sequences

The amount of literature pertinent to crop rotations, intercropping, or other crop sequences is voluminous. Thus, the literature reported covers those attributes that can be utilized in management.

Clegg (1982) reported that grain yield of sorghum grown after soybean was equal to the grain yield of continuous sorghum fertilized with 76 kg N ha^{-1}. Total nitrogen uptake by sorghum was significantly increased by rotation and estimates of residual soil nitrogen as nitrate after soybean were 50 to 60 kg N ha^{-1} (Gakale and Clegg, 1987).

Crop rotation systems using soybean, sorghum, maize, and red clover (*Trifolium pratense*) were studied by Varvel and Peterson (1990). Rotations with legumes reduced the amount of applied nitrogen needed for optimum maize and sorghum yields. This is favorable for the farmer because it reduces the possibility of leaching residual N and reduces production costs.

Comparisons of growing cotton and sorghum in rotation or in continuous culture (Bilbro, 1972) showed continuous cotton to yield higher. However, sorghum yielded more when grown in rotation with cotton.

A series of rotations using sorghum, oat, fescue (*Festuca arundinacea*) grass, and sweetclover (*Melilotus sp.*) were studied by Adams (1974). Results indicated crop rotations can significantly improve water intake and soil loss for 20 months. Crop yields were improved for up to 4 years. However, in both instances the benefits depended upon the crops and crop sequence in the rotation.

The objective of a study by Peterson et al. (1990) was to determine the effects of previous crop on row crops under rainfed conditions. Maize was the most sensitive to the previous crop as indicated by grain yield variability. Yields of maize, soybean, and sorghum were 74, 25, and 10% higher following fallow as compared to continuous monoculture. In years of average rain, maize-soybean was the highest yielding sequence as compared to years of both above and below normal precipitation where the sorghum-soybean was the highest yielding rotation sequence.

Holderbaum et al. (1990) concluded that fall-seeded legumes into maize could reduce the amount of applied nitrogen needed and provide soil protection in winter months. Similarly, vetch (*Vicia* sp.) legumes were interseeded in mid-September into maize and sorghum as a cover crop (Blevins et al., 1990). The legumes were killed prior to planting the cereal crops the next spring. Grain yield increases were equivalent to those from application of 75 kg N ha^{-1} for maize and 130 kg N ha^{-1} for sorghum. The higher N equivalent for sorghum was due to a longer growth period for the legume before it was killed. They concluded that with these increased grain yields, use of the winter cover crop could be an important cropping strategy.

Crookston et al. (1991) observed first crop maize or soybean after a sequence of soybean or maize, respectively, resulted in 15 to 17% superior yields. Therefore, they concluded for Minnesota that at least a sequence of three and possibly more crops be incorporated into cropping strategies.

Maranville et al. (1971) were able to double-crop sorghum after wheat. In this system, water at planting was critical since the wheat had depleted the soil moisture. Sweetcorn can be successfully double-cropped after canning peas (*Pisum sativum*) in Wisconsin (Ndon and Harvey, 1981). However, precautions need to be taken with respect to weed control and tillage.

Use of crop rotations, intercropping, or other cropping sequences can be used in various cropping strategies. Maximum light interception can be achieved with use of crops tailored for different parts of the season or by more than one crop in the field at the same time. There is often a benefit from nutrient cycling in more complex patterns. Leguminous crops generally contribute to increased nitrogen availability for a succeeding crop. Cover crops used for erosion control during the off-season can also contribute to increased fertility, since fewer nutrients are lost with soil runoff. Choice of the crops in a complex cropping system may provide benefits due to the rotation effect of growing two or more crops in the proper sequence. Other positive consequences of diversity or rotations are less well understood.

Irrigation

If available, irrigation can be used to reduce the effect of water stress. However, water available for irrigation is declining and becoming more expensive. Priority is being directed towards water to be used for human consumption, industry, recreation, and natural wildlife habitats. Productivity per unit of water used is low in agriculture compared to some other industries. Also in certain regions underground stored water has been exhausted by deep well pumping and slow recharge.

Maize is probably the field crop most responsive to irrigation. Although maize yields were increased with irrigation, Bitzer et al. (1983a) concluded that yields were not influenced as much by total amount of water the crop received, as by distribution of the water.

When several irrigation levels were applied to soybean, the highest yields occurred under the water regime that did not allow stress to occur (Doss et al., 1974). The most responsive period to apply water was during pod fill. Yield increased linearly with increased water in this stage, indicating this was the critical period for irrigation. Kadhem et al. (1985) reported that the maturity stages in soybean R = 3.5 and R = 4.5 represent the portion of the reproductive period most responsive to irrigation.

Water applied to soybean double-cropped after wheat increased yields of soybean 700 kg ha^{-1} over nonirrigated soybean (Daniels and Scott, 1991). Sorghum responded similarly in Nebraska (Maranville et al., 1971).

Yield reduction of pearl millet was dependent on the time of stress onset in relation to flowering (Mahalakshmi et al., 1988). Peak use of water by wheat coincided with the period from heading to completed head extension (Rickman et al., 1978). Water was needed during the grain filling period, especially in soils with restricted rooting depths (< 150 cm).

Research indicates that water needs to be applied at the inception of stress and at growth stages such as flowering and grainfill when the crops are susceptible to stress. Thus, with planning, application of water when crop needs are critical will reduce the amount of water needed and increase yields. Conservation of water will be even more important in the near future, and efficient water use is a critical dimension of sustainable production systems.

Harvest/Quality

Harvest can be influenced by many variables. A proper harvest time and conditions are necessary to insure a quality product. This may be influenced by the climate, but also by cultural practices. Awareness of the influence of cultural practices on quality factors can aid in decisions on harvest methods and timing of harvest.

Crop maturity varies significantly among varieties in most crops (Nelson et al., 1991b,c). Genotype selection can aid in obtaining grain sorghum and maize types that have a lower moisture at harvest. This gives the producer the opportunity to choose varieties that will mature earlier or dry faster to insure low moisture at harvest and adequate test weight.

Irrigation increased kernel breaking susceptibility (KBS) in maize. In contrast, increased nitrogen decreased KBS. Kernel density (KD) was increased or decreased with irrigation or nitrogen depending on the type of maize hybrid (Kniep and Mason, 1989). In all 3 years of the study, irrigation increased yield and decreased protein content. Increased nitrogen applications increased protein content. Opaque-2 maize had considerably higher lysine, although total grain yield was not as high as normal maize (Kniep and Mason, 1991).

Although total yield of soybean is the factor that contributes mostly to its economic return, extra value of soybean, based on the oil and protein content of the seed, can be estimated. These factors indicate premiums could be used in buying soybean based on quality (Nelson et al., 1991a).

Growing sorghum after soybean in rotation hastens maturity and increases protein content and total protein (Gakale and Clegg, 1987). Proper application of nitrogen also will improve the protein quality and total protein.

Consideration of quality factors such as protein and oil in the economic structure would result in decisions that may not depend on maximum yields. Input levels would be determined by their availability and management control increasing the efficiency of their use. To achieve more sustainable yields and economic returns, grain quality and alternative products from the grain crops need to be considered.

Models

As early as 1920, Mooers (1920) recognized the variability of different maize varieties in relation to plant population. From that research, he described mathematical formulas (models) for predicting the plant population to use. He concluded the best planting practice was to use a row width that permits satisfactory use of tillage equipment and that allowed for the determined number of stalks to be as widely spaced as possible.

Knowledge of the water status of the soil is important for making decisions for crop choice and water application if irrigation is available. A soil water assessment model for corn, wheat, sorghum, and soybean grown on reasonably flat terrain, using near-real time weather data accurately estimated soil water during the growing season (Robinson and

Hubbard, 1990). This can be used as a management tool for the efficient use of water.

A model to predict the development of a maize canopy has been presented by Retta et al. (1991). Accurate estimation of phenology is important because most diseases and insects are dependent on stage of growth of a crop. Sensitivity of this model to light and water also indicated it could be modified to mimic weed competition. This could allow a producer to estimate the economic threshold of weeds to reduce crop loss.

A management model for budgeting and scheduling crop production operations is available (Chen, 1986). Models for chemical runoff and leaching and estimating surface residue can be obtained (Clark, 1991). PRE-AP provides information on the rate chemical compounds run off or leach into the groundwater. This model is designed around a model called GLEAMS (Groundwater Loading Effects of Agricultural Management Systems). Information from models can be presented graphically. Growers also can be provided surface residue information of 21 crops due to tillage and decomposition with a residue management decision support program called RESMAN. The program analyzes the grower's tillage practices and makes recommendations to obtain the final desired residue coverage.

Many models are available that can be used for decision input for crop management. Various programs using these models are listed in farm, computer, and professional magazines or furnished by educational and agricultural industries. Information furnished by these programs can substantiate and guide a producer in decision making. There is still a need to develop integrative models that allow combinations of fields, crops, and producer objectives on a whole-farm basis. The fine-tuning of production practices for specific fields and farms, and to help each family realize their unique goals, can contribute to a more sustainable agriculture.

CONCLUSIONS

High crop yields from a hectare of land require an intense level of management. Today crop production management is even more complex. Environmental concerns of chemical contamination of water, residual chemical contamination of food and feed, soil erosion, water availability, and many more are entered into the production equation. Although maximum yields may be desired, many times maximum crop yields are not the most profitable (Colville, 1966). With the many factors to consider in our crop production systems, it becomes difficult to determine the emphasis that should be given to each factor.

This complicated process can be less difficult if we can depend on computers to aid in decision making. Today personal computers are relatively inexpensive and have increased speed, memory, and internal storage for data analysis, prediction, and graphics. These will allow the use of more complex programs needed for tomorrow. Networking is available for access to the information data bases needed to predict crop growth and yield, moisture use, fertilizer requirements based on soil test and previous crops, probability statistics, and financial costs. These tools, coupled with greater information on the production history of the farm, should aid in management decisions for efficient crop production. More efficient resource use can lead to reduced production costs, less impact on the off-farm environment, and a more sustainable food production system.

REFERENCES

Adams, J. E. 1974. Residual effects of crop rotations on water intake, soil loss, and sorghum yield. *Agron. J.* 62:299-403.

Albus, W. L., J. Weigel, and H. Sayfikar. 1988. The effect of planting date on irrigated corn in southeastern North Dakota. *N.D. Farm Res. Agric. Exp. St.* 47:22-25.

Alessi, J. and J. F. Power. 1974. Effects of plant population, row spacing and relative maturity on dryland corn in the northern plains. *Agron. J.* 66:316-319.

Bell, A. A. 1982. Plant pest interaction with environmental stress and breeding for pest resistance: plant diseases. In: *Breeding Plants for Less Favorable Environments*. John Wiley & Sons, New York. pp. 335-364.

Bilbro. J. D. 1972. Yield probabilities for cotton and grain sorghum grown under dryland conditions of the Texas high plains. *Agron. J.* 64:140-142.

Bird, G. W., T. Edens, F. Drummond, and E. Groden. 1990. Design of pest management systems for sustainable agriculture. In: *Sustainable Agriculture in Temperate Zones*. C. A. Francis, C. B. Flora, and L. D. King (Eds.). John Wiley & Sons, New York. pp. 55-110.

Bitzer, M. J., C. G. Poneleit, and T. Martin. 1983a. Effect of irrigation, row width, and plant population on corn yields. *K. Agric. Exp. Stn. Prog. Rep.* p. 33.

Bitzer, M. J., J. H. Herbek, D. Pilcher, T. W. Pfeiffer, and C. R. Tutt. 1983b. Row width studies for full-season soybeans. *K. Agric. Exp. Stn. Prog. Rep.* p. 46.

Blevins, R. L., J. H. Herbek, and W. W. Frye. 1990. Legume cover crops as a nitrogen source for no-till corn and grain sorghum. *Agron. J.* 82:769-772.

Blum, A. 1985. Breeding crop varieties for stress environments. *Crit. Rev. Plant Sci.* 2:199-238.

Board, J. E., B. G. Harville, and A. M. Saxon. 1990. Narrow-row seed-yield enhancement in determinate soybean. *Agron. J.* 82:64-68.

Boquet, D. J. 1990. Plant population density and row spacing effects on soybean at post-optimal planting dates. *Agron. J.* 82:59-64.

Brun, L. J., E. T. Kanemasu, and W. L. Powers. 1972. Evapotranspiraton from soybean and sorghum fields. *Agron. J.* 64:145-148.

Campbell, G. S. 1977. *An Introduction to Environmental Biophysics.* Springer-Verlag, New York.

Chang, J.-Hu. 1968. *Climate and Agriculture—An Ecological Survey.* Aldine Publishing, Chicago.

Chen, L. H. 1986. Microcomputer model for budgeting and scheduling crop production operations. *Trans. Am. Soc. Agric. Eng.* 29:908-911.

Christiansen, M. N. and C. F. Lewis (Eds.). 1982. *Breeding Plants for Less Favorable Environments.* John Wiley & Sons, New York.

Clark, B. 1991. Tech talk. *Ag. Consultant*, August, p. 12.

Clark, R. B., J. W. Maranville, and H. J. Gorz. 1978. Phosphorus efficiency of sorghum grown with limited phosphorus. In: *Plant Nutrition 1978*. A. R. Ferguson, R. L. Bielski, and I. B. Ferguson (Eds.). *Proc. 8th Int. Colloq. Plant Anal. Fert. Prob., Auckland, New Zealand*, pp. 93-99.

Clegg, M. D. 1972. Light and yield related aspects of sorghum canopies. In: *Sorghum in the Seventies*, N. G. P. Rao and L. R. House (Eds.). Oxford and IBH Publishing, New Delhi. pp. 279-301.

Clegg, M. D. 1982. Effect of soybean on yield and nitrogen response of subsequent sorghum crops in eastern Nebraska. *Field Crops Res.* 5:233-239.

Clegg, M. D., W. W. Biggs, J. D. Eastin, J. W. Maranville, and C. Y Sullivan. 1974. Light and transmission in field communities of sorghum. *Agron. J.* 66:471-476.

Colville, W. L. 1966. Environment and maximum yield of corn. In: *Maximum Crop Yield—Challenge.* ASA Spec. Publ. No. 9:21-26.

Crookston, R. K., J. E. Kurle, P. J. Copeland, J. H. Ford, and W. E. Lueschen. 1991. Rotational cropping sequence affects yield of corn and soybean. *Agron. J.* 83:108-113.

Dalton, L. G. 1967. A positive regression of yield on maturity in sorghum. *Crop Sci.* 7:271.

Daniels, M. B. and H. D. Scott. 1991. Water use efficiency of double-cropped wheat and soybean. *Agron. J.* 83:564-570.

Dao, T. H. and H. T. Nguyen. 1989. Growth response of cultivars to conservation tillage in a continuous wheat cropping system. *Agron. J.* 81:923-929.

Deibert, E. J. 1989. Soybean cultivar response to reduced tillage systems in northern dryland areas. *Agron. J.* 81:672-676.

Devine, T. E. 1982. Genetic fitting of crops to problem soils. In: *Breeding Plants for Less Favorable Environments.* M. N. Christiansen and C. F. Lewis (Eds.). John Wiley & Sons, New York. pp. 143-173.

Doss, B. D., R. W. Pearson, and H. T. Rogers. 1974. Effect of soil water stress at various growth stages on soybean yield. *Agron. J.* 66:297-299.

Edwards, C. A., R. Lal, P. Madden, R. H. Miller, and G. House (Eds.). 1990. *Sustainable Agricultural Systems.* Soil and Water Conservation Society, Ankeny, IA.

Elmore, R. W. 1990. Soybean cultivar response to tillage systems and planting date. *Agron. J.* 82:69-73.

Fehr, W. R. 1987. *Principles of Cultivar Development, Theory and Technique.* Vol. 1. Macmillan, New York.

Fortin, M. C. and F. J. Pierce. 1990. Developmental and growth effects of crop residues on corn. *Agron. J.* 82:710–715.

Francis, C. A. (Ed.) 1986. *Multiple Cropping Systems.* Macmillan, New York.

Francis, C. A. 1990. Breeding hybrids and varieties for sustainable systems. II. In: *Sustainable Agriculture in Temperate Zones.* C. A. Francis, C. B. Flora, and L. D. King (Eds.). John Wiley & Sons, New York. pp. 24–54.

Francis, C. A., C. B. Flora, and L. D. King (Eds.). 1990. *Sustainable Agriculture in Temperate Zones.* John Wiley & Sons, New York.

Fukai, S. and M. A. Foale. 1988. Effects of row spacing on growth and grain yield of early and late cultivars. *Aust. J. Exp. Agric.* 28:771–777.

Gakale, L. P. and M. D. Clegg. 1987. Nitrogen from soybean for dryland sorghum. *Agron. J.* 79:1057–1061.

Griffith, D. R., S. D. Parsons, and J. V. Mannering. 1990. Mechanics and adaptability of ridge-planting for corn and soya bean. *Soil Tillage Res.* 18:113–126.

Harper, L. A., J. E. Giddens, G. W. Langdale, and R. R. Sharpe. 1989. Environmental effects on nitrogen dynamics in soybean under conservation and clean tillage systems. *Agron. J.* 81:623–631.

Hattendorf, M. J., M. S. Redelfs, B. Amos, L. R. Stone and R. E. Gwin, Jr. 1988. Comparative water use characteristics of six row crops. *Agron. J.* 80:80–85.

Holderbaum, J. F., A. M. Decker, J. J. Meisinger, F. R. Mulford and L. R. Vough. 1990. Fall-seeded legume cover crops for no-tillage corn in the humid east. *Agron. J.* 82:117–124.

Imholte, A. A. and P. R. Carter. 1987. Planting date and tillage effects on corn following corn. *Agron. J.* 79:746–751.

Jenkins, J. N. 1982. Plant pest interactions with environmental stress and breeding for pest resistance. In: *Breeding Plants for Less Favorable Environments.* M. N. Christiansen and C. F. Lewis (Eds.). John Wiley & Sons, New York. pp. 365–374.

Kadhem, F. A., J. E. Specht, and J. H. Williams. 1985. Soybean irrigation serially timed during stages R1 to R6. II. Yield component responses. *Agron. J.* 77:299–304.

Kniep, K. R. and S. C. Mason. 1989. Kernal breakage and density of normal and opaque-2 maize grain as influenced by irrigation and nitrogen. *Crop Sci.* 29:158–163.

Kniep, K. R. and S. C. Mason. 1991. Lysine and protein content of normal and opaque-2 maize grain as influenced by irrigation and nitrogen. *Crop Sci.* 31:177–181.

Lauer, J. G. and J. R. Partridge. 1990. Planting date and nitrogen effects on spring malting barley. *Agron. J.* 82:1083–1088.

Leopold, A. C. and P. E. Kriedemann. 1975. *Plant Growth and Development,* 2nd ed. McGraw-Hill, New York.

Mahalakshmi, V., F. R. Bidinger, and G. D. P. Rao. 1988. Timing and intensity of water deficits during flowering and grain-filling in pearl millet. *Agron. J.* 80:130–135.

Makeham, J. P. and L. R. Malcom. 1981. *The Farming Game*. Gill Publications, Armidale, NSW, Australia.

Maranville, J. W., R. B. Clark, and W. M. Ross. 1980. Nitrogen efficiency in grain sorghum. *J. Plant Nutr*. 2:577–589.

Maranville, J. W., H. W. Wittmus, M. D. Clegg, and J. D. Eastin. 1971. Using sorghum for double cropping. *Farm Ranch and Home Quarterly* (spring).

Marshall, B. and R. W. Willey. 1983. Radiation interception and growth in an intercrop of pearl millet/groundnut. *Field Crop Res*. 7:141–160.

Martin, J. H., W. H. Leonard, and D. L. Stamp. 1976. *Principles of Field Crop Production*, 3rd ed. Macmillan, New York.

Mooers, C. A. 1920. Plant rates and spacing for corn under southern conditions. *Agron. J*. 12:1–22.

Munawar, A., R. L. Blevens, W. W. Frye, and M. R. Saul. 1990. Tillage and cover crop management for soil water conservation. *Agron. J*. 82:773–777.

Ndon, B. A. and R. G. Harvey. 1981. Influence of herbicides and tillage on sweet corn double cropped after peas. *Agron. J*. 73:791–795.

Nelson, L. A., R. W. Elmore, R. S. Moomaw, G. W. Hergert, and R. N. Klein. 1991a. Nebraska soybean variety tests, 1990. Ext. Circ. 90–104. University of Nebraska-Lincoln, Institute of Agriculture and Natural Resources, Agricultural Research Division, and Cooperative Extension. 49 pp.

Nelson, L. A., R. W. Elmore, P. T. Nordquist, R. N. Klein and D. D. Baltensperger. 1991b. Nebraska grain sorghum hybrid tests, 1990. Ext. Circ. 90–106. University of Nebraska-Lincoln, Institute of Agriculture and Natural Resources, Agricultural Research Division, and Cooperative Extension, 37 pp.

Nelson, L. A., R. S. Moomaw, R. W. Elmore, P. T. Nordquist, R. N. Klein and D. D. Baltensperger. 1991c. Nebraska corn hybrid tests, 1990. Ext. Circ. 90–105. University of Nebraska-Lincoln, Institute of Agriculture and Natural Resources, Agricultural Research Division, and Cooperative Extension.

Nielsen, K. F. 1971. Roots and root temperatures. In: L. W. Carson (Ed.) *The Plant Root and Its Environment*. University Press of Virginia, Charlottesville, VA. pp. 293–333.

Parker, M. B., W. H. Marchant, and B. J. Mullinis, Jr. 1981. Date of planting and row spacing effects on four soybean cultivars. *Agron. J*. 73:759–762.

Peterson, T. A., C. A. Shapiro, and A. D. Flowerday. 1990. Rainfall and previous crop effects on crop yields. *Am. J. Altern. Agric*. 5:33–37.

Quisenberry, J. E. 1982. Breeding for drought resistance and plant water use efficiency. In: *Breeding Plants for Less Favorable Environments*. M. N. Christiansen and C. F. Lewis (Eds.). John Wiley & Sons, New York. pp. 193–212.

Rees, D. J. 1986. Crop growth and development and yield in semi-arid conditions in Botswana. I. The effects of population density and row spacing on *Sorghum bicolor*. *Exp. Agric*. 22:153–167.

Retta, A., R. L. Vanderlip, R. A. Higgins, L. J. Moshier, and A. M. Feyerherm. 1991. Suitability of corn growth models for incorporation of weed and insect stresses. *Agron. J*. 83:757–765.

Rickman, R. W., R. R. Allmaras, and R. E. Ramig. Root-sink descriptions of water supply to dryland wheat. *Agron. J*. 70:723–728.

Ritchie, J. T. and E. Burnet. 1971. Dryland evaporative flux in a subhumid climate: II. Plant influences. *Agron. J.* 63:55–62.

Robinson, J. M. and K. G. Hubbard. 1990. Soil water assessment model for several crops in the high plains. *Agron. J.* 82:1141–1148.

Roder, W., S. C. Mason, M. D. Clegg, and K. R. Kneip. 1989. Yield-soil water relationships in sorghum-soybean cropping systems with different fertilizer regimes. *Agron. J.* 81:470–475.

Rubel, D. 1935. The replaceability of ecological factors and the law of the minimum. *Ecology*, 16:336–341.

Sasser, J. N. 1982. Plant pest interactions with environmental stress and breeding for pest resistance: Nematodes. In: *Breeding Crops for Less Favorable Environments*. M. N. Christiansen and C. F. Lewis (Eds.) John Wiley & Sons, New York. pp. 375–390.

Schulze, L. D. 1971. The Effect of Plant Population on Sorghum (*Sorghum bicolor* L.). p. 78, M.S. Thesis, University of Nebraska.

Shatz, B. G., A. A. Schneiter, and J. C. Gardner. 1990. Effect of plant density on grain sorghum production in North Dakota. *N.D. Farm Res. N.D. Exp. Stn.*, 47:15–17.

Thomas, G. A., J. Standley, H. M. Hunter, G. W. Blight, and A. A. Webb. 1990. Tillage and crop residue management affect Vertisol properties and grain sorghum growth over seven years in the semi-arid sub-tropics. 3. Crop growth, water use and nutrient balance. *Soil Tillage Res.* 18:389–407.

Tetio-Kagho, F., and F. P. Gardner. 1988. Responses of maize to plant population density. II. Reproductive development, yield, and yield adjustments. *Agron. J.* 80:935–940.

Unger, P. W. 1978. Straw mulch effects on soil temperature and sorghum germination and growth. *Agron. J.* 78:858–864.

Varvel, G. E. and T. A. Peterson. 1990. Residual soil nitrogen as affected by continuous, two-year, and four-year crop rotation systems. *Agron. J.* 82:958–962.

Villar, J. L., J. W. Maranville, and J. C. Gardner. 1989. High density sorghum production for late planting in the central great plains. *J. Prod. Agric.* 2:333–338.

Wilhelm, W. W., H. Bouzersour, and J. F. Power. 1989. Soil disturbance-residue management effect on winter wheat growth and yield. *Agron. J.* 81:581–588.

Zandstra, H. C. 1977. Cropping systems research for the Asian farmer. In: *Proc. Symp. Cropping Systems Res. and Dev. for the Asian Rice Farmer*. IRRI, Los Banos, Laguna, Philippines.

Zweifel, T. R., J. W. Maranville, W. M. Ross, and R. B. Clark. 1987. Nitrogen fertility and irrigation influence on grain sorghum nitrogen efficiency. *Agron. J.* 79:419–422.

6

Pest Management–Weeds

Frank Forcella and Orvin C. Burnside

TABLE OF CONTENTS

INTRODUCTION

Weeds continue to plague agriculture today as they have done in the past, but the time and effort devoted to weed management by producers has decreased considerably. This savings in labor requirements can be attributed largely to increased use of chemical weed control. In addition, chemical weed control has enhanced crop yields not only through better weed suppression, but also by permitting important management modifications, e.g., changing corn culture from checked hill planting to high-density, narrow rows. However, chemical weed control is not a panacea for weed management. Herbicides suffer a number of disadvantages with respect to sustainable agriculture. Although many of these disadvantages may be more perceived than real, society is becoming adamant in dictating that alternative forms of weed control be implemented.[48]

Weed management methods can be divided into at least four overlapping categories: biological, cultural, physical, and chemical control. The first three methods, of course, are the major alternatives to chemical control. Combination of two or more of these methods is referred to as integrated weed management. Weed management always has been integrated (Figure 1); however, from the advent of agriculture to the early 20th century, physical weed control (hand-weeding and hoeing, and subsequently tillage) dominated over other methods. During most of this

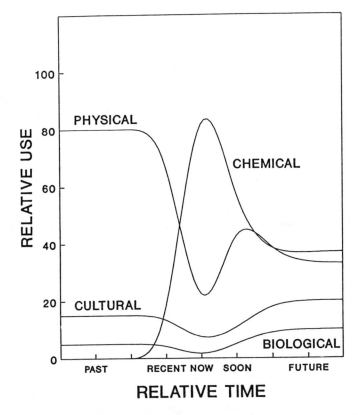

FIGURE 1.   Relative use of four weed control methods in the past, present, and future in the United States.

time, cultural and biological control of weeds probably occurred more fortuitously than purposefully. With the development of herbicides, chemical control quickly dominated weed management methods. Consequently, all alternative methods decreased in importance, although physical control still was used more widely than cultural control, which in turn ranked higher than biological control. This ranking remains so today (Figure 1).

What will be the future for weed management? Over the next 10 to 20 years we expect chemical weed control to decrease in importance, due primarily to social and environmental concerns. Simultaneously, cultural and biological weed control will increase to levels of importance greater than those they experienced prior to the herbicide revolution. Physical weed control will experience a resurgence of importance because it is the easiest to substitute for chemical control. Renewed soil degradation due to excessive tillage subsequently may lessen the importance of physical weed control, perhaps stimulating use of new and safer herbicides. In any event, sustainable weed management in the future, which is the only type of weed management possible, will have a more equitable distribution or integration of control categories (Figure 1). The final rankings may be: chemical $\approx$ physical > cultural > biological control. The rationale for these rankings is:

1. Chemicals will continue to provide a "quick fix" for weed problems. Fortunately, new chemicals will be safer and will be used more judiciously in a truly integrated weed management program.
2. Both traditional and novel mechanical implements will be important aids in future weed management. Their use will be coupled with knowledge of weed seedling ecology.
3. Crops, including smother crops, engineered for high competitive ability will compliment physical control.
4. Biocontrol agents, notably plant pathogens and their metabolites, will be used more likely in the management of specific weeds.

Future weed management also will be information intensive. Farm managers will increasingly integrate weed, soil, weather, and control information with their management experience. Rapid information processing will be necessary, and consequently computer models, decision aids, and expert systems will be used routinely for economical and environmental optimization of weed management systems. Although today these techniques are in their infancy, we expect rapid progress in these information systems within 10 to 20 years.

Weed science will experience in the near future a rapid evolution toward a greater understanding of weed biology and ecology. Research in physiology, biochemistry, and genetics of herbicide/plant interactions will continue, but in a diminished role. Bioengineering of more efficient crops will occur. Agricultural engineers will place increased emphasis on mechanical control implements and directed herbicide applicators. Greater cooperation will occur among economists, sociologists, and weed scientists. And finally, all relevant disciplines must not only continue exploring their current fields of interest, but must integrate their research into total farm management models.

We will amplify and rationalize the above speculations through the reviews, analyses, and ideas presented. Examples from the North American Corn Belt will be emphasized, since this area is one of the world's largest recipients of herbicides.

WEED SANITATION

Every farm manager should be a weed sanitation official, with a working knowledge of (1) those weed species likely to become problems, (2) the manner in which weed propagules are likely to spread to and within farms, and (3) the economic consequences of ineffective sanitation. Analogously, weed sanitation personnel of every county, state, region, and country should have similar knowledge or awareness.

Over 2000 plant species have been considered weeds in the United States, and 1400 species are considered important enough to be placed on the "Composite List of Weeds."[141] Holm et al.[63] estimated that at least 1200 additional weed species have not yet entered the country. More recently, Rapoport[113] calculated there to be 26,000 species of weeds and potential weeds worldwide. Which of these weeds, or other species not yet considered as weeds, will enter the United States? Of those that do enter the country, which will become economically important and which will become mere roadside waifs? Writing for the U.S. Department of Agriculture Animal and Plant Health Inspection Service, Westbrooks[142] indicates that weed control should be the last resort of weed management. Weed management strategies should begin with: (1) prevention of weed propagules from entering commerce; (2) exclusion when propagules are detected at ports of entry; (3) detection, containment, and eradication of incipient colonies; and, lastly, (4) perpetual control of widespread species.

Recent research on weed invasions has attempted to provide objective criteria by which plants can be screened for their potential threat as weeds prior to colonization.[54,55] Once colonization has occurred, species distributions can be monitored and categorized objectively as to which has the greatest potential for subsequent spread.[46] Potential ecological range studies have considerable value in this regard, and techniques employed in Australia[104] and the United States[106] are at the forefront of this endeavor. On smaller scales of weed invasion, economic impacts[3] and eradication strategies[153] have been studied.

Although weed invasions are recognized as a multiscaled problem,[142] agencies responsible for intervention along that scale are not especially well coordinated. In any event, final responsibility for responding to agricultural weed invasions remains with the individual farm manager.

Weeds Invading Farms

Weeds have innumerable mechanisms of spread; however, by far the most important for agriculture is as contaminants of crop seeds and livestock feeds. Despite weed seed contamination of commercial crop seed being much less than 1% (a common allowance), even such low allowances are ample for successful colonization of weed species. Thorough examination of purchased seed is necessary. Purchased hay may originate from over 1000-km distance, especially during regional droughts, and contain new species of noxious weeds.[113] Seeds of many weed species are not killed by passage through intestinal tracts, and with an intestinal retention time of 5 to 7 days, unconfined livestock may void weed seeds great distances from points of ingestion. An acceptable treatment combination for contaminated livestock feeds would be ensiling, feeding ensilage to confined livestock, and composting manure.[8]

Livestock disperse weed propagules independent of feed stuffs, e.g., by adhesion to fur. Because livestock may be transported long distances, particularly during regional droughts,[131] the animals and their transport vehicles may carry weed species alien to the livestock destination.

Farm equipment, especially tillage implements, are responsible for short-range transport of weed propagules because of adherence to equipment of rhizomes, roots, and seed-laden soil. Harvest equipment easily carry and readily disseminate weed seeds. Custom combines may transport weed seeds thousands of kilometers from Mexico to Canada as they follow the wheat harvest.

Surface irrigation water is a common source of weed seeds.[39] Irrigation water becomes increasingly contaminated with weed seeds as it progresses through canals.[149] Screens on irrigation systems may trap some seed.[110] Cooperation among landowners who share irrigation/drainage systems is important for weed sanitation.

Other human-mediated sources of propagules of new weed species are ballast; sand, gravel, and soil fill materials; grass sods; nursery stock; and escaped cultivated plants.

Weed Laws

The first seed law in the United States that regulated weed seed content was passed by Connecticut in 1921 to prohibit the sale of forage-grass seed containing Canada thistle (*Cirsium arvense*). Many nations have adopted legislation dealing directly with weed sanitation. In the U.S., both federal and state laws have a significant impact on weed prevention, but improvements are necessary.[15,142]

Seed laws have been enacted by the United States government as well as each of its states. Considerable uniformity exists among these laws because of the work of a joint committee of the International Crop Improvement Association, Association of American Seed Control Officials, Association of Official Seed Analysts, and American Seed Trade Association. Seed laws in the United States have been important in reducing the spread of weed species as well as in increasing the use of adapted high quality crop seeds. Crop seed certification standards, which have been developed by individual state crop improvement associations, have set high and widely emulated standards for sanitation of commercial crop seeds.

The Federal Noxious Weed Law was passed in 1974. Enforcement and compliance are expensive; consequently, the law has not been fully effective. The National Association of State Departments of Agriculture drafted a model uniform state weed law based on laws from California, Kansas, and Minnesota. In general, seed and preventive laws have been only moderately effective, because they lack widespread support and adequate incentives.[15,142] Thus, many weeds that could be contained become widespread. Unfortunately, once a weed species becomes widespread, eradication is then not feasible.[15,153]

Quarantine of relatively small areas infested with particularly noxious weeds has occurred. In North and South Carolina, quarantine and eradication of witchweed (*Striga asiatica*) and itchgrass (*Rottboellia cochinchinesis*) have been attempted with some success, but also at a very high cost.

CULTURAL WEED CONTROL

Cultural weed control includes use of management techniques such as sowing date, seeding rate, row spacing, crop species and variety selection, crop rotation, and smother crops.

Sowing Date

Sowing date has distinct but differential effects on crop/weed interactions. The differences arise when comparing cool- and warm-season plants, and because of the difficulty in separating the effects on weed control of sowing date and its immediately preceding seedbed preparation.

Early sowing of cool-season crops, such as spring wheat or lupins, enhances the ability of the crop to compete with weeds and decreases weed competition and fecundity.[105,109] This apparently occurs because cool-season crops germinate and commence vigorous vegetative growth before many common summer annual weeds.

In contrast, early sowing of warm-season crops, such as soybean, often is penalized because of severe weed competition. This occurs because low soil temperatures retard crop germination and thereby eventually permit simultaneous emergence of both crop and weeds. With delayed sowing, most of the current year's population of annual weed seedlings already have emerged and are killed by seedbed preparation.[49] This phenomenon may be illustrated with early- and late-sown 'Evans' soybean in central Minnesota (Forcella, unpublished data). In the absence of herbicides, yields of early- (May 1) and late-planted (June 1) soybean, averaged over 2 years (1987 and 1988), were 1.5 and 2.2 Mg/ha in adjacent subplots. Soybean yield in both early- and late-sown weed-free plots averaged 2.6 Mg/ha. Midseason topgrowth of green and yellow foxtail (*Setaria viridis* and *S. glauca*) associated with these yield reductions was 0.56 (early sowing) and 0.14 Mg/ha (late sowing). Also in Minnesota, Gunsolus[60] noted decreased yields and increased giant foxtail (*S. faberi*) densities with early- (May 12) as opposed to late-sown (June 2) soybeans. Similarly, in an Ontario study, soybean yield, averaged over 3 years, was reduced 45% by jimsonweed (*Datura stramonium*) in early (mid-May) sowings, but only 30% in late (early June) sowings.[139] In Wisconsin, corn yields and wild-proso millet (*Panicum miliaceum*) control increased when corn was planted in mid-May rather than late April.[62]

Successes in weed control due to sowing date probably would be transitory if delayed sowing was attempted over a number of years. This would occur because of the abundance of both warm- and cool-season weed species whose varied physiologies may preadapt them to continuous delayed-sowing environments, e.g., the protracted emergence of barnyard grass (*Echinochloa crus-galli*) in late spring and summer may preadapt it to delayed-sowing crops.

Seeding Rate

Increased seeding rates of solid-seeded crops (e.g., small-grain cereals, peas, etc.) reduces weed growth and increases associated crop yields.[25,80] In contrast, in row crops, where row widths are greater than 50 cm, increasing within-row density of crop plants above that normally recommended has little influence on weed growth and/or weed-crop competition.[42] This distinction based upon row-width probably occurs for several reasons, but primarily because of light interception. In row crops, abundant light penetrates crop canopies during the first several weeks after planting regardless of within-row crop density. Weeds between the rows may capture such light directly, whereas weeds within the row still may capture some of this light tangentially. On the other hand, increased seeding rates of solid-seeded crops tends to broadly maximize light interception by the crop while minimizing that available to weeds.

Row-Spacing

In row crops, row widths are the result of a compromise farmers make with weeds. Equidistant crop spacing within and between rows (low rectangularity) theoretically results in maximum yield.[45] Spatial arrangement of high-density crops, such as wheat, does not affect weed competition.[84] However, spatial arrangement does affect weeds in low-density row crops, but it simultaneously inhibits interrow cultivation.

In the absence of herbicides in row crops, the spacing between rows has a greater effect on yield than within-row density,[42] i.e., weed competition is less with low than with high crop rectangularity. Reduced row spacing increases light interception by the crop canopy and lessens the quantity of light incident on leaves of any weeds that may be present. Accordingly, narrow row-spacing will increase the ability of the crop to compete with weeds and often reduce the crop's requirement for herbicides.

In the absence of either chemical or mechanical weed control, reducing soybean row width from the conventional 76-cm spacing to 25 cm reduced the biomass of a mixture of several common annual weeds by 52%.[16] Using the same soybean row spacings as above, Legere and Schreiber[76] showed that both biomass and leaf area of redroot pigweed (*Amaranthus retroflexus*) was halved in narrow rows. Equations developed for sicklepod (*Cassia obtusifolia*) and common cocklebur (*Xanthium strumarium*) in soybean reveal almost identical reductions of weed biomass in 25-cm compared to 76-cm row spacings.[107]

Reductions of both weed seed production and biomass are important in weed management. Narrow row (20 cm) soybean reduced pitted morning glory (*Ipomoea lacunosa*) seed production by 90% in comparison to that in 76-cm-wide rows.[65]

Historically, herbicide rate requirements have been determined only after innumerable field tests across broad regions. Rates for row crops normally were determined for plantings with conventional 76-cm row widths. Narrow rows should decrease herbicide rate requirements, and the relationship between row spacing and herbicide requirement is roughly proportional. That is, if row spacing is reduced by 50%, then the requisite rate of an appropriate herbicide may be halved. Burnside and Colville[16] provided one of the earliest examples of such a reduced rate requirement using preemergence herbicide applications in soybean (Figure 2). In the absence of both tillage and herbicide, soybean yields increased linearly as row width decreased. With a chloramben application of 1.1 kg/ha (about one third the recommended rate), wide-row soybean yields were appreciably lower than those at chloramben rates of 2.2 and 4.5 kg/ha. However, as row width decreased to 50 and 25 cm, soybean yields at all chloramben application rates were equivalent. Recently, these results were duplicated using postemergence herbicide applications on soybean in western Minnesota (Forcella, unpublished data). 'Evans' soybean, planted in 76-cm rows (interrows cultivated) and 38-cm rows (no cultivation), received postemergence applications of sethoxydim at either 0.30 or 0.15 kg ai/ha (full and half recommended rates). Green

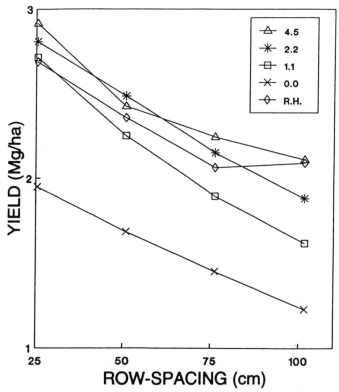

FIGURE 2. Soybean yields as influenced by row spacing, rates (kg a.i./ha) of chloramben, and rotary hoeing (R.H.) in eastern Nebraska. Chloramben application rates were approximately 0, 50, 100, and 200% of labeled recommendations. (Data adapted from Burnside and Colville [1964]).

and yellow foxtail control was similar (>85%), and soybean yield was identical (2.6 Mg/ha) in all treatments.

Herbicide rate reductions with decreased row spacings are applicable to crops other than soybean. Weed control and crop yields of both narrow row sunflower (25 cm) and corn (38 cm) receiving 25 to 33% of their recommended herbicide application rates were equivalent to that of their conventional (76-cm rows) counterparts receiving full herbicide rates plus interrow cultivation.[53]

Decreased row widths not only permit decreases in herbicide application rates, but also decreased frequency of postemergence "tillage." In soybean, for example, chloramben plus rotary hoeing in 20- to 40-cm rows substituted for chloramben plus 2 to 3 interrow cultivations in 61- to 102-cm-wide rows.[108]

With respect to weed control, there are limits as to crop row widths. For example, no differences in wild-proso millet control were observed between wide (108 cm) and narrow row (76 cm) corn.[62] In this case, the "narrow" rows exceeded the minimum row width within which crop canopies can usurp space quickly and shade competing weeds.

Crop and Variety Selection

The effects of crop and variety selection on weed/crop interference depend upon the competitive abilities of the crop and weed. Competitive ability is a function of numerous factors, the most important being the development rate and duration of the crop canopy, which includes rapidity of emergence, height growth, and lateral expansion. The importance of rapid root development also must be important, but this aspect of crop ecology is less well documented. A quantitative function that integrates these factors would be a useful tool in weed management.

Two examples from Colorado illustrate the interactions of crop canopy development and duration. Schweizer et al.[122] compared weed control in four crops: barley, corn, pinto bean, and sugarbeet. After moderate herbicide application, barley had the fewest weeds ($<1/m^2$) and corn the most ($15/m^2$). The primary reasons for this were probably early emergence and dense canopy of barley relative to the row crops. In this case, these two factors probably superceded the importance of the corn canopy's height and duration advantage. Interestingly, these results are contradicted by an earlier, but similar experiment, where barley, corn, and sugarbeets were compared.[121] After moderate herbicide application, the fewest number of weeds occurred in sugarbeet ($0.1/m^2$), followed by corn ($0.3/m^2$), and barley ($30/m^2$). The reason barley was the "worst" crop in the 1984 study and the "best" crop in the 1988 experiment with respect to weed escapes relates to weed species composition. In the 1984 study, early-germinating lambsquarters (*Chenopodium* spp.) accounted for 30% of the seedbank population. Emergence of lambsquarters would have coincided with that of barley, making it competitive and more difficult to control. In the 1988 study, late-germinating barnyard grass (*Echinochloa crus-galli*) and redroot pigweed together comprised 85% of the seedbank. These late-emerging species would have been suppressed by the early-sown barley crop.

Another illustration is from Georgia. Eight crops were compared for their ability to yield under two general weed management systems, cultivation only and chemical control.[57] The crops were cabbage, cowpea, corn, cucumber, peanut, snapbean, soybean, and turnip. Yields of most crops were low in the absence of herbicides. Only cowpea and turnip yielded equally under both weed management systems. Turnip was planted in late winter (February) and would have developed an extensive canopy by the time potential weeds such as Texas panicum (*Panicum texanum*) and tall morning glory (*Ipomoea purpurea*) germinated.

Although cowpea was planted in early summer (June), its rapid canopy growth apparently conferred its high competitive ability.

Differences in competitive ability with weeds also exist within crop species, although these differences have yet to be fully exploited agronomically. This topic has been reviewed recently.[17,18] Although soybean has been studied most intensively in this regard (e.g., Reference 117), intraspecific variation for competitive ability is known for at least 18 crop species. Several traits have been identified that are believed to confer differential competitive ability. These traits include seed weight, emergence rate, growth rate, canopy growth rate, canopy diameter, canopy density, leaf area index, axillary branching, tiller number, height, indeterminate growth, bush habit, maturity, yield, fecundity, vernalization response, nitrogen absorption, and potassium utilization. Unfortunately, whether these traits specifically confer competitive ability is unknown because of the problem of pleiotropy. However, through the experimental use of isogenic crop lines, one trait, leaf area expansion rate, was shown to be associated unequivocally with crop tolerance to weeds.[47]

One of the most obvious examples of differing crops' influences on weed control is that of winter wheat and spring wheat. Here, the former effectively eliminates populations of spring annual weeds such as Russian thistle (*Salsola iberica*),[152] and the latter reduces populations of winter annual weeds such as downy brome (*Bromus tectorum*).[91] The mechanism by which this is achieved is simply disruption of the weeds' life cycles by crops with seasonally contrasting life cycles.

Crop Rotation

Annual crop rotation decreases weed populations. For example, Johnsongrass (*Sorghum halapense*) populations in Ohio increased linearly and rapidly where corn was planted continuously for 5 years, despite repeated herbicide applications.[7] In contrast, where corn was rotated annually with soybean, Johnsongrass populations either remained stable or decreased slowly. Similarly in Alabama, annual sowing of soybean over 4 years resulted in an increased sicklepod seedbank.[10] Conversely, annual corn-soybean rotation caused a slow but downward trend in the size of the sicklepod seedbank.

Separating the effects of crop rotation and associated herbicide rotation on weed populations is difficult. However, some studies do permit this distinction. In North Carolina, broadleaf signalgrass (*Brachiaria platyphylla*) is a problem weed that is tolerant to alachlor, a standard grass herbicide used in corn (C), peanut (P), and soybean (B) production. Population density and seed production of broadleaf signalgrass and large crabgrass (*Digitaria sanguinalis*) were compared over 3 years with and without herbicides in four crop rotations: C-P-C, P-C-P, C-B-C, and B-C-B.[69] In the absence of herbicides, the two rotations dominated by corn (C-P-C and C-B-C) reduced density and seed production of signalgrass relative to that of P-C-P and B-C-B. Unfortunately, large crabgrass behaved oppositely to broadleaf signalgrass, and competition from the former weed, rather than corn, may explain the observed decrease in broadleaf signalgrass populations. Similarly, MacHoughton (as reported in Reference 60), found that continuous cropping of corn, soybean, or wheat greatly increased populations of giant foxtail, velvetleaf (*Abutilon theophrasti*), or both. In contrast, in a corn/wheat/soybean rotation, giant foxtail populations increased only slightly and velvetleaf populations declined.

A possibly important, but unstudied, aspect of crop rotations is the selection of the initial crop to begin the rotation. Selection of this crop may have to be matched with initial weed species composition and density and acceptable intensity of weed management. A fortuitous example of this is the rotation study from Colorado[120] using a full factorial of cropping sequences with the following four crops: barley (B), corn (C), pinto beans (P), and sugar beets (S). When weed management levels (as indexed by frequency of herbicide applications over 4 years) were high, all crop sequences greatly reduced densities of weeds escaping control (Figure 3). However, with low herbicide input levels, the crop sequence S-B-C-P restricted weed escapes considerably. The sequences B-C-P-S and P-S-B-C were similar and intermediate in their restriction of weeds escaping control. In contrast, the sequence C-P-S-B permitted high numbers of weeds to escape control. The primary weed was barnyard grass, and its lack of control in corn during the first year of the experiment may have preconditioned poor control in subsequent crops at the lower herbicide input levels.

Smother Crops

Abundant literature exists regarding smother or cover crops. Shear[123] summarized this research for corn. Unfortunately, only a small fraction of these studies dealt specifically with weed control. In many of those that did, smother crops did not reduce weed/crop competition.[9,11] Although in other situations smother crops aided in controlling weeds, crop yields also were reduced.[38,101] Where smother crops controlled weeds effectively and simultaneously maintained high crop yields, they exerted their effects on weeds by either inhibiting germination allelochemically, impeding seedling emergence physically, and/or competing with emerged plants.

Smother crop life cycles must be matched to those of the crop and cropping system. For example, perennial sods, which normally occur in

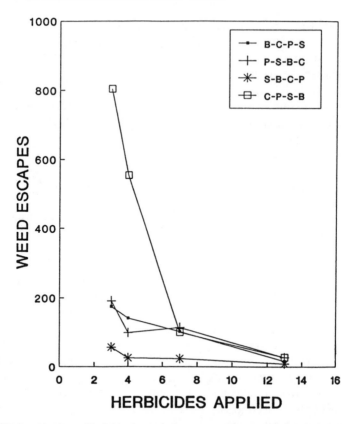

FIGURE 3. Number of individual weed plants escaping control as influenced by the number of herbicides applied over four years in four crop rotations in central Colorado. Abbreviations are barley (B), corn (C), pinto bean (P), and sugarbeet (S). (Data adapted from Schweizer et al. [1988]).

pastures being rotated to row crops in the Corn Belt, are not effective cover crops. Their suppression requires timely herbicide applications, and even then regrowth often interferes with crop development.[13] For row crops, annual smother crops, especially winter-hardy cereals, appear more effective for the necessary combination of weed suppression and crop tolerance.[143]

One of the best smother crops known in the Corn Belt is winter rye. When sown in autumn, this species has high winter-hardiness and luxuriant springtime growth, and when killed with a herbicide just prior to crop planting, abundant allelochemicals leach from its tissues to suppress weeds.[111] Winter rye in combination with a spring application of glyphosate or paraquat has been a highly successful form of weed management in vegetables,[111] corn,[123] and soybean.[137,138] However, because winter rye uses much soil water prior to crop emergence, high yields in soybean/ winter rye systems arise only with receipt of more than 0.7 m annual precipitation.[137,138]

Another potential smother crop for corn is winter rape (*Brassica napus*). When autumn-sown and killed with 2,4-D in spring immediately after no-till corn planting, winter rape provided excellent annual weed control while maintaining high corn yields (Table 1). Unfortunately, winter rape is not hardy in the northern Corn Belt; for survival in that area it requires insulation by early and heavy snowfalls.

Burndown herbicides typically are necessary to suppress even annual smother crops.[123] However, delayed crop-sowing dates may alleviate needs for herbicides, at least in regions with long growing seasons. In Louisiana, for example, burndown herbicides were not needed to control annual legume smother crops, because these crops were senescing at the late date of soybean and sorghum planting.[59] In Wisconsin, mowing winter rye at anthesis provided highly effective weed control in soybean, but this resulted in later-than-usual soybean planting dates and low soy-

Table 1.   Effect of a Winter Rape (*Brassica napus* 'Dwarf Essex') Cover Crop on Corn Development, Height, Yield, and Weed Control at the USDA-ARS Swan Lake Experiment Station, Stevens Co., MN, 1986[a]

Treatment	Corn > 1st leaf June 2 (%)	Corn height June 20 (cm)	Weed control June 18 (%)	Corn yield (kg/ha)
Conventional Fall moldboard Alachlor + cyanazine PPI Field cultivator Seedbed-sown corn	58 b	32 a	95 a	7736 b
Winter cover crop Fall moldboard Fall-sown rape No-till corn 2, 4-D POST	43 a	36 b	99 a	7673 b
Weedy check Fall moldboard Field cultivator Seedbed-sown corn	38 a	37 b	0	3208 a

[a]F. Forcella, unpublished data. Values in columns in table lacking common letters differ significantly according to ANOV and LSD (p = 0.05).

Table 2. Eight Tillage Systems Commonly Used in the North-Central and Northwestern Corn Belt of the U.S. The Use of a Specific Tillage System Depends upon Soil Texture, Soil Water Content in Autumn, the Previous Crop, and Idiosyncrasies of the Farm Manager

Tillage System	Soil	Previous Crop
Fall MBD[a] + spring DSK/HRW	Heavy	Corn (or soybean)
Spring MBD + Packer	Light/wet	Corn (or soybean)
Fall CSL + spring DSK/HRW	Heavy	Corn, soybean, wheat
Spring CSL + DSK/HRW	Light/wet	Corn, soybean, wheat
Fall DSK + DSK/HRW	Heavy	Wheat
Spring DSK/HRW	Light/wet	Wheat
Till-plant (ridge-till)	Heavy/wet	Corn, soybean
No-till	Light	Soybean, wheat

[a]Abbreviations: tandem disk (DSK), harrow (HRW), moldboard plow (MBD), chisel (CSL). In the northern Corn Belt, a field cultivator ("digger") often is substituted for a tandem disk.

bean yields.[32] However, because herbicides were not used in this system, profitability was equivalent to conventional soybean production.

In time, smother crops may be developed specifically to reduce dependence on herbicides in row crops. DeHaan et al.[30] provided a futuristic example of a genetically tailored smother crop. By mechanically and chemically altering the density, height, and longevity of *Brassica hirta* in corn, they developed a blueprint of an ideal spring-sown smother crop: 1000 seeds per square meter, 10 cm tall, and 4- to 8-week lifespan. They subsequently genetically manipulated dwarf *Brassica* populations to develop a facsimile of the ideal smother crop. Initial field-testing on corn, in the absence of herbicides, appeared highly encouraging.[30]

PHYSICAL WEED CONTROL

Physical weed control includes tillage, as well as other forms of physical disturbance of weed plants, such as burning, electrocution, and solarization.

Tillage Systems

Tillage can be divided into tillage systems, with at least eight commonly used in the Corn Belt (Table 2). Their use depends primarily upon the previous crop, autumn soil water content, soil texture, fuel price, and idiosyncrasies of the farm manager. Several examples of differential weed control and competition have been compared between and among tillage systems in the following paragraphs.

Soil tillage, at least in row crop production, also can be classified into three forms: primary, secondary, and tertiary. Primary tillage is the main tillage operation performed on the soil, and in the Corn Belt it typically is synonymous with moldboard plowing or, increasingly, with chisel plowing. Secondary tillage, performed after primary tillage and immediately before crop sowing, typically is accomplished with tandem disks or field cultivators plus harrows or their equivalents. Tertiary tillage is performed after planting and may have two components: (1) rotary hoeing, rod weeding, or harrowing, and (2) interrow cultivation.

Primary Tillage

Primary tillage has been the subject of considerable research in weed management.[146] From 1985 to 1990, over 400 citations dealing with weeds and tillage are listed in the information retrieval system, AGRICOLA. These tillage publications may be divided into two overlapping categories: differential weed species composition, and differential herbicide efficacy and persistence.

Several authors have compared tillage systems with respect to promotion or inhibition of weed species. Generally, annual grass weeds proliferate in no-till (NT) systems and are restricted in conventional till (CT) systems (Table 3). In contrast, annual broadleaf weeds tend to proliferate in CT and are inhibited by or indifferent to NT. Common lambsquarters is an example of an exception to this generalization; this weed appears to survive regardless of tillage system. As expected, perennial weeds are inhibited by the extra soil disturbance in CT (Table 3).

A reduced tillage (RT) system characterized as fall-chisel (chisel plowing in autumn followed by disking/harrowing in spring) is becoming widely adopted in the Corn Belt. Several weeds (e.g., giant foxtail) recently have been shown to be promoted by a fall-chisel tillage system (Table 3). This probably occurs because RT distributed 75% of weed seeds below the soil surface but above 0.05 m depth,[126] precisely the burial depth that maximizes giant foxtail seedling emergence.[87] In contrast, CT distributes less than 10% of weed seeds above 0.05 m, and NT distributes nearly all seeds on the soil surface, both of which function to inhibit germination and emergence of species such as giant foxtail.

Till-plant (TP), otherwise known as ridge-till, systems appear to be highly effective in inhibiting many weed species. When the ridges are truncated at planting a large proportion of the buried weed seeds, which had been proximal to crop rows, are moved into interrows where they subsequently can be controlled with cultivation.[144] In corn/soybean rotations, >80% of weed seeds in crop rows at planting were moved to

Table 3. Affinity of Weed Species to Tillage Systems

Weed Species	Tillage System[a]				Refs.
	CT	RT	TP	NT	
Annual grass					
Barnyard grass	•			−	144
	−			+	112
Fall panicum	−			+	112
Giant foxtail	−	+	−	+	12
	−	•		+	14
	•	+	−	+	68
	−			+	112
Green foxtail	−		•		52
	−	−		+	150
Itchgrass	+			−	5
Annual broadleaf					
Horseweed		−		+	11
Lambsquarters	•	•		•	14
	•		•		122
	•	+	•	•	68
	•	•			135
	+			−	112
Redroot pigweed	•	+		•	14
	+			•	57
	+	−			31
	+			−	112
Sicklepod	+			−	6
	+			−	116
	+			−	10
Velvetleaf	+	•	−	•	12
	+	•		•	14
	•	+		•	56
	•			•	29
Perennial					
Canada thistle	−			+	34
Foxtail barley	−	•		+	33
	−	−		+	150
Quackgrass	−	+		+	86

[a]Abbreviations: conventional tillage (CT), reduced tillage (RT), till plant (TP), no tillage (NT), promotion (+), indifference (•), and inhibition (−).

interrows, reducing short-term needs for herbicides while maintaining high crop yields.[51,52] However, in continuous corn, smaller percentages of weed seeds are moved to interrows.[144,51]

Although certain weed species tend to decrease in importance when tillage is abandoned entirely, some weeds will proliferate in NT. Thus, the necessity of herbicide use may increase in these systems, which, in turn, decreases the potential role that NT will play in agriculture's sustainable future.

Another criticism of NT is that efficacy of preemergence herbicides may be low in these systems due to volatilization, photodecomposition, or residue retention. However, recent research comparing preemergence herbicide efficacy across tillage systems does not support this contention (Table 4). Indeed, associations of herbicide efficacy and tillage systems appears random. Examples from Table 4 follow: (1) efficacies of most formulations of alachlor appear indifferent to tillage system, but a granular formulation actually performs better in NT than in CT or RT (Table 4); (2) a tank mix of alachlor plus metribuzin is more effective in CT and NT than in RT, whereas efficacy of alachlor plus chloramben is higher in

Table 4. Tillage System Effects on Herbicide Efficacy

Herbicides and Application Time	Tillage System[a]			Refs.
	CT	RT	NT	
Preemergence				
Alachlor				
l	•	•	•	68
m	•	•	•	68
g	•	•	+	68
Alachlor + metribuzin	+	−	+	56
Alachlor + chloramben	−	−	+	56
Pendimethalin + chloramben	−	+	−	56
Clomazone	+	•	−	130
	•		•	90
	+		•	90
Imazethapyr				
w	+		•	89
d	•		+	89
Imazaquin				
w	+		•	89
d	•		+	89
Postemergence				
Sethoxydim	•		•	58
Fluazifop	•		•	58

[a]Abbreviations as in Table 3, plus the following: liquid formulation (l), microencapsulated formulation (m), granular formulation (g), dry growing season (d), wet growing season (w). Symbols as in Table 3.

NT than in CT or RT; and in contrast, (3) pendimethalin plus chloramben were most effective in RT.

Differential efficacy of preemergence herbicides with tillage system may be environmentally dependent. For example, in wet years both imazaquin and imazethapyr are more effective in CT than in NT, but in dry years the reverse is true. The higher levels of soil water in NT than CT in dry years[140] may explain this alternation of preemergence efficacy with tillage system.

As might be expected, efficacies of herbicides applied postemergence, such as sethoxydim and fluazifop, are indifferent to tillage system.[58]

Secondary Tillage

Secondary tillage is used when primary tillage leaves unacceptably large clods of soil or levels of residue on the soil surface. It is used to prepare seedbeds, typically involving disking (or its equivalent) to 5- to 10-cm soil depth and harrowing.

Because disking distributes weed seeds evenly in a layer near the soil surface[23,126] from which almost any seedling can emerge, most weed species respond positively to seedbed preparation. However, the timing of seedbed preparation effects weed populations considerably. Many weed species in the northern climates, for example, emerge primarily in one or a few flushes in spring.[102,127] Significant crop/weed competition can be expected if seedbeds are prepared and crops planted prior to emergence flushes of important weed species. However, if seedbed preparation is delayed until after major flushes of weed emergence, subsequent weed seedling populations are small, mechanical weed control efficacy is high, less herbicide is required, and crop/weed interactions are reduced greatly.[49]

Tertiary Tillage

The two main forms of tertiary tillage are rotary hoeing for both row and drilled crops, and interrow cultivation for row crops. Rod weeding and spiketooth harrowing also are forms of tertiary tillage. Because of high efficacy in weed control, shallow interrow cultivation in row crops is almost universal. Four concerns associated with interrow cultivation are: (1) delayed control until crops are tall enough (25 cm) to withstand some soil burial, (2) root pruning and shoot damage, (3) ineffective within-row weed control, and (4) tractor fuel costs of about $5/ha.

Rotary hoeing is an inexpensive ($2/ha fuel cost) but more controversial method of mechanical postemergence weed control. It consists of rapidly spinning tynes that flail weed seedlings both within and between rows. The weed seedlings are controlled by either shredding or uprooting, or both. Leaves of crop seedlings also may be shredded, but because crop seeds were sown more deeply, their seedlings are not uprooted. Speed of the rotary hoe and size of weed seedlings are extremely important for efficacy. Indeed, a main conclusion from Gunsolus'[60] review of the agronomic literature on rotary hoeing in soybean and corn is that timeliness is the key element for efficacy. The first rotary hoeing must occur after weed seed germination but before weed seedling emergence. One or two subsequent passes at 5-day intervals may also be necessary. Unfortunately, information on rotary hoeing is difficult to glean from published literature; and where it is readily available, the results may be contradictory. For example, in a series of reports by Moomaw and Robison[92-94] from Nebraska, rotary hoeing was found to reduce weed growth appreciably in 1 of 3 years in soybean, 1 of 2 years in corn, and 2 of 2 years in sorghum. Other Nebraska studies indicated that rotary hoeing is as effective as herbicides in maintaining high crop yields.[16] More recently, rotary hoeing was reported to reduce weed seedling populations, but had little effect on crop yields.[119] In any event, rotary hoeing efficacy, either alone or when used with herbicides, is greater in dry than in wet years.[60] Variability in the results of rotary hoeing experiments means that the ultimate value of this inexpensive and environmentally appealing form of weed management, and then only remains questionable for sustainable agriculture.

Farmers, engineers, and entrepreneurial inventors routinely develop new agricultural implements. With increased emphasis on physical weed control, we can expect new useful implements to be developed. A potential example is the within-row, row crop, cultivator with "spyders, spinners, and torsion weeders" being promoted by a California inventor. Improvements in traditional cultivators are increasing weed control efficacy under high crop residue levels. Unfortunately, most new inventions have limited utility, but a small proportion will prove practical.

Burning

Burning of crop residues in late summer or autumn is an old and effective means of reducing some weed species in winter wheat and rice. In Britain, seed populations of slender foxtail (*Alopecurus myosuroides*) were reduced by 71% by burning, and seedling populations in the crop were reduced by 72%.[100] Similarly, wheat residue burning in Oregon reduced downy brome populations.[115] Although inexpensive ($< \$1/\text{ha}$) and effective, extensive burning of crop residues unfortunately produces hazardous levels of smoke. For this reason, and because of potential losses of valuable soil organic matter, residue burning will not be a sustainable form of weed management.

Flaming and electrocution are other forms of burning weeds. Neither method has been widely adopted, nor are they likely to be adopted extensively in the future due to high energy costs.[135]

Solarization

Heating of wet soil surfaces to temperatures greater than 65°C kills nondormant weed seeds and greatly reduces weed seedling populations.[40] Soil heating generally has been accomplished by covering soil with sheets of clear plastic. For significant effects to occur, the plastic must remain on the soil surface for 1 week to control species such as redroot pigweed, days must be clear and sunny, and the soil must be wet. Coupled with the expense of plastic sheeting as well as its environmental persistence, solarization remains a nonsustainable form of weed management, and then only in high-value horticultural crops.

A novel and more sustainable form of solarization is the use of lenses that reradiate and concentrate sunlight onto a narrow band. Within a few seconds, temperatures along this band can reach 290°C, resulting in 100% control of cotelydon-stage seedlings of redroot pigweed.[67]

BIOLOGICAL CONTROL

Phytophageous Insects

A pessimist might summarize the biological control literature in weed science by stating "never has so much been written about so little." Research concerning the biological control of weeds is dominated by entomologists, and this is reflected by the information reported in biocontrol publications. The emphasis in these publications is insect demographics. Control of weeds is mentioned only rarely, probably because weed control by insects has been achieved only rarely.[70] After the initial and extraordinary successes of control of Klamath weed (*Hypericum perforatum*) and prickly pear (*Opuntia* spp.) by insects, no dramatic success stories have appeared. At best, success has been marginal (e.g., see Reference 66). For weeds in crops, there are no examples of effective control by insects.

Plant Pathogens

Successful control of weeds in crops by microorganisms may hold more promise than that by insects. Perhaps the first widespread success of fungi used for biocontrol of a crop weed was that of *Puccinia chondrillina* on rush skeletonweed (*Chondrilla juncea*) in Australian wheat fields.[26,27]

More recently, repetitive (annual) and innundative applications of pathogenic fungi (mycoherbicides) have shown promise for weed control in crops. The two major achievements have been with the commercialization of *Phytophora palmivora* for control of stranglervine (*Morrenia odorata*) and *Colletotrichum gloeosporioides* for control of northern jointvetch (*Aeschynomene virginica*).[129] Smith[124] briefly summarized mycoherbicide research for other weed species. Another development involving purposeful manipulation of microbes is the growing interest in weed seed pathogens. Crop environments may be managed to promote such pathogens and/or the organisms may be applied directly to flowering weeds.[72]

Probably because fungi require long periods of relatively high humidity to become infectious, most successful mycoherbicide research has been concentrated in the humid southeastern United States. Technological advances may solve such problems, e.g., invert emulsion carriers maintain high humidities for fungal spore germination on weed leaf surfaces,[28] and bran/fungi formulations retain spores on or near leaf surfaces.[95] Nevertheless, dramatic successes for many important weeds in crops do not appear forthcoming. Admittedly, however, mycoherbicide research is still in its infancy.

CHEMICAL WEED CONTROL

Herbicides are generally synthetic organic chemicals that kill plants or inhibit their normal growth. In the past, inorganic chemicals, such as various salts, ash, smelter wastes, and metallic compounds, were used as soil sterilants. Beginning in the late 19th century, copper salts were used in controlling broadleaf weeds in cereals. However, not until the 1940s were "true" organic herbicides developed, i.e., MCPA and 2,4-D. Herbicides such as these ushered in a new era of efficiency and effectiveness in crop production.

Since 1940 there has been a frenzy of activity in the area of herbicide development. Indeed, the Weed Science Society of America listed over 160 currently used herbicides in 1991. These chemicals account for 85% of the 210 million kilograms of pesticides used in row crops and small grain crops in the United States during 1990.[133] Insecticides accounted for only 14% of pesticides used, while the remaining 1% is accounted for by fungicides, nematicides, etc. Although farmers applied herbicides to less than 10% of corn, cotton, and wheat acreage in 1952, they now treat more than 90% of their land with herbicides.[103]

The advent of herbicides came at a time when world agriculture was entering an era of increased mechanization, altered production techniques, expanded acreage per farmer, reduced labor supplies, increased use of off-farm inputs, and increased crop yields and quality, all while per unit production costs were being reduced. Although mechanical and cultural weed control practices had sufficed in the past, modern agriculture demanded improved, more economical, and labor-saving weed control methods. Herbicides met those requirements. Farmers quickly found that they were able to farm nearly twice as much land with equivalent labor, and their weed control was better. Consequently, profitability and farm competitiveness were improved. Herbicides not only changed weed control on farms, but also in industrial, transport, forest, aquatic, and recreational areas worldwide.

Individuals using herbicides should be familiar with various classification schemes. The differing schemes are used to better categorize herbicides into groups with similar properties such as toxicity, use, persistence, volatility, weed spectrum controlled, chemical family, etc. Various classification schemes are discussed briefly below.

Herbicide Families

Herbicide chemists have divided these compounds into families based on chemical similarities. The traditional families (followed by an example) include: acid amide (alachlor), aliphatic carboxylic (dalapon), aromatic carboxylic (2,4-D), benzonitrile (bromoxynil), carbamate (EPTC), triazine (atrazine), dinitroaniline (trifluralin), diphenyl ether (acifluorfen), phenol (dinoseb), urea (linuron), uracil (bromacil), and miscellaneous (glyphosate). Two recently discovered families of herbicides are sulfonyl urea (chlorsulfuron) and imidazilinone (imazethapyr). New herbicidal compounds are discovered routinely, and new herbicide families can be expected in the future.

Knowledge of herbicide families is useful not only for organic chemists, but also for farm managers. Two practical examples follow: (1) Annually repetitious use of any of several carbamate herbicides leads to a profusion of carbamate-degrading soil microorganisms. Once selected, these microbes can degrade EPTC quickly enough to reduce efficacy significantly. (2) Annually repetitious use of a triazine herbicide selects for weed populations with cross-resistance to other triazine herbicides.

Application Time

"Preplant" herbicides are applied prior to planting. They may be unincorporated or mechanically incorporated into the soil to either activate them or to prevent their loss by volatilization or photodecomposition. Preplant foliar treatments are generally nonselective, and they may kill plants by contact or be translocated within the plants. Preplant soil treatments may vary in lengths of residual activity after application. Generally, preplant incorporated herbicides are the chemicals of choice for farm managers who are adverse to risk.

"Preemergence" selective herbicides are applied after planting but before emergence of weeds or crops. Although some of these herbicides may require shallow mechanical incorporation for activation, most require incorporation by rain or irrigation within a few days of application for high efficacy.

"Postemergence" herbicides are foliar herbicides applied after weed emergence. These are are generally selective herbicides that are absorbed and translocated within the plants and may be applied broadcast, directed, or banded. Most postemergence herbicides that were developed relatively recently are expensive to apply (> $20/ha) and, consequently, they are used judiciously. Older chemicals, such as 2,4-D, are inexpensive ($5/ha) and are used widely. Because postemergence herbicides are used only when a weed problem is apparent, they probably have a more secure future in sustainable agriculture than prophylactic preplant and preemergence herbicide applications.

Persistence

Herbicides may be classified according to their residual activity. "Nonresidual" herbicides lose their phytotoxicity almost immediately upon contact with soil. The well-known compound, glyphosate, is an example. "Residual" herbicides have varying degrees of persistence. Popular herbicides used in corn production, such as alachlor, metolachlor, and cyanazine, have residual activities lastings about 1 month, almost exactly the length of time necessary for a crop to be weed free without incurring yield losses. Long-residual herbicides include atrazine, chlorsulfuron, and trifluralin, which can retain toxicity to some plant species for at least 1 year. Such herbicides reduce crop rotation options and are less acceptable for sustainable agriculture.

Herbicide Selectivity, Type, and Formulation

Because farm managers typically desire to grow one crop species and eliminate competing species, most present herbicides show varying degrees of selectivity. Nonselective herbicides kill or damage all vegetation, and they may be applied to soil or foliage. In contrast, selective herbicides control undesirable weeds while not seriously affecting the crop. Selectivity may be due to differential application, retention, absorption, adsorption, translocation, metabolism, accumulation at site of action, or mode of action of the herbicide among weeds and crops. Selective herbicides are applied as contact foliar sprays, systemic foliar sprays, or soil treatments that act on germinating seeds or roots.

Herbicides vary markedly in biological activity, and thus dosage is an important criteria for selectivity. Recently developed herbicides show activity on weeds at very low rates (g/ha instead of kg/ha). Although use of these herbicides will reduce total environmental loads of synthetic chemicals, whether this solves the problem of environmental contamination is questionable.

Herbicide formulation also affects selectivity. Because of targeting difficulties, dust formulations have been eliminated. Granular formulations eliminate drift, are not retained by crop canopies, and fall through residues to reach the soil surface. Following rain or mechanical incorporation, they are effective against weed seedlings.

Protective barriers or seed coatings of activated charcoal have been used to improve herbicide selectivity. Various formulations, adjuvants, and wetting agents also affect herbicide phytotoxicity.

Herbicide Selection

For herbicide development, large numbers of compounds were screened in the past against only a few major crops and weed species to ascertain selectivity. Currently, tissue culture techniques can be used to evaluate a new herbicide for crop selectivity on numerous crop biotypes quickly and efficiently.

Gene transfer technology can be used to produce crops tolerant to both old and new herbicides. This technology should result in greater herbicide selectivity in the future. However, because of the potential for misuse of such crops and associated chemicals (i.e., repetitious application), creating conditions for rapid development of resistant weed species, adoption of corn, soybean, or wheat cultivars engineered for herbicide resistance may be inhibited. More sustainable targets for the extraordinary capabilities of genetic engineers might be herbicide-tolerant, low-acreage specialty crops.

Herbicide Application

Soil or foliar applications affect selectivity and will be dictated by characteristics of the herbicide and crop-weed situation. Herbicides applied to the soil often attack weeds at an earlier and more sensitive stage, and they may provide increased safety to the crop. In contrast, foliar applications have the advantages of requiring lower rates and often being translocated within the plant so perennial weeds can be controlled effectively. Complete spray coverage is less critical for translocated than contact herbicides.

A number of conventional and special sprayers are designed to improve herbicide selectivity and phytotoxicity.[79] Special sprayers and spray applications include rotary atomization, electrostatic atomizers, herbicide application through irrigators, granular applications, controlled release formulations, herbicide-fertilizer combinations, weed-wipers, injection into soil, herbicide-treated crop seed, application based on geographical information system (GIS)-mediated soil maps and weed problems, and application activation based on electronic/sensory detection of weed plants. This latter weed-activated spraying has been shown to save over 90% of the herbicide vs. broadcast application for fallow land spraying in Australia.[43]

State of Weed-Crop Growth

Generally, the younger a weed plant the easier it is to control with herbicides. Foliar translocated herbicides are best applied to vigorously growing annuals less than 10-cm tall; or to autumn growth of perennials, since there is then a large leaf surface for herbicide absorption plus active translocation to roots. In the event of stomatal uptake of foliar-applied herbicides, morning or evening applications are best. Contact herbicides sprayed during early crop growth may kill emerged broadleaf weeds with exposed growing points, but not crops such as corn or peas whose growing points are still below ground at that time. Selective weed control in crops depends upon the differential between crop tolerance and weed susceptibility at time of application.

Environmental Concerns

The general public has expressed environmental concerns with current herbicide use that include food safety, farmworker safety, cancer risks, birth defects, wildlife mortality, protection of endangered species, surface and ground water contamination, and herbicide-resistant weeds. Not surprisingly, the issue of ground water contamination has alarmed the farming community because this water resource provides 97% of drinking water for rural populations.[2]

Herbicide-tolerant weeds were first discussed in the 1950s and reported in 1970.[61,118] Since then there has been considerable research in this area, both because of an increasing threat to agricultural production from such weeds, and the potential benefits from developing herbicide-tolerant crops. There are now at least 40 dicot and 15 monocot plant species with biotypes resistant to triazine herbicides.[64] Also, at least 44 weed species have been reported to have biotypes resistant to one or more of 15 other herbicides. This list of herbicide-tolerant weeds will continue to grow, especially where there have been repeated and prolonged exposures of weeds to herbicides with the same mode of action. Extremely high efficacies of recent herbicides (e.g., chlorsulfuron) may accelerate selection for tolerant weed biotypes. Efficacies of all weed control methods are transitory, however, as there are no permanent solutions to weed control except through integrated weed management, as discussed below.

New Herbicides

Chemical industry personnel, primarily, as well as a few public scientists, continue to discover new compounds with herbicidal activity. Because of stringent registration regulations, new commercial herbicides are invariably safer for public health than the "average" older compounds.

From an organic chemist's perspective, exciting developments in herbicide development are occurring. Naturally occurring compounds are being tested and used as blueprints for the "biorational design" of herbicides.[36] These natural compounds are being isolated from higher plants as well as microorganisms. A related novelty is that of "photodynamic" herbicides.[132] One of the more interesting photodynamic compounds is the rather simple amino acid, δ-aminolevulinic acid (ALA). In conjunction with modulators, ALA causes plants to oversynthesize tetrapyrroles within their cells, which, when irradiated, produce harmful singlet oxygen.[83] Lastly, through classical herbicide screening, two extremely phytotoxic families of herbicides have been developed recently: imidazilinones and sulfonyl ureas. Herbicides within these families attain exceptionally high efficacies at rates as low as a few grams per hectare. Curiously, the site of action of both families is identical, despite their differing chemistries.

INTEGRATED MANAGEMENT

Integrated weed management (IWM) is the most feasible approach for weed control in sustainable agriculture, since no single control method can be relied upon as a panacea for weed management. Each control method (Figure 4) will provide some fraction of total weed control. The "sum" of these fractions should provide dependable and adequate control of the weed vegetation present in any given situation. IWM does not mean that specific fractions of each of the four basic types of weed control (Figure 4) must be used monotonously. Instead, IWM should place annually varying emphasis on the control techniques best suited to the weed problem at hand. Few individuals will have the necessary information and training to select the appropriate mix of control techniques skillfully for effective IWM. Consequently, weed management in sustainable agriculture will depend heavily on weed/crop management models.

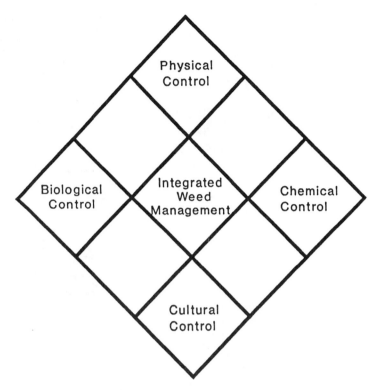

FIGURE 4. Idealized relationships among the major categories of weed control. Integrated weed management is central to all categories. In sustainable agricultural fields, combinations of control techniques must be employed and varied annually, as overreliance on any one technique may lead to unwanted environmental and/or agronomic problems.

Weed/crop models are relatively new to weed management. However, they are being developed rapidly, and we expect their adoption and use to increase equally fast in the near future. Weed management models vary in their abilities and usefulness with respect to sustainable agriculture. They can be divided superficially into three categories: weed threshold and population models, herbicide decision models, and bioeconomic models.

Weed Threshold and Population Models

The initial problem weed management modeling faced was determination of the weed thresholds. Thresholds represent levels of weed pressure at which crop yields are decreased sufficiently to justify intervention, typically in the form of a postemergence herbicide application.[24] A second problem concerned the choice of the most appropriate measure of weed pressure (e.g., density, cover, or biomass) to be used in management models. Considerable progress has occurred during the past decade in relating crop yield losses to various measures of weed abundance during the early stages of crop growth. Although this work has been worldwide in scope, it has been led by British and Dutch weed scientists.

A major hinderence to development of a general theory for threshold models was an appropriate equation that described the effect of weed pressure on crop yield reduction. Although not universal, the rectangular hyperbola equation[20] has received wide acceptance. The equation is as follows:

$$Y = Y_{wf}\{1 - I*D/[100 (1 + I*D/A)]\}$$

where Y = crop yield, Y_{wf} = weed-free crop yield, I = percentage yield loss per unit weed density, D = weed density, and A = asymptotic percentage yield loss at high weed densities. Weed density generally is considered to be the most practical measure of weed pressure.[99,150]

Continuing problems with this equation, as well as any other, are fluctuations of its coefficient values. For example, for bedstraw (*Galium aparine*) in winter wheat at the same site, I was equal to 4.1 per plant per square meter in 1987 but only 1.2 per plant per square meter in 1988.[150] Fluctuations in coefficient values also may occur because of differing soil types,[44] rainfall,[97] soil fertility,[71] emergence times,[21] weed species diversity,[4] crop densities,[84] and background herbicide levels.[1] More dynamic equations that are environmentally sensitive will be required in the future. Toward this end, attempts are underway to simulate crop-weed interactions using more complex procedures.[73,147] Unfortunately, a general modeling paradigm is that as model complexity increases, utility, or user-friendliness, diminishes. Nevertheless, weed management in sustainable agriculture will require explicit development of species-specific and process-oriented equations describing the effects of weeds on crop yields.

The first weed/crop modeling endeavors that coupled threshold models with demographic models of individual weed species were for slender foxtail, *Alopecurus myosuroides*,[35] and wild oats, *Avena fatua*,[22] in winter wheat. Ultimate goals of these models are weed control and economics. In practice, they are extremely useful tools for exploratory agronomic research but are not sophisticated enough for actual farm-based weed management. What these models lack are decision rules providing multiple control options for multispecies weed assemblages.

Herbicide Decision Models

Herbicide decision models[98] are useful aids that permit farm managers and others to select herbicides according to the crop and expected spectrum of weed species. An example of this type of model is WEEDIR (Weed Control Directory), developed by Minnesota Extension Service programmers[88] and based on annually-updated, expert advice of extension weed scientists.[37] Pessimists might conclude that such decision aids perpetuate high chemical useage and therefore detract from the ideals of integrated weed management in sustainable agriculture. The utility of these models, however, is in their ability to select only the most appropriate herbicides for the task at hand. At the very least this should result in fewer misapplications of inappropriate herbicides, or less poor timing and overapplication of all herbicides. In addition, herbicide decision aids can act as submodels in more complex multiple-decision bioeconomic models.

Multiple-Decision Bioeconomic Models

Bioeconomic models relate, in varying degrees, weed control decision aids, weed density thresholds, and fiscal information. At present, these extremely valuable models are at three levels of comprehensiveness. The first is typified by HERB,[148] a decision model for applying postemergence herbicides to soybean in North Carolina. This model allows users to "input" the postplant seedling densities of any of 76 weed species, which were previously ranked as to their competitive ability. A "total competitive load" is then calculated and potential yield reductions determined. Herbicide recommendations based on the eight most important weed species are then "output" along with financial information. Although this model has been used in the field with positive results (i.e., predicted vs. actual soybean yield loss; personal communication of H.D. Coble, May 6, 1991), it has not been compared economically with farmers' decisions.

A second-level bioeconomic model has been developed for irrigated corn in Colorado.[77] The Colorado model has profited greatly from collaboration of weed scientists with agricultural economists. Although the model places increased emphasis on weed ecology, it relies solely on herbicides for weed control. Preplant and postemergence weed control decisions made by the model are driven primarily by preplant weed seedbank estimates, but also by post-plant seedling densities, herbicide costs, expected grain yields, and commodity prices. The model has been field-tested under experimental conditions as well as in farmers' fields. Results have been encouraging, with maintenance of high gross profit margins while achieving considerable reductions (50 to 75%) in herbicide usage as compared to standard farmer practices.

A third-level bioeconomic model is currently under development. This model, WEEDSIM, is devoted to the North American Corn Belt,[128] a region that annually receives a disproportionately high load of the continent's herbicides. The specific goal of WEEDSIM is to reduce dependence on herbicides. Consequently, the model's recommended weed control options include not only herbicides, but mechanical strategies as well. Although initial development and programming of WEEDSIM has been conducted by agricultural economists, a regional team of weed scientists (NC-202 Committee of the Consortium of Agricultural Experiment Stations from the North Central States) from all of the Corn Belt states participates in conceptual structuring, parameter estimation, and data acquisition.

WEEDSIM currently relies heavily on weed biology and crop-weed ecology. After weed seedbank data are input, the model determines various species-specific factors: seedling emergence rates, seedling mortalities due to varying control options, competitive effects on corn and soybean yields, seed production of escaping weeds, and seedbank mortalities, as well as all relevant costs and profits. The model was successfully field-tested in 1991 and in 1992 for continuous corn and corn/soybean rotations.

Planned improvements for WEEDSIM include providing optional entry of current daily environmental data and projections of near-term temperature and rainfall (i.e., cool/wet, warm/wet, cool/dry, and warm/dry). This is being done to more adequately estimate (1) proportional seedbank emergence, which varies with spring temperatures;[50] (2) efficacies of both mechanical[49] and chemical control options; (3) competitive abilities;[82] (4) weed seed production; and (5) crop yields and losses.[96]

SUMMARY

Weed management systems in the future will differ greatly from those employed today. These future management systems will be less dependent upon material inputs and more dependent upon information and modeling. Operating these systems will require greater understanding of weed ecology and detailed knowledge of each cropped field. Weather information, including past records, present readings, and both short- and long-term forecasts, will play crucial roles in future weed management systems. Mimicking the "crop reports" of today's media (e.g., percent of fields planted, harvested, etc.), daily weed management information will be disseminated in the future. Examples include (1) expected emergence percentages for annual weeds on predominant soil types during planting time; (2) dates to expect maximum benefit for timely seedbed preparation; (3) expected crop yield losses due to specific weed densities based upon recent and anticipated weather; (4) treatment dates for maximum perennial weed control, and (5) expected net income based upon a variety of management choices.

Herbicides will play a continuing but diminished role in future weed management systems. New chemicals will be safer environmentally, more active against weeds, more selective to the crop, and applied less repetitiously. Engineering of crop biotypes resistant to new herbicides will continue, especially for minor crops. Threat of evolution of weed biotypes resistant to herbicides for which major crops could be engineered will dampen enthusiasm for widespread genetic manipulation in these crops. In contrast, genetic engineering and breeding efforts will increase to develop crops that are more competitive with weeds. Greater use of mechanical and cultural control methods will occur, but will vary with the crop species grown. Biological weed control in crops using plant pathogens, especially in the form of mycoherbicides, also will increase.

ACKNOWLEDGMENTS

We thank Douglas Buhler, Bruce Maxwell, and Alan Tasker for their comments on earlier versions of this paper. Several fellow weed scientists supplied useful information and references. This paper represents a joint contribution of USDA-ARS, the University of Minnesota, and Minnesota Agricultural Experiment Station, *Minnesota Science Journal*, Ser. No. 19, 275.

REFERENCES

1. Adcock, T. E. and Banks, P. A., Effects of preemergence herbicides on the competitiveness of selected weeds, *Weed Sci.* 39, 54, 1991.
2. Anonymous. National water survey 1986: hydrologic events and groundwater quality, Water Supply Paper 2325, U.S. Geological Survey, Washington, D.C., 1988.
3. Auld, B. A. and Tisdale, C. A., Impact assessment of biological invasions. In: *The Ecology of Biological Invasions*, Groves, R. H. and Burdon, J. J., Eds., Cambridge University Press, New York, 1986, 79.
4. Austin, M. P., Groves, R. H., Fresco, L. M. F., and Kaye, P. E., Relative growth of six thistle species along a nutrient gradient with multispecies competition, *J. Ecol.* 73, 667, 1985.
5. Ayeni, A. O., Duke, W. B., and Akobundu, I. O., Weed interference in maize, cowpea and maize/cowpea intercrop in a subhumid tropical environment. I. Influence of cropping season, *Weed Res.* 24, 269, 1984.
6. Banks, P. A., Tripp, T. N., Wells, J. W., and Hammel, J. E., Effects of tillage on sicklepod (*Cassia obtusifolia*) interference with soybean (*Glycine max*) and soil water use, *Weed Sci.* 34, 143, 1985.
7. Bendixen, L. E., Corn (*Zea mays*) yield in relation to johnsongrass (*Sorghum halepense*) population, *Weed Sci.* 34, 449, 1986.

8. Blackshaw, R. E. and Rode, L. M., Effect of ensiling and rumen digestion by cattle on weed seed viability, *Weed Sci.* 39, 104, 1991.

9. Bridgemohan, P. and Brathwaite, R. A. I., Weed management strategies for the control of *Rottboellia cochinchinensis* in maize in Trinidad, *Weed Res.* 29, 433, 1989.

10. Bridges, D. C. and Walker, R. H., Influence of weed management and cropping systems on sicklepod (*Cassia obtusifolia*) seed in the soil, *Weed Sci.* 33, 800, 1985.

11. Brown, S. M. and Whitwell, T., Influence of tillage on horseweed, *Conyza canadensis, Weed Technol.* 2, 269, 1988.

12. Buhler, D. D. and Daniel, T. C., Influence of tillage systems on giant foxtail, *Setaria faberi*, and velvetleaf, *Abutilon theophrasti*, density and control in corn, *Zea mays, Weed Sci.* 36, 642, 1988.

13. Buhler, D. D. and Mercurio, J. C., Vegetation management and corn growth and yield in untilled mixed-species perennial sod, *Agron. J.* 80, 454, 1988.

14. Buhler, D. D. and Oplinger, E. S., Influence of tillage systems on annual weed densities and control in solid-seeded soybean (*Glycine max*), *Weed Sci.* 38, 158, 1990.

15. Burnside, O. C., Progress and potential for nonherbicidal control of weeds through preventive weed control, in *International Conference on Weed Control*, Holstun, J. T., Ed., Food and Agriculture Organization, UNESCO, Rome, 1970, 464.

16. Burnside, O. C. and Colville, W. L., Soybean and weed yields as affected by irrigation, row spacing, tillage, and amiben, *Weeds.* 12, 109, 1964.

17. Callaway, M. B., Crop varietal tolerance to weeds: a compilation, *Plant Breed. Biom. Publ. Ser. No. 1990-1*, Cornell University, Ithaca, NY, 1990.

18. Callaway, M. B. and Forcella, F., Crop tolerance to weeds. In: *Crop Improvement for Sustainable Agriculture*, C. A. Francis, and Callaway, M. B., Eds. University of Nebraska Press, 1993, 105.

19. Cavers, P. B. and Kane, M., Responses of proso millet (*Panicum miliaceum*) seedlings to mechanical damage and/or drought treatments, *Weed Technol.* 4, 425, 1990.

20. Cousens, R., A simple model relating crop yield loss to weed density, *Ann. Appl. Biol.* 107, 239, 1985.

21. Cousens, R., Brain, P., O'Donovan, J. T., and O'Sullivan, A., The use of biologically realistic equations to describe the effects of weed density and relative time of emergence on crop yield, *Weed Sci.* 35, 720, 1987.

22. Cousens, R., Doyle, C. J., Wilson, B. J., and Cussans, G. W., Modeling the economics of controlling *Avena fatua* in winter wheat. *Pestic. Sci.* 17, 477, 1986.

23. Cousens, R. and Moss, S. R., A model of the effects of cultivation on the vertical distribution of weed seeds within the soil, *Weed Res.* 30, 61, 1990.

24. Cousens, R., Wilson, B. J., and Cousens, G. W., To spray or not to spray: the theory behind the practice, *Proc. Brit. Crop Prot. Conf. Weeds.* 1, 671, 1985.

25. Cudney, D. W., Jordan, L. S., Holt, J. S., and Reints, J. S., Competitive interactions of wheat (*Triticum aestivum*) and wild oats (*Avena fatua*) grown at different densities, *Weed Sci.* 37, 538, 1989.

26. Cullen, J. M., Evaluating the success of the programme for the biological control of *Chondrilla juncea* L., *Proc. Int. Symp. Biol Conf. Weeds.* 4, 117, 1976.

27. Cullen, J. M., Kable, P. F., and Catt, M., Epidemic spread of a rust imported for biological control. *Nature.* 244, 462, 1973.

28. Daigle, D. J., Connick, W. J., Quimby, P. C., Evans, J., Trask-Morrell, B., and Fulgham, F. E., Invert emulsions: carrier and water source for the mycoherbicide, *Alternaria cassiae, Weed Technol.* 4, 327, 1990.

29. Defilice, M. S., Witt, W. W., and Barrett, M., Velvetleaf (*Abutilon theophrasti*) growth and development in conventional and no-tillage corn (*Zea mays*), *Weed Sci.* 36, 609, 1988.

30. DeHaan, R. L., Wyse, D. L., Ehlke, N. J., and Putnam, D. H., Development of a *Brassica* sp. smother plant for weed control in corn, *Proc. North Central Weed Sci. Soc.* 45, 27, 1990.

31. Dessaint, F., Chadoeuf, R., and Barralis, G., Etude de la dynamique d'une communaute adventice. III. Influence a long terme des technique culturales sur la composition specifique du stock semencier, *Weed Res.* 30, 319, 1990.

32. Doll, J. D. and Bauer, T. L., Use of rye to control weeds in soybean, *Proc. N. Central Weed Sci. Soc.* 45, 118, 1990.

33. Donald, W. W., Primary tillage for foxtail barley (*Hordeum jubatum*) control, *Weed Technol.* 4, 318, 1990.

34. Donald, W. W., Management and control of Canada thistle, *Cirsium arvense, Rev. Weed Sci.* 5, 193, 1990.

35. Doyle, C. J., Cousens, R., and Moss, S. R., A model of the economics of controlling *Alopecurus myosuroides* Huds. in winter wheat, *Crop Prot.* 5, 143, 1986.

36. Duke, S. O., Naturally occurring chemical compounds as herbicides, *Rev. Weed Sci.* 2, 15, 1986.

37. Durgan, B. R., Gunsolus, J. L., and Becker, R. L., Cultural and Chemical Weed Control in Field Crops, Minn. Ext. Serv. AG-BU-3157, St. Paul, 1990.

38. Echtenkamp, G. W. and Moomaw, R. S., No-till corn production in a living mulch system, *Weed Technol.* 3, 261, 1989.

39. Egginton, G. E. and Robins, W. W., Irrigation water as a factor in the dissemination of weed seeds, Col. Agric. Exp. Stn. Bull. 253, 1920.

40. Egley, G. H., Weed seed and seedling reductions by soil solarization with transparent polyethylene sheets, *Weed Sci.* 31, 404, 1983.

41. Egley, G. H., High temperature effects on germination and survival of weed seeds in soil, *Weed Sci.* 38, 429, 1990.

42. Felton, W. L., The influence of row spacing and plant population on the effect of weed competition in soybean (*Glycine max*), *Aust. J. Exp. Agric.* 16, 926, 1976.

43. Felton, W. L., Use of weed detection for fallow weed control, *Proc. Great Plains Cons. Till. Symp.* Bismark, ND, 1990, 241.

44. Firbank, L. G., Cousens, R., Mortimer, A. M., and Smith, R. G. R., Effects of soil type on crop-weed density relationships between winter wheat and *Bromus sterilis, J. Appl. Ecol.* 27, 308, 1990.
45. Fischer, R. A. and Miles, R. E., The role of spatial pattern in the competition between crop plants and weeds: A theoretical analysis, *Math. Biosci.* 18, 335, 1973.
46. Forcella, F., Final distribution is related to rate of spread in alien weeds, *Weed Res.* 25, 181, 1985.
47. Forcella, F., Tolerance of weed competition associated with high leaf-area expansion rate in tall fescue, *Crop Sci.* 27, 146, 1987.
48. Forcella, F., Importance of pesticide alternatives to sustainable agriculture, *Nat. Forum* 68, 41, 1988.
49. Forcella, F., Weed seedling emergence models are practical tools for weed control, WSSA Abstr. 31, 122, 1991.
50. Forcella, F., Prediction of weed seedling densities from soil seed reserves, *Weed Res.* 32, 29, 1992.
51. Forcella, F. and Lindstrom, M. J., Movement and germination of weed seeds in ridge-till crop production systems, *Weed Sci.* 36, 56, 1988a.
52. Forcella, F. and Lindstrom, M. J., Weed seed populations in ridge and conventional tillage, *Weed Sci.* 36, 500, 1988b.
53. Forcella, F., Westgate, M. E., and Warnes, D. D., Effect of row spacing on herbicide and cultivation requirements for row crops, *Am. J. Altern. Agric.,* 7, 161, 1992.
54. Forcella, F. and Wood, J. T., Colonization potential of alien weeds are related to their native distributions: Implications for plant quarantine, *J. Aust. Inst. Agric. Sci.* 50, 35, 1984.
55. Forcella, F., Wood, J. T., and Dillon, S. P., Characteristics distinguishing invasive weeds within *Echium* (bugloss), *Weed Res.* 26, 351, 1986.
56. Freed, B. E., Oplinger, E. S., and Buhler, D. D., Velvetleaf control for solid-seeded soybean in three corn residue management systems, *Agron. J.* 79, 119, 1987.
57. Glaze, N. C., Dowler, C. C., Johnson, A. W., and Sumner, D. R., Influence of weed control programs in intensive cropping systems, *Weed Sci.* 32, 762, 1984.
58. Glenn, S., Peregoy, R. S., Hook, B. J., Heimer, J. B., and Wiepke, T., *Sorghum halapense* (L.) Pers. control with foliar-applied herbicides in conventional and no-tillage soyabeans, *Weed Res.* 26, 245, 1986.
59. Griffin, J. L. and Dabney, S. M., Preplant-postemergence herbicides for legume cover-crop control in minimum tillage systems, *Weed Technol.* 4, 332, 1990.
60. Gunsolus, J. L., Mechanical and cultural weed control in corn and soybeans, *Am. J. Altern. Agric.* 5, 114, 1990.
61. Harper, J. L., The evolution of weeds in relation to the resistance to herbicides, *Proc. Brit. Weed Cont. Conf.* 3, 179, 1956.
62. Harvey, R. G. and McNevin, G. R., Combining cultural practices and herbicides to control wild-proso millet (*Panicum miliaceum*), *Weed Technol.* 4, 433, 1990.

63. Holm, L. H., Plucknett, D. L., Pancho, J. V., and Herberger, J. P., *The World's Worst Weeds—Distribution and Biology*, University of Hawaii Press, Honolulu, 1977.

64. Holt, J. S. and LeBaron, H. M., Significance and distribution of herbicide resistance, *Weed Technol.* 4, 141, 1990.

65. Howe, O. W. and Oliver, L. R., Influence of soybean (*Glycine max*) row spacing on pitted morninglory (*Ipomoea lacunosa*) interference, *Weed Sci.* 35, 185, 1987.

66. Huffaker, C. B., Hamai, J., and Nowierski, R. M., Biological control of puncturevine, *Tribulus terrestris*, in California after twenty years of activity of introduced weevils, *Entomophaga.* 28, 387, 1983.

67. Johnson, D. W., Krall, J. M., Delaney, R. H., and Pochop, L. O., Response of monocot and dicot weed species to fresnel lens concentrated solar radiation, *Weed Sci.* 37, 797, 1989.

68. Johnson, M. D., Wyse, D. L., and Lueschen, W. E., The influence of herbicide formulation on weed control in four tillage systems, *Weed Sci.* 37, 239, 1989.

69. Johnson, W. C. and Coble, H. D., Crop rotation and herbicide effects on the population dynamics of two annual grasses, *Weed Sci.* 34, 452, 1986.

70. Julien, M. H., Kerr, J. D., and Chan, R. R., Biological control of weeds: an evaluation, *Prot. Ecol.* 7, 3, 1984.

71. Konesky, D. W., Siddiqi, M. Y., and Glass, A. D. M., Wild oat and barley interactions: varietal differences in competitiveness in relation to phosphorus supply, *Can. J. Bot.* 67, 3366, 1989.

72. Kremer, R. J., Management of weed seed banks with microorganisms, *Ecol. Appl.* 3, 42, 1993.

73. Kropff, M. J., Modelling the effects of weeds on crop production, *Weed Res.* 28, 465, 1988.

74. Lawson, H. M., Competition between annual weeds and vining peas grown at a range of population densities: effects on the crop, *Weed Res.* 22, 27, 1982.

75. Lawson, H. M. and Topham, P. B., Competition between annual weeds and vining peas grown at a range of population densities: effects on the weeds, *Weed Res.* 25, 221, 1985.

76. Legere, A. and Schreiber, M. M., Competition and canopy architecture as affected by soybean (*Glycine max*) row width and density of redroot pigweed (*Amaranthus retroflexus*), *Weed Sci.* 37, 84, 1989.

77. Lybecker, D. W., Schweizer, E. E., and King, R. P., Economic analysis of four weed management systems, *Weed Sci.* 36, 846, 1988.

78. Lybecker, D. W., Schweizer, E. E., and King, R. P., Weed management decisions in corn based on bioeconomic modeling, *Weed Sci.* 39, 124, 1991.

79. McWhorter, C. G., and Gebhardt. M. R., *Methods of Applying Herbicides*, Weed Science Society of America, Champaign, IL, 1987.

80. Martin, R. J., Cullis, B. R., and McNamara, D. W., Prediction of wheat yield loss due to competition by wild oats (*Avena* spp.). *Aust. J. Agric. Res.* 38, 487, 1987.

81. Marwat, K. B. and Nafziger, E. D., Cocklebur and velvetleaf interference with soybean grown at different densities and planting patterns, *Agron. J.* 82, 531, 1990.
82. Maxwell, B. A., Neighborhood approach as a basis for simulating weed competition in corn, *Proc. N. Central Weed Sci. Soc.* 46, 106, 1991.
83. Mayasich, J. M., Mayasich, S. A., and Rebeiz, C. A., Response of corn (*Zea mays*), soybean (*Glycine max*), and several weed species to dark-applied photodynamic herbicide modulators, *Weed Sci.* 38, 10, 1990.
84. Medd, R. W., Auld, B. A., Kemp, D. R., and Murison, R. D., The influence of wheat density and spatial arrangement on annual ryegrass, *Lolium rigidum* Gaudin, competition, *Aust. J. Agric. Res.* 36, 361, 1985.
85. Menges, R. M., Weed seed population dynamics during six years of weed management systems in crop rotations on irrigated soil, *Weed Sci.* 35, 328, 1987.
86. Merivani, Y. N. and Wyse, D. L., Effects of tillage and herbicides on quackgrass in a corn-soybean rotation, *Proc. N. Central Weed Cont. Conf.* 39, 97, 1984.
87. Mester, T. C. and D. D. Buhler, Effects of soil temperature, seed depth, and cyanazine on giant foxtail (*Setaria faberi*) and velvetleaf (*Abutilon theophrasti*) seedling development, *Weed Sci.* 39, 204, 1991.
88. Miller, D., Petersen, P., and Hoefer, F., WEEDIR: A Weed Control Decision Aid Program, Minn. Ext. Serv. AG-CS-2163, St. Paul, 1990.
89. Mills, J. A. and Witt, W. W., Effect of tillage systems on the efficacy and phytotoxicity of imazaquin and imaethypyr in soybean (*Glycine max*), *Weed Sci.* 37, 233, 1989.
90. Mills, J. A., Witt, W. W., and Barrett, M., Effects of tillage on the efficacy of clomozome in soybean (*Glycine max*), *Weed Sci.* 37, 217, 1989.
91. Moomaw, R. and Martin, A., Crop rotation for weed control, *Proc. Univ. Nebraska Crop Prot. Clinics* 1986, 165, 1986.
92. Moomaw, R. S. and Robison, L. R., Broadcast or banded chloramben with tillage variables in soybeans, *Weed Sci.* 20, 502, 1972.
93. Moomaw, R. S. and Robison, L. R., Broadcast or banded atrazine plus propachlor with tillage variables in corn, *Weed Sci.* 21, 106, 1973.
94. Moomaw, R. S. and Robison, L. R., Broadcast or banded atrazine + propachlor with tillage variables in grain sorghum, *Agron. J.* 65, 274, 1973.
95. Morris, M. J., A method for controlling *Hakea sericea* Schrad. seedlings using the fungus *Colletotrichum gloeosporoides* (Penz.) Sacc. *Weed Res.* 29, 449, 1989.
96. Mortensen, D. A. and Bauer, T. A., Comparison of techniques for measuring weed interference in soybean, *Proc. North Central Weed. Sci. Soc.* 46, 1991.
97. Mortensen, D. A. and Coble, H. D., The influence of soil water content on common cocklebur (*Xanthium strumarium*) interference in soybeans (*Glycine max*), *Weed Sci.* 37, 76, 1989.
98. Mortensen, D. A. and Coble, H. D., Two approaches to weed control decision-aid software, *Weed Technol.* 5, 445, 1991.
99. Mortimer, A. M., Sutton, J. J., and P. Gould, On robust weed population models, *Weed Res.* 29, 229, 1989.

100. Moss, S. R., Influence of tillage, straw disposal system and seed return on the population dynamics of *Alopecurus myosuroides* Huds. in winter wheat, *Weed Res.* 27, 313, 1987.

101. Neilson, J. C. and Anderson, J. L., Competitive effects of living mulch and no-till management systems on vegetable productivity, *West. Soc. Weed Sci. Res. Prog. Rep.* 1989, 148, 1989.

102. Ogg, A. G. and Dawson, J. H., Time of emergence of eight weed species, *Weed Sci.* 32, 327, 1984.

103. Osteen, C. D. and Szmedra, P. J., Agricultural pesticide use trends and policy issues, *U.S.D.A. Econ. Res. Serv. Rep. No. 622*, 1989.

104. Panetta, F. D. and Dodd, J., Bioclimatic prediction of the potential distribution of skeleton weed *Chondrilla juncea* L. in Western Australia, *J. Aust. Inst. Agric. Sci.* 53, 11, 1987.

105. Panetta, F. D., Gilbey, D. J., and D'Antuono, M. F., Survival and fecundity of wild radish (*Raphanus raphanistrum* L.) plants in relation to cropping, time of emergence and chemical control, *Aust. J. Agric. Res.* 39, 385, 1988.

106. Patterson, D. T., Musser, R. L., Flint, E. P., and Eplee, R. E., Temperature responses and potential for spread of witchweed (*Striga lutea*) in the United States, *Weed Sci.* 30, 87, 1982.

107. Patterson, M. G., Walker, R. H., Colvin, D. L., Wehtje, G., and McGruire, J. A., Comparison of soybean (*Glycine max*)-weed interference from large and small plots, *Weed Sci.* 36, 836, 1988.

108. Peters, E. J., Gebhardt, M. R., and Stritzke, J. F., Interrelations of row spacings, cultivations and herbicides for weed control in soybeans, *Weeds.* 13, 285, 1965.

109. Peterson, D. E., Foxtail Competition in Spring Wheat and *Ustilago neglecta* for Biological Control of Yellow Foxtail, Ph.D. thesis, North Dakota State University, Fargo, 1987.

110. Pugh, W. J. and Evans, N. A., Weed seed and trash screens for irrigation water, *Col. Agric. Exp. Stn. Bull. 522S*, 1964.

111. Putnam, A. R., Vegetable weed control with minimal herbicide inputs, *HortSci.* 25, 155, 1990.

112. Putnam, A. R., DeFrank, J., and Barnes, J. P., Exploitation of allelopathy for weed control in annual and perennial cropping systems, *J. Chem. Ecol.* 9, 1001, 1983.

113. Rapoport, E. H., Tropical versus temperate weeds: A glance into the present and future. In *Ecology of Biological Invasions in the Tropics*, Ramakrishnan, P.S., Ed., International Scientific Publications, New Delhi, 1991, 41.

114. Ramesl, R. E. and Wicks, G. A., Use of winter wheat (*Triticum aestivum*) cultivars and herbicides in aiding weed control in an ecofallow corn (*Zea mays*) rotation, *Weed Sci.* 36, 394, 1988.

115. Rasmussen, P. E. and Rohde, C. R., Stubble burning effects on winter wheat yield and nitrogen utilization under semiarid conditions, *Agron. J.* 80, 940, 1988.

116. Robinson, E. L., Langdale, G. W., and Stuedmann, J. A., Effect of three weed control regimes on no-till and tilled soybeans (*Glycine max*), *Weed Sci.* 32, 17, 1984.
117. Rose, S. J., Burnside, O. C., Specht, J. E., and Swisher, B. A., Competition and allelopathy between soybeans and weeds, *Agron. J.* 76, 523, 1984.
118. Ryan, G. F., Resistance of common groundsel to simazine and atrazine, *Weed Sci.* 18, 614, 1970.
119. Schmitt, R. M., Lueschen, W. E., and Gonsolus, J. L., Rotary hoeing used in conjunction with herbicides, *Proc. N. Central Weed Sci. Soc.* 45, 111, 1990.
120. Schweizer, E. E., Lybecker, D. W., and Zimdahl, R. L., Systems approach to weed management in irrigated crops, *Weed Sci.* 36, 840, 1988.
121. Schweizer, E. E. and Zimdahl, R. L., Weed seed decline in irrigated soil after rotation of crops and herbicides, *Weed Sci.* 32, 84, 1984.
122. Schweizer, E. E., Zimdahl, R. L., and Mickelson, R. H., Weed control in corn (*Zea mays*) as affected by till-plant systems and herbicides, *Weed Technol.* 3, 162, 1989.
123. Shear, G. M., Introduction and history of limited tillage, in *Weed Control in Limited-Tillage Systems*, Wiese, A. F., Ed., Weed Science Society of America, Champaign, IL, 1985, 1.
124. Smith, R. J., Biological control of northern jointvetch (*Aeschynomene virginica*) in rice (*Oryza sativa*) and soybean (*Glycine max*) — a researcher's view, *Weed Sci.* 34, Suppl. 1, 17, 1986.
125. Stahlman, P. W. and Miller, S. D., Downy brome (*Bromus tectorum*) interference and economic thresholds in winter wheat (*Triticum aestivum*), *Weed Sci.* 38, 224, 1990.
126. Staricka, J. A., Burford, P. M., Allmaras, R. R., and W. W. Nelson, Tracing the vertical distribution of simulated shattered seeds as related to tillage, *Agron. J.* 82, 1131, 1990.
127. Stoller, E. W. and Wax, L. M., Periodicity of germination and emergence of some annual weeds, *Weed Sci.* 21, 574, 1973.
128. Swinton, S. M., A Bioeconomic Model for Weed Management in Corn and Soybean, Ph.D. thesis, University of Minnesota, St. Paul, 1991.
129. Templeton, G. M., Smith, R. J., and TeBeest, D. O., Progress and potential of weed control with mycoherbicides, *Rev. Weed Sci.* 2, 1, 1986.
130. Thelen, K. D., Kells, J. J., and Penner, D., Comparison of application methods and tillage practices on volatilization of clomazone, *Weed Technol.* 2, 323, 1988.
131. Thomas, A. G., Gill, A. M., Moore, P. H. R., and Forcella, F., Drought feeding and the dispersal of weeds, *J. Aust. Inst. Agric. Sci.* 50, 103, 1984.
132. Towers, G. H. N. and Arnason, J. T., Photodynamic herbicides, *Weed Technol.* 2, 545, 1988.
133. Uri, N., Gill, M., Vesterby, M., Bull, L., Taylor, H., and Delvo, H. W., Agricultural resources: inputs, situation and outlook report, *U.S.D.A. Econ. Res. Serv. AR-20*, Washington, D.C., 1990.
134. Vangessel, M. J. and Renner, K. A., Effect of soil type, hilling time, and weed interference on potato (*Solanum tuberosum*) development and yield, *Weed Technol.* 4, 299, 1990.

135. Vigneault, C., Benoit, D. L., and McLaughlin, N. B., Energy aspects of weed electrocution, *Rev. Weed Sci.* 5, 15, 1990.
136. Wallace, R. W. and Bellinder, R. R., Potato (*Solanum tuberosum*) yields and weed populations in conventional and reduced tillage systems, 3, 590, 1989.
137. Warnes, D. D., Eberlein, C. V., Ford, J. H., and Lueschen, W. E., Effect of precipitation and management on weed control and crop yields with a winter rye cover crop system, *Agron. Abstr.* 81, 296, 1989.
138. Warnes, D. D., Ford, J. H., Eberlein, C. V., and Lueschen, W. E., Effects of a winter rye cover crop system and available soil water on weed control and yield in soybean, in Nat. Conf. on Cover Crops for Clean Water, Jackson, TN, April 9 to 11, 1991, 149–150.
139. Weaver, S. E., Factors affecting threshold levels and seed production of jimsonweed (*Datura stramonium* L.) in soyabeans [*Glycine max* (L.) Merr.], *Weed Res.* 26, 215, 1986.
140. Webber, C. L., Kerr, H. D., and Gebhardt, M. R., Interrelations of tillage and weed control for soybean (*Glycine max*) production, *Weed Sci.* 35, 830, 1987.
141. Weed Science Society of America, Composite list of weeds, *Weed Sci.* 32, Suppl. 2, 1984.
142. Westbrooks, R. G., Plant protection issues. I. A commentary on new weeds in the United States, *Weed Technol.* 5, 232, 1991.
143. Weston, L. A., Cover crop and herbicide influence on row crop seedling establishment in no-tillage culture, *Weed Sci.* 38, 166, 1990.
144. Wicks, G. A. and Somerholder, B. R., Effect of seedbed preparation on distribution of weed seed, *Weed Sci.* 19, 666, 1971.
145. Wicks, G. A., Smika, D. E., and Hergert, G. W., Long-term effects of no-tillage in a winter wheat (*Triticum aestivum*)–sorghum (*Sorghum bicolor*)–fallow rotation. *Weed Sci.* 36, 384, 1988.
146. Wiese, A. F., *Weed Control in Limited-Tillage Systems*, Monogr. No. 2, Weed Science Society of America, Champaign, IL, 1985.
147. Wilkerson, G. C., Jones, J. W., Coble, H. D., and Gunsolus, J. L., SOY-WEED: a simulation model of soybean and common cocklebur growth and competition, *Agron. J.* 82, 1003, 1990.
148. Wilkerson, G. C., Modena, S. A., and Coble, H. D., HERB: decision model for postemergence weed control in soybean, *Agron. J.* 83, 413, 1991.
149. Wilson, R. G., Dissemination of weed seeds by surface irrigation water in western Nebraska, *Weed Sci.* 28, 87, 1980.
150. Wilson, B. J. and Wright, K. J., Predicting the growth and competitive effects of annual weeds in wheat, *Weed Res.* 30, 201, 1990.
151. Wrucke, M. A. and Arnold, W. E., Weed species distribution as influenced by tillage and herbicides, *Weed Sci.* 33, 853, 1985.
152. Young, F. L., Russian thistle (*Salsola iberica*) growth and development in wheat (*Triticum aestivum*), *Weed Sci.* 34, 901, 1986.
153. Zamora, D. L., Thill, D. C., and Eplee, R. E., An eradication plan for plant invasions, *Weed Technol.* 3, 2, 1989.

7

Management of Arthropod Pests

Joseph E. Funderburk and Leon G. Higley

TABLE OF CONTENTS

INTRODUCTION

Indiscriminate use of chemical pesticides began shortly after World War II. Within two decades, many target and nontarget pests were genetically resistant, and some pesticides were no longer effective. Contamination of soil and water with long-lived pesticides affected bird populations and other wildlife, with some species even nearing extinction. From this disaster, an ecologically compatible methodology of pest control was developed called integrated pest management (IPM). Stringent laws also were passed in the U.S. banning pesticides that contaminated the environment or posed unnecessary health risks and forcing industry to develop new and safer pesticides.

Because of known resistance of many arthropod pests to insecticides, applied entomologists became involved in the development of IPM concepts and programs. A multidisciplinary approach is required for successful development of IPM programs. In addition to ecology, expertise in crop physiology and crop production is important. Management strategies directed at arthropods can influence problems from other pests, emphasizing the need for additional expertise in and understanding of weed, disease, and nematode management. Finally, economic and sociologic factors must be understood for successful implementation. The objective of IPM, as described by Huffaker and Smith (1980), is "the development of improved, ecologically oriented pest management systems that optimize, on a long-term basis, costs and benefits of crop protection." In view of the current emphasis placed on the development of sustainable agriculture, note should be made of the previous importance placed on sustainability in the theory and practice of IPM. Individual control tactics employed against pests are sustained through appropriate integration of multiple control tactics. The use of control tactics is regulated by public agencies to prevent pollution and depletion of natural resources. Economic sustainability also is a major consideration in IPM programs.

IPM programs have been developed and implemented in many agricultural production systems with favorable economic and ecological results. IPM theory is based on sound scientific principles, with refinements in theory resulting from a dynamic process of intensive research. Problems, however, sometimes result in the practice of IPM.

Most of the problems result from the large amount of knowledge and expertise needed to develop, implement, and continually improve an IPM program. Although simple solutions are sometimes possible when managing pest problems, there is a tendency to rely too heavily on strategies developed from inadequate knowledge and without sufficient consideration of all consequences of their use.

More alarming is the expectation on the part of government and industry that new innovative control tactics will be a panacea in solving pest problems. Such a "silver bullet" attitude regarding pesticides directly led to the pesticide crisis, and this cycle may be repeating itself in planned uses of genetically engineered crops. Moreover, looking for silver bullets results in inadequate support for more appropriate broad-based approaches to management such as are provided by IPM programs. Indeed, insufficient support for implementation of IPM programs continues to be a major impediment to sustainable pest management. Government may espouse such a practice through ignorance or as a short-term solution to crisis pest problems. Industry may be interested in short-term, rather than sustained economic returns. Regulatory practices based on political criteria rather than valid scientific factors also may force industry into misuse of a technology.

In this chapter, we discuss ecologically based, practical approaches to management of arthropod pests, beginning with a brief overview of relevant aspects of arthropod ecology. The scientific basis for discussion is derived primarily from the vast literature published on the theory and practice of IPM. Examples from numerous crops are included to illustrate points, but particular attention has been given to soybean, which is a major commodity grown in many different geographical regions. Our rationale is an attempt to give perspective to the similarities and differences associated with development of management systems for an individual crop in different agroecosystems. Currently, emphasis is being placed on ensuring that IPM programs are compatible with sustainable agricultural systems. Several important problems associated with management of arthropod pests in sustainable agricultural systems are highlighted, and potential ways to rectify these problems are discussed.

ECOLOGY OF ARTHROPOD COMMUNITIES IN AGROECOSYSTEMS

Development of ecologically based management systems for arthropod pests begins with knowledge of their ecology, and increased understanding of pest ecology improves the ability to develop ecologically and economically desirable ways to prevent avoidable crop damage. Ecosystems are communities of organisms and their physical environment interacting as an ecological unit. As Crossley et al. (1987) point out, most are managed, so the distinction between natural and artificial ecosystems is unclear. Agroecosystems are artificial ecosystems managed for production of food, fiber, and other resources. As is true of all ecosystems, agroecosystems are open, exchanging biotic and abiotic elements with other ecosystems. The vegetation growing in such places as farm ponds, hedgerows, wooded areas, drainage ditches, roadways, and crop fields within an agroecosystem provides niches for arthropod species. As with other ecosystems, biotic and abiotic elements are exchanged between habitats within an agroecosystem.

Crops are grown in areas away from their centers of origin, thereby making available new niches for use by arthropods. The center of origin of soybean is Asia, but about half of the total area worldwide is now planted in North America. Numerous pests have adapted and inhabit soybean in North America. However, entire guilds that cause severe economic losses in Asia are completely lacking (Kogan and Turnipseed,

1987). Because niche occupancy is a dynamic process, phytophagous species may eventually adapt to fill these nonsaturated feeding niches.

There are many noncrop plant species that provide niches within an agroecosystem for pest and innocuous arthropod species and their natural enemies. Certain weed species are able to survive both tillage operations and herbicides and are tolerated in crop fields. Other plants are adapted to utilize crop fields during fallow periods. Plant associations also become established in drainage ditches, hedgerows, roadways, wooded areas, and other noncrop areas. Many natural enemy species require food sources in the form of pollen, nectar, or innocuous arthropods that are not provided by crop plants. Alternative food requirements provided to natural enemies by noncrop plants have been reviewed by Altieri et al. (1977) and Altieri and Whitcomb (1979). Other arthropods require habitat of noncrop vegetation that allows for survival at critical times when crops are not being produced. By providing additional niches, noncrop vegetation greatly influences the diversity and abundance of arthropod species within an agroecosystem.

The biotic exchange between agroecosystems and between different habitats within an agroecosystem can be illustrated by selecting examples from soybean. The velvetbean caterpillar (*Anticarsia gemmatalis*) and the green cloverworm (*Plathypena scabra*) migrate northward from southern overwintering areas each year, thereby causing outbreaks in production centers of the northern U.S. where survival year round is not possible (Herzog and Todd, 1980, Buntin and Pedigo, 1983). Populations of the corn earworm (*Helicoverpa zea*) and the southern green stink bug (*Nezara viridula*) migrate into soybean fields each year from adjoining habitats of crops or wild vegetation (Stinner et al., 1982; Panizzi and Slansky, 1985). Established populations of lesser cornstalk borers (*Elasmopalpus lignosellus*) frequently survive preplant tillage practices, feed on partially buried weed or crop residues, and transfer to seedling soybeans (All, 1980).

Many indigenous natural enemies of soybean pests occur in all production areas. In some production areas, additional exotic natural enemies have been purposely introduced into the agroecosystem. Changes in cropping sequence and soil tillage operations prevent continuous survival during the year of most natural enemy species directly within crop fields, and their ecology within an agroecosystem is greatly influenced by the abundance and spatial arrangement of noncrop vegetation. For example, many species of predaceous ants do not nest in soybean fields, but forage in from adjacent weedy hedgerows (Whitcomb et al., 1972). Weeds directly within soybean fields also provide niches for natural enemy species and can result in an increase in the diversity and abundance of the associated arthropod community. Both broadleaf and grassy weeds within fields increase predator populations (Altieri, 1981; Shelton and Edwards, 1983).

Habitats of plants within ecosystems are fluctuating and sometimes are disparate resources for herbivorous species. This is especially true in agroecosystems where management activities greatly influence the diversity and abundance of vegetation in crop and noncrop habitats. The amount of utilization of an available niche by adapted species is initially determined by the process of finding plant hosts. Abundance and dispersion patterns of plant hosts greatly influences successful location and selection by herbivores. The influences of spatial arrangement and combinations of plant species on herbivore abundance are discussed and reviewed by Denno and McClure (1983a,b).

Morphology and biochemistry of plant hosts determine final acceptance and suitability to herbivore populations. These factors are inherited, but expression is sometimes influenced by environment. According to coevolutionary theory, a plant species produces a succession of defenses against herbivore populations. However, some herbivore species are able to adapt genetically and utilize the plant's resources. This process is explained in Janzen (1968). Another adaptive strategy for plant species is to tolerate some injury from herbivores and carry on normal physiological function with little or no reduction in overall productivity.

Microclimatic conditions within the niche of an arthropod species are altered by the abundance, spatial arrangement, vegetative condition, and reproductive condition of the plants providing the habitat. Changes of microclimate directly influence development, survival, and reproduction of herbivore species. Similar effects of changes in microclimate on the biologies of natural enemy species result in indirect effects on the population dynamics of herbivore populations. For example, planting soybean in narrow rows at high seeding rates increases mortality of lepidopterous herbivores by fungal pathogens, as does alterations in soybean plant maturity and growth resulting from changes in planting date (Sprenkel et al., 1979).

Many abiotic and biotic factors combine to affect herbivore populations, and the population dynamics of individual species are complex. The number of parameters affecting population dynamics of local populations of the simplest herbivore species is quite large (Gilbert, 1989). Population dynamics of herbivore populations in relation to IPM is discussed in Rabb and Guthrie (1970) and Geier et al. (1973). Research on the population dynamics of soybean pests is extensive. Despite this broad base of ecological knowledge, the ability to make reliable forecasts about population dynamics does not exist for any individual pest species (Kogan and Turnipseed, 1987).

Because the transformation of natural ecosystems into managed agroecosystems results in reductions in plant species diversity and increases in the abundance of crop species, arguments have been made that instability of herbivore populations is a by-product and that outbreaks of pests are more likely. Letourneau (1990) reviews and discusses the relevant literature. In actuality, outbreaks of some pest species are decreased by reductions in vegetation diversity. Therefore, indiscriminate diversification of vegetation within an agroecosystem cannot be expected to automatically reduce pest outbreaks. Although there are many species of phytophagous arthropod species within an agroecosystem, populations of only a limited number have the potential to reach outbreak densities. Effects of planned combinations of plants on the likelihood of outbreaks of these pests can be understood, and the information practically used in management programs.

PLANT RESPONSE TO PEST INJURY

Understanding the ecology of arthropod pests in agroecosystems is essential to their management. However, an equally important consideration is the relationship between those arthropod pests and their host plants. Plant responses to pest injury are important determinants of how we can minimize the impact of that injury. The nature of the injury and physiological responses of crops to injury set the boundaries within which management actions are possible. Consequently, understanding the physiological responses of the plants to pest injury is an essential undertaking for developing IPM programs.

Despite the importance of the topic, surprisingly few reviews of plant responses to arthropod injury are available. Bardner and Fletcher (1974), Fenemore (1982), Pedigo et al. (1986), and Welter (1989) review aspects of plant response to arthropod injury. This limited number of reviews on plant response to arthropod injury reflects the fact that this area has not received the attention it merits. Indeed, many workers have pointed out the need for additional research emphasis on this topic (e.g., Bardner and Fletcher, 1974; Boote, 1981; Pedigo et al., 1981, 1986; Wintersteen and Higley, 1992).

In examining relationships between plants and pests, it is important to distinguish between injury and damage. Injury is a stimulus causing a change in physiological processes, usually a deleterious change. Typically, arthropod injury is associated with feeding, such as leaf feeding, root feeding, phloem removal, etc. In contrast, damage is a measurable reduction in crop yield, utility, or fitness arising out of injury. Injury is a stimulus from pest activities that gives rise to damage, which is the plant response to those injuries (Pedigo et al., 1986).

Although details on relationships between specific types of arthropod injury and corresponding plant responses are uncertain in many instances, the theoretical relationship between injury and damage has been known for 30 years. In 1961, Tammes described the damage curve which represents the theoretical relationship between injury and damage (Figure 1). The damage curve describes the relationship between increasing injury and corresponding changes in yield (or other plant responses of interest). In the damage curve, damage is represented by crop yield loss. Subsequently, Pedigo et al. (1986) described specific portions of the damage curve and presented a terminology for plant responses corresponding to different portions of the curve (Figure 1). These responses were: tolerance—no damage per unit injury; overcompensation—negative damage per unit injury (a yield increase for some level of injury); compensation—increasing damage per unit injury; linearity—maximum damage per unit injury; desensitization—decreasing damage per unit injury; and inherent impunity—no damage per unit injury. These six terms and their corresponding portions of the damage curve describe the theoretical relationships between injury and damage. Not all plants display this entire array of responses to injury; however, all potential responses to injury are encompassed by the damage curve and these individual portions of that curve.

Plant responses to injury depend on a number of important factors. Specifically, these are the type of injury, the timing of injury, the plant part injured, the intensity of injury, and environmental factors (Pedigo et al., 1986). Each of these factors is an important determinant of how a crop will respond to arthropod injury. The injury type is important because it specifies which physiological processes are impaired by the

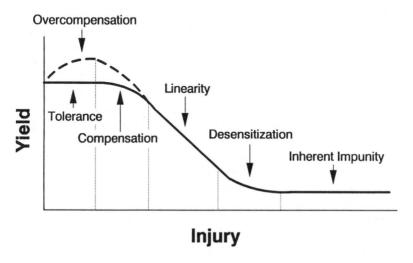

FIGURE 1.   The damage curve—the theoretical relationship between yield and injury.

pest activities. Arthropod injury to crops can fit into one of at least seven different categories: stand reduction, leaf mass consumption, assimilate removal, turgor reduction, fruit destruction, architecture modification, and phenological disruption (Boote, 1981, Pedigo et al., 1986). Injury by a single pest can fall into many of these categories. For instance, seed-corn maggot injury to soybean produces stand reduction, architecture modification, and phenological disruption (Higley and Pedigo, 1991). Because plant responses to injury are very much a function of physiological processes affected by injury, the type of injury is an important variable in understanding those responses.

Injury early in plant development, such as to seedling plants, can have serious repercussions; however, injury to young plants often is of less significance because of greater potential for compensation. Commonly, arthropod injury has the most serious impact on crop yields when it occurs during reproductive development of the plant. For example, the impact of defoliation on soybean yields is much more severe when defoliation occurs during reproductive stages than during vegetative growth (Teigen and Vorst, 1975). The timing of pest injury poses particular challenges to pest management in that management of the same pest may differ greatly, depending when the injury by that pest occurs during crop growth.

Direct injury, injury to yield-producing organs, typically has more serious consequences than indirect injury, injury to non-yield-producing organs. Distinctions in injury to different ages of plant parts also are possible. Some insects preferentially feed on younger leaves, for example, and plant responses to these tissues are likely to be more severe than to the same levels of injury made to older, less productive leaves.

As the damage curve indicates, greater amounts of injury (greater injury intensity) cause greater reductions in yield. After sufficient injury, crops may be insensitive to further injury, but typically this only happens after unacceptable losses already have been incurred. For arthropods, the intensity of injury is directly tied to their numbers. Consequently, managing most arthropod injury has focused on reducing arthropod numbers in an effort to decrease the intensity of injury.

Environmental effects on crop response to injury fall into two categories: abiotic and biotic effects. Abiotic factors are temperature, light intensity, CO_2 concentrations, pollutants, soil factors, and water availability. Of these, water availability is probably the most important single biotic factor that alters crop responses to injury. For example, the impact of defoliating insects is much more severe on water-stressed plants than on unstressed plants (Hammond and Pedigo, 1982). Among biotic factors that influence crop responses, arthropods, diseases, weeds, and intraspecific competition within the crop all are important variables. Changes in plant populations, for example, may alter responses of a crop to a given arthropod pest. Similarly, changes in cultural practices can lead to changes in intraspecific competition, disease occurrence, soil conditions, and water availability, all of which can have an important impact on crop responses to pest injury. For these reasons, changing cultural practices necessitate a re-examination of relationships between arthropod pests and their associated crop hosts.

This overview of crop responses to arthropod injury illustrates the complexity of these relationships. Details on all aspects of these relationships are valuable in devising effective management programs for arthropod pests. Unfortunately, the complexity of relationships between arthropod pests and crops has prevented their thorough examination for most crop pest situations. Indeed, relatively little is known about the specific physiological responses of plants to most types of arthropod pest injury. In the absence of such understandings, IPM tends to rely heavily on reducing the intensity of injury; in other words, controlling pest numbers. As we will discuss, this emphasis on controlling pest numbers has seriously constrained IPM programs and has even led to failures in management. Ongoing research efforts to identify general models of crop response to different types of pest injury offer the promise of providing more comprehensive understandings of relationships between pest injury and crop response (Wintersteen and Higley, 1992). Such understandings are essential for improving existing IPM programs and for directing management efforts into new areas.

IPM PROGRAMS

Two fundamental questions must be answered when dealing with pest problems. First, is any action needed; second, if needed, what action should be taken? The simplistic approach of pest control ignores these questions and seeks to kill as many pests as possible, without any detailed evaluation of the situation and without considering externalities. In contrast, IPM programs are designed to evaluate an array of factors when making management decisions about pests. Specifically, IPM has three broad objectives. The first is to maintain profitability, or economic soundness, when managing pests. Pest management actions should only be taken if economically justified. The second is to minimize selection pressure on pest populations from management tactics. Pest resistance develops when a given tactic, such as the use of a pesticide, puts undue selection pressure on a pest population. The third is to maintain environmental quality, specifically by minimizing the impact of management tactics on the environment.

Historically, IPM programs have emphasized the first of these objectives, ensuring that management actions are economically justified.

More recently, other criteria, such as environmental quality and reducing resistance, have received equal emphasis in IPM programs. Although formal mechanisms exist for examining the economic justification of a management action, corresponding guidelines for considering environmental safety and minimizing selection pressure have not been as forthcoming. Nevertheless, by calling for management actions only when economically justified, IPM programs have greatly reduced the use of pesticides, thereby improving environmental safety and diminishing the prospect of pest resistance development.

The foundation of IPM programs is the ability to make objective decisions about managing pests. In 1959, Stern et al. presented the concept of the economic injury level (EIL), which is a decision tool for making objective management decisions. The EIL is that level of pest injury necessary to justify management action against the pest. Because it is difficult to measure pest injury directly, EILs usually are described in terms of pest densities, with pest density acting as an index of injury. Because Stern et al. (1959) did not explicitly show how to calculate EILs and associated variables, the application of their ideas was delayed. Indeed, the first calculated EIL for a pest species did not appear until the early 1970s (Stone and Pedigo, 1972). The EIL incorporates economic factors such as cost of control and value of the commodity as well as biological information on injury and damage relationships between pests and crop plants.

Stated simply, the EIL is that level of injury where control costs equal benefits of the control. For pesticides, control costs include cost of the pesticide and cost of application; benefits are the yield loss that is prevented by use of a pesticide. In its most general form, the EIL equation is

$$EIL = C/VD(ti)IK$$

where C equals control cost, V equals value of the commodity, D(ti) equals damage (yield loss) as a function of total injury (ti), I equals injury per pest, and K equals the proportion of injury prevented by the control tactic. More simply, the EIL can be expressed as:

$$EIL = control\ cost/(commodity\ value * yield\ loss\ per\ pest)$$

By integrating economic factors regarding crop value and management costs, as well as biological information on relationships between pest injury and yields, the EIL provides a powerful approach for making management decisions regarding pests. Although establishing EILs requires substantial research, particularly with respect to establishing yield losses per unit injury and injury per pest, the EIL is an essential tool for rationalizing pest management decisions.

Besides the substantial research requirements for establishing EILs, there are a number of other serious limitations to their development and use (Pedigo et al., 1986; Poston et al., 1983). In particular, the EIL only addresses economic criteria in pest management decisions and does not directly address the important issues of maintaining environmental quality or of minimizing selection pressure from a management tactic on a pest. Nevertheless, the use of EILs has been extremely important in improving environmental safety and minimizing the development of resistance, because the use of EILs eliminates unnecessary pesticide use. Considerable research efforts are under way to expand both the availability and usefulness of EILs. In particular, recent work has provided a mechanism for incorporating considerations of environmental risks directly into the economic injury level. Higley and Wintersteen (1992) propose one method for developing environmental EILs, which assigns a dollar value to environmental risks for various pesticides and incorporates these costs into the EIL equation. These environmental EILs provide one approach for combining both economic and environmental considerations in pest management decisions. Additionally, the EIL provides a framework for other approaches to maintaining environmental quality in pest management (Pedigo and Higley, 1992).

Actions against pests require an evaluation of the pest in the situation, a decision on whether management action is warranted or not, and a decision regarding the choice of appropriate management tactics. Evaluating pest problems involves both appropriate identification of the pest involved and quantification of the intensity of pest injury, usually by estimating pest densities. This information is necessary for using EILs. The EIL is the basis for making management decisions on pests; however, it is not used directly. If we wait until a pest population reaches the EIL to make a decision, we will have accrued unnecessary yield losses. Consequently, it is better to decide to take action while a pest population is well below the EIL, but when we are still reasonably certain it will reach the EIL. The economic threshold (ET) is used to identify this point.

The ET is a pest or injury level set below the EIL that indicates when management action should be taken (Pedigo et al., 1986). The ET is predictive. It predicts when a pest population is going to reach the EIL. Consequently, the practical tool used for making management decisions regarding pests is the ET, which is based on a calculated EIL. In part, the ET represents the solution to a dilemma. On the one hand, if a pest is going to reach the EIL, we want to control it as early as possible to avoid unnecessary losses. Alternatively, the earlier we take action, the more uncertain we are that pest numbers actually will reach the EIL. The ET represents a compromise between these two objectives. Often the ET is set as a percentage of the EIL (e.g., 80%); however, techniques do exist for actually calculating an ET (Pedigo et al., 1986, 1989).

The choice of a management tactic is much more problematic. The more lead time available for responding to a pest problem, the greater the array of options for managing that problem. When a pest problem can be anticipated very early, such as before the crop has been planted, a variety of management tactics can be used to prevent or minimize pest injury. When rapid action is needed against a pest population, pesticides often are the only available tactic. Such differences in our ability to anticipate pest problems lead to two types of management actions, therapeutic and preventive. Therapeutic actions are those used to remedy or ameliorate an existing problem. Preventive tactics are those used to avoid potential pest problems. When pest problems can be anticipated, usually it is best to use preventive management tactics. In contrast, when sudden unpredictable pest problems occur, therapeutic actions must be used. Although therapeutic and preventive actions provide two approaches toward managing pest problems, a challenge to IPM programs is that our choice of individual management tactics often is limited, particularly in alternatives to pesticides. Consequently, one important research activity in IPM is to identify alternative tactics and to refine existing tactics.

Successful preventive management tactics have advantages over alternatives. They provide relatively inexpensive management of pest problems and may not require as much detailed information on pest status in a crop. In contrast, therapeutic tactics require detailed information on developing pest populations and rely on the use of EILs and ETs in making decisions on management actions. Both therapeutic and preventive tactics are essential for IPM programs, but individual tactics often are limited to only one type of action.

Therapeutic Tactics

Many, perhaps most, pest problems cannot be anticipated. Fluctuations in pest populations and changes in host susceptibility to injury both contribute to variability in the occurrence of pest problems. Occasional problems are unavoidable; consequently, it is essential that tactics to minimize the impact of these problems are available and are properly employed. Indeed, because pests are an inherently destabilizing factor in crop production, methods to minimize their impact are needed to provide more stable production systems.

For therapeutic tactics to be successful they must be properly employed against pests. Production systems must be monitored for pests, and, as problems develop, pest densities must be quantified. Monitoring pests can represent a significant economic cost, given the substantial labor required for sampling. Additionally, all sampling involves some uncertainty. Consequently, when developing sampling programs for arthropod pests, the need for large numbers of samples (necessary for providing precise information on pest populations) must be weighed against the economic cost of obtaining those samples. A vast literature exists on approaches and techniques for sampling arthropod pests (e.g., see Kogan and Herzog [1980] and Southwood [1978]). Much work in this area has been directed toward developing more efficient techniques for sampling arthropods. In particular, approaches such as sequential sampling can provide more rapid information on pest populations with no sacrifice in accuracy of the estimate. Establishing a sampling program for an arthropod pest depends very much on circumstances specific to the crop and ecological factors associated with the pest. Questions such as proper timing of a sampling program, choice of appropriate sampling technique, desired level of precision, and similar considerations all come into play in the formation of a sampling program. Ultimately, it is information obtained through sampling arthropod pests that is necessary for use with ETs and EILs in making management decisions.

Relatively few therapeutic management tactics other than pesticides are available. This reliance on pesticides for therapeutic management reflects the suitability of pesticides for rapid management actions and their cost effectiveness in this role. Ecologically, there is no strong argument against the use of pesticides as therapeutic management tactics. For occasional use against sudden pest outbreaks, pesticides can be used rapidly with correspondingly rapid effects on pest populations. No other management tactic offers such flexibility in use and fast action. Problems with pesticide use frequently develop when pesticides are used not in a therapeutic role but, instead, are used preventively. Repeated use of pesticides contributes to pest resistance as well as to environmental contamination. The goal of most IPM programs is to preserve pesticide use for those situations where pesticides are most applicable and to minimize any deleterious impacts of that use.

Besides cost, the key features of pesticides that are important for IPM are toxicity and persistence. Ideally, pesticides should be specific and relatively nonpersistent. Obviously, compounds must persist long enough to impact pest populations, but they should not persist so long as to pose a threat to the environment. Similarly, the best pesticides are those that selectively kill the target pests with little impact on natural enemies and nontarget organisms. In practice, few pesticides meet both of these goals. Indeed, given the substantial development and registration costs associated with producing a pesticide, most chemical companies look for broad spectrum pesticides that can be used for a wide variety of pest species, rather than for highly specific pesticides that may have a more limited market.

The availability of pesticides is not only limited by economic constraints in terms of development and registration of pesticides, but also is limited by their modes of action. Relatively few unique mechanisms for insecticidal action against pests have been discovered (Beeman, 1982). As a pest becomes resistant to one of these methods, options for managing that pest become further constrained. Initially, pesticides were developed with broad spectrum activity and long persistence. The chlorinated hydrocarbons, which include compounds like DDT, had both of these properties and, because of their serious hazards to the environment, are largely discontinued. Other compounds employing different modes of action include the carbamate, organophosphate, and pyrethroid classes of insecticides. In addition to these, recent developments exploit unique features of insect behavior and physiology in an effort to provide acceptable pest mortality with minimal impact on nontarget species. Despite these efforts, therapeutic management tactics for arthropod pests still must draw from a limited pool of acceptable pesticides. Consequently, there is a strong incentive to try to conserve these pesticides and avoid their misuse. Without such therapeutic tactics, many pest problems would be unmanageable.

Besides pesticides, a few other techniques are available as therapeutic tactics. One approach is inundative releases of natural enemies. This involves the mass rearing of natural enemies, such as parasitoids, and their release where a pest problem is occurring. Because there is a time lag between natural enemy release and declines in pest populations, inundative releases work best when they can be made before injurious pest populations have developed (van Lenteren and Woets, 1988). Another promising approach is the use of entomopathogens, which involves applying a pathogen to a pest population and allowing the subsequent disease to suppress that population. An important approach in some crops is to alter harvest times to control pests. For example, early alfalfa cuttings often are used as a management tactic for alfalfa insect pests. One other therapeutic approach involves combining tactics. The cultural practice of trap cropping, growing a section of a crop that is highly attractive to insect pests, can be used to draw pests away from the important crop, with the pest population in the trap crop later treated with insecticide. Hokkanen (1991) provides a review of uses of trap cropping.

Preventive Tactics

Decisions by individual producers to employ cultural control and host plant resistance in their IPM programs usually are made before crop planting and without specific knowledge that the target pest or pests will reach outbreak population densities. The same is true for decisions involving biological control tactics designed to enhance natural enemy populations. From an operational standpoint, decisions to employ such preventive tactics are based, at least nominally, on risk assessment and EILs.

Preventive application of an insecticide or acaracide is not compatible with IPM philosophy, but is considered for very economically important pests in which outbreak populations cannot be feasibly detected by scouting or reliably controlled with rescue insecticide application. The southern corn rootworm (*Diabrotica undecimpunctata howardii*) is such a pest in peanuts produced along the Atlantic Coast of the U.S. Only fields with soils of high organic matter or clay content, however, are at high risk, and preventive insecticide application is recommended only in such fields (Brandenburg and Herbert, 1991). When integrated with a partially resistant cultivar, reduced rates of insecticide are adequate.

Crop fields are suitable habitat for a wide range of indigenous beneficial arthropod species. These beneficial insects, mites, and spiders form a complex fauna. Disease organisms also kill arthropod pests. The impacts of this complex fauna and flora of beneficial organisms on arthropod pest populations are rarely well understood, but there is no doubt that indigenous natural enemies are very important in suppressing arthropod pest populations in most production situations. Consequently, conservation of natural enemy populations is a primary consideration in IPM programs. There are many predators, parasites, and diseases of soybean pests in most agroecosystems. Kogan and Turnipseed (1987) review the literature and discuss the importance of natural enemies in soybean IPM programs in different ecological regions. Insecticide applications are greatly reduced where conservation of natural enemies is carefully considered in chemical control recommendations of soybean. For example, fungicide applications in soybean IPM programs in Ecuador are geared so as not to interfere with natural velvetbean caterpillar mortality from an important fungal disease (Stansly and Orellana, 1990).

True plant resistance is usually considered to be under genetic control, which operates through purposeful manipulation of the defensive mechanisms of plants against herbivore populations. Traditionally, resistant plant cultivars have been developed by selecting from natural resistance mechanisms. Recently, molecular techniques have been developed which allow for artificial incorporation of resistant mechanisms into a plant's defenses. Genetically engineered crops have been successfully developed that express the *Bacillus thuringiensis* toxin (Gould, 1988).

Mechanisms of plant resistance are nonpreference, antibiosis, and tolerance (Painter, 1951). Nonpreference and antibiosis involve both plant and arthropod pest characteristics. Nonpreference operates by impairing normal pest behavior. Antibiosis impairs the pest's metabolic processes, which can affect survival, development, and behavior. Tolerance is a plant response in which satisfactory yields are obtained in spite of the pest's injury. Antibiosis has been most widely sought after by plant breeders developing resistant cultivars.

Plant breeders have focused on developing resistant crop cultivars for some of the most economically severe pests. At least thirteen genes that confer resistance in wheat to Hessian fly (*Mayetiola destructor*) have been identified (Gallun, 1977; Hatchett et al., 1981). Resistant cultivars have been developed possessing one resistance-conferring gene. In many geographical production centers, Hessian fly biotypes have evolved that are unaffected by the antibiotic resistance factors conferred by individual wheat resistance genes. Plant breeders have responded by recommending the sequential use of single resistance genes (Gallun, 1977; Gallun and Khush, 1980).

Length and orientation of pubescence hair are characteristics affecting the preference of potato leafhopper (*Empoasca fabae*) populations for soybean (Broersma et al., 1972; Turnipseed, 1977). Pubescence charac- teristics imparting resistance have been incorporated into commercial soybean cultivars, and these varieties have been successfully used for decades in IPM programs to prevent economic damage from this pest. Several soybean plant introductions with antibiosis and nonpreference genes for the leaf-feeding guild also have been identified (Sullivan, 1985). These plant introductions have been used to develop resistant commer- cial cultivars, which have not been widely accepted. The reason is that most leaf-feeding pests are easily controlled with inexpensive insecti- cides, and susceptible cultivars with slightly higher yields are preferred by producers.

Cultural control involves manipulation of a pest's requisites within the agroecosystem to reduce rates of pest increase and damage. The favora- bility of the environment for the pest is reduced by altering the diversity, abundance, and spatial arrangement of the vegetation or by modifying the practices used to produce crops. Development of cultural control requires an understanding of pest ecology as it relates to outbreaks in a crop. Cultural control tactics usually are designed to exploit a vulnerabil- ity in the pest's life cycle.

Changing the diversity, abundance, and spatial arrangement of the vegetation within the agroecosystem is a major planning decision with other potential economic and ecologic ramifications. Consequently, these types of cultural controls usually have been employed against very economically damaging, difficult-to-control pests. Spider mites (*Te- tranychus* spp.) have a broad host range and overwinter on wild plant hosts present in field margins (Brandenburg and Kennedy, 1982). These hosts and early spring crop hosts such as corn serve as nurseries for population increase and dispersal into adjacent peanut fields. Managing spider mites is difficult because of problems in scouting, rapid reproduc- tive rates, high reproductive potential, and mite sensitivity to pesticides and environmental conditions. Crop rotation of peanuts away from corn reduces infestations. In addition, field borders are not mowed in the summer to avoid dispersal into the field.

Damage to crucifers from the cabbage maggot (*Delia radicum*) and the cabbage aphid (*Brevicoryne brassicae*) is greatly reduced by between-row interplantings of beans, grass, clover, or spinach (Coaker, 1980). Cabbage aphid damage declines because of reduced immigration and increased predation; cabbage maggot populations are reduced because of reduced oviposition on the crucifer crop. Early maturing soybean varieties planted near main soybean plantings attract the bean leaf beetle (*Ceratoma trifurcata*) and southern green stink bug (*Nezara viridula*) (Newsom and Herzog, 1977). Insecticide applications in the early maturing trap crop control the target pests and prevent their spread to the main crop. Natural enemy populations also are conserved in the main crop. Reductions in spread of bean pod mottle virus, which is vectored by the bean leaf beetle, also occurs. Although this trap cropping system is very effective, it has not been widely accepted by producers in regions where these pests are serious economic problems. These pests are easily controlled with insecticides, and growers are unwilling to accept potential reductions in yield in the soybean trap crop.

Crop rotation can be an effective management practice against arthropod pest species with long generational cycles. Rotation of host and nonhost crops has been used successfully for managing soil arthropod pests that become established during one cropping season, overwinter within the field, and are present in an injurious life stage during the next cropping season. Whitefringed beetles (*Graphognathus* spp.) can be very damaging to soybean. Rotating between seasons with a grass crop such as corn or grain sorghum reduces problems (Ottens and Todd, 1979). Eggs of western corn rootworms (*Diabrotica virgifera virgifera*) and northern corn rootworms (*D. barberi*) are almost exclusively laid in corn (Levine and Oloumi-Sadeghi, 1991). In the Midwest, rotation with a noncorn crop such as soybean has been used to manage the pests in corn. However, a genetic mechanism of extended egg diapause of more than 1 year has been documented in some northern corn rootworm populations. In instances where crop rotation has been heavily relied upon as a control tactic, a proportion of rootworm populations have extended egg diapause, and heavy damage to corn sometimes occurs.

The selection of which crops to grow within the agroecosystem also is very important in determining overall injury from arthropod pests. Outbreaks of soybean looper (*Pseudoplusia includens*) are common in soybean grown in agroecosystems containing cotton (Burleigh, 1972; Wuensche, 1976). Adult feeding on cotton nectar increases oviposition, mating frequency, and longevity (Jensen et al., 1974). Discontinuing cotton production within an agroecosystem would reduce damage, but is not utilized because of the economic importance of cotton to producers.

Although cultural controls that operate by modifying the vegetation within the agroecosystem are employed, required changes impact the economics of farming operations, which frequently impedes adoption by producers. Herzog and Funderburk (1986) discuss these problems and review the literature on this subject. Cultural control practices, in which the methods used to plant, maintain, or harvest a crop are altered, usually require less drastic changes in production practices and are frequently employed in IPM programs. Risks for seedcorn maggot (*Delia platura*) injury in soybean in the Midwest are great when green organic matter of rye, wheat, or alfalfa is incorporated into the soil, but low when only corn or soybean residues are present (Hammond and Stinner, 1987). In situations where soybean must be planted into a field containing green organic matter, planting with no-tillage practices virtually eliminates the damage potential. Economic damage from velvetbean caterpillar in the Southeastern U.S. can be evaded by planting early maturing soybean varieties and/or by planting early (Herzog, 1979).

Most cultural control practices are employed specifically to prevent or reduce damage from a particular pest. Recent research is revealing that production practices can be tailored, based on ecologic and economic evaluations, to ameliorate overall damage from pests, while still optimizing crop production. In soybean fields double-cropped with winter wheat in the Southeastern U.S., biological control from indigenous populations of bigeyed bugs (*Geocoris* spp.) and damsel bugs (*Nabis* and *Raduviolus* spp.) can be increased by using disk tillage rather than no-tillage (Funderburk et al., 1988). Soil fertility levels of phosphorus also affect pest and natural enemy populations in this double-cropped system (Funderburk et al., 1991). Avoiding overfertilization, in addition to reducing the cost of the fertilizer, reduces the likelihood of pest outbreaks in soybean from southern green stink bugs and velvetbean caterpillars.

Government Regulation and IPM

Regulatory control tactics are used to prevent the introduction of pests from a region of occurrence to a region of nonoccurrence. Pests usually have few natural enemies in introduced regions, and severe outbreaks can result if populations become established. Regulatory control involves inspections and quarantines of imported hosts to prevent introductions of exotic pest populations. Eradication of most indigenous pests has proved unfeasable for ecological and environmental reasons. However, eradication of an accidentally introduced pest can be successful if pursued aggressively before populations are established over a wide area in large densities. Knipling (1979) discusses regulatory control and eradication efforts. These functions are government responsibilities and are appropriately handled by public agencies.

The Environmental Protection Agency (EPA) is empowered to regulate pesticides throughout the U.S. The EPA considers public health and environmental safety before approving pesticides for use. Registration requires studies to understand a pesticide's effects on laboratory animals and to evaluate its behavior in the environment. Approved pesticides are available to producers as long as they are used according to label requirements. Monitoring and enforcement are conducted by the EPA, the Food and Drug Administration (FDA), and individual state agencies. Although these organizations are legally empowered only in the U.S., there is no doubt that the net effects on the environment and public health worldwide have been very positive. Many countries wish to avoid environmental problems and to ensure public health, and follow EPA regulations within their own country (e.g., Stansly and de Sevilla, 1990). Fear of losing export markets is another impetus for care in pesticide usage in foreign countries.

A highly controversial aspect of the regulatory process involves pesticide residues on food. Residue tolerances established by EPA are developed from worst-case scenarios and have highly conservative built-in safety factors. Random testing in the U.S. conducted by the FDA and state agencies continually indicates that domestic and imported foods do not contain unsafe pesticide residues. Although there is no certainty about some aspects of this issue, available scientific evidence indicates that the food supply produced with pesticides is safe and healthy. Beall et al. (1991) discuss the pesticide registration process and related aspects of public health and environmental safety.

Agricultural enterprises are small businesses, and pesticide usage involves microeconomic decision making by producers. Pesticides are available for use by growers at their discretion, so long as they are used according to label requirements. In most cases, it is legal for growers to use a pesticide in ways not recommended in IPM programs. Unnecessary usage and overusage sometimes results in development of resistance by individual arthropod pests to one or more insecticides. Roush and McKenzie (1987) discuss and review the scientific literature on development of resistance in arthropod populations. Private industry and regulatory agencies are beginning to consider sustainability of pesticides when developing labels. For example, proposed labels for abamectin (Merck and Co., Inc., Rahway, New Jersey) and cyromazine (Ciba-Geigy, Greensboro, North Carolina) against leafminer (*Liriomyza trifolii*) on lettuce and celery require alternating the pesticides as a resistance management technique.

Federal and state agencies in the U.S. are involved in protection of soil and water resources. Conservation of natural resources in relation to sustainable agriculture is described elsewhere in this book; however, some aspects relate directly to IPM programs. As previously discussed, protection of natural resources from pesticide contamination is a regulatory function of EPA. Additionally, soil tillage and irrigation practices sometimes are purposely modified to control arthropod pests, and the effects of such practices on possible depletion of natural resources must be considered in IPM programs.

IPM AND SUSTAINABLE AGRICULTURE

The goal of sustainable agriculture is to have agricultural systems that can be maintained through time without extraordinary inputs and without significant impacts on the environment. Specifically, sustainable systems must provide for stability in production and in producer profits, while also maintaining environmental stability. Pests, including arthropod pests, present a particular challenge to sustainable agriculture because the effect of pests on agricultural systems is intrinsically destabilizing. These disruptive effects of pests fall into two areas, the direct effect of pests on production and the effects of tactics used to manage pests. The direct effect of pests on production is to stress crops and livestock resulting in reduced yields, reduced quality, and increased susceptibility to other stressors. The effects of management tactics include potential environmental disruption, such as by contamination from pesticides or increased soil erosion from cultural control tactics, and the development of pests that cannot be readily controlled (because of pest resistance to management tactics, arising through misuse of those tactics). Any pest problem represents an economic challenge to producers, either through crop losses from the pests or through costs of managing the pests. In the extreme, pests may routinely limit or prevent production of certain commodities. Alternatively, pests may only occasionally impact a production system, but even an infrequent pest problem may be sufficient to reduce or eliminate profits and disrupt production. Consequently, it is impossible to have a truly sustainable agricultural system without having sustainable solutions for pest management.

IPM was developed to provide sustainable solutions to pest problems. The concept and practices of IPM predate those of sustainable agriculture and have no doubt served as a model for sustainable agriculture; therefore, IPM is both compatible and essential to sustainable agriculture. However, it is important to recognize that no tactic or approach, including IPM, can completely eliminate the impact of pests on agricultural systems. Thus, IPM is not a panacea for solving all pest problems; rather, it is the best approach for providing economically and environmentally sound approaches for reducing the deleterious effects of pests.

The specific goals of IPM fit well with the objectives of sustainable agriculture. Overall, IPM provides reliable procedures for managing pests, which is an obvious requirement for maintaining agricultural productivity. More specifically, IPM programs address issues of economic and environmental sustainability, as well as providing for the conservation of management tactics. Economic sustainability is provided through IPM by undertaking management actions only when they are economically justified. Criteria such as the EIL and ET were developed for making management decisions that consider the economic merits of action. Environmental sustainability is provided by the use of procedures to minimize unnecessary pesticide use (such as EILs and ETs), by the use of multiple tactics and of nonchemical tactics, and by the development of decision tools that explicitly consider environmental factors (such as environmental EILs). Conservation of management tactics is provided through IPM programs by using multiple tactics and by using individual tactics in ways to minimize selection pressure on pests, thereby reducing the likelihood of pest resistance developing.

The therapeutic approach for managing pests in IPM programs has been successful for many crops in many production situations. Inputs of pesticides are reduced, thereby reducing costs and potential environmental contamination. Reductions in pesticide usage also help to conserve indigenous natural enemies and, when pest populations must be suppressed chemically, pesticides frequently can be selected that reduce negative effects on natural enemies. Highly selective biological pesticides are available for some pests. Overall, the therapeutic approach reduces the risk for development of insecticide resistance in pest populations through tolerance of subeconomic pest densities and maximization of natural mortality from abiotic and biological control factors. In concept, the therapeutic approach is the core around which a complete IPM program incorporating preventive control tactics is built.

The therapeutic approach can not be used to manage all pests. Density of some pests can not be economically or reliably estimated in scouting programs. Or, a management tactic may not be available that rapidly and reliably suppresses an outbreak population. Such problems are typical when managing soil insect pests (Cheshire et al., 1989). The therapeutic approach also is not applicable for severe pests in which populations remain above economic densities. Overreliance on an individual management tactic frequently occurs under these circumstances. As previously shown, arthropod populations can develop resistance to insecticidal controls, cultural controls, and host plant resistance. Development of resistance to biological controls such as *B. thuringiensis* also can occur (Tabashnik et al., 1990).

In situations where the therapeutic approach can not be employed, development of resistance to control tactics by arthropod pests continues to be a significant problem. Georghiou and Mellon (1983) review examples of pesticide resistance to arthropods, and Gallun and Khush (1980) review adaptation of arthropod pests to plant resistant varieties. Development of resistance sometimes can be avoided in these situations by reducing the selection pressure on arthropod populations for individuals unaffected by the resistance factor. For pesticidal control, this can be accomplished, as previously described, by alternating pesticides with different modes of action. This resistance management strategy is described in Roush and McKenzie (1987). Theories have been advanced for ways to prevent adaptation of arthropod pests to resistant crop varieties. Sequential deployment or mixed plantings of varieties with different antibiotic resistance factors seem preferable to releasing a variety containing all of the resistance factors (Gould, 1986). Resistance management is an extremely important issue for genetically engineered crops, such as those containing the *B. thuringiensis* toxin.

Because of movement between agroecosystems of arthropod pests, resistance management must be employed over widespread geographical areas to be successful. For example, soybean looper populations in soybean in the Southeast U.S. are resistant to some insecticides despite a successful IPM program on soybean in the region (Leonard et al., 1990). The soybean looper does not overwinter in the region, and resistant populations are believed to have developed as a result of exposure to heavy pressure from insecticides in overwintering areas of extreme southern Florida and the Caribbean. Some form of legal mandate obviously would be required to ensure that adequate numbers of producers participated in a resistance management program over such a widespread geographical area. In this case, the problem would be confounded by the fact that the insecticide-resistant populations might be developing from populations outside the U.S. EPA historically has focused on public health and environmental safety, but the proposed label requirements for abamectin and cyromazine previously discussed are an indication that pesticide resistance management may receive greater emphasis.

For pests where the therapeutic approach is not an effective management strategy, employing preventive control tactics that do not result directly in pest mortality is a resistance management strategy. For example, tolerance as a host plant resistance factor allows for greater pest populations without economic loss. In some situations, this can result in greater natural enemy populations and increased biological control throughout the agroecosystem. Selection pressure for individuals not affected by the resistance factor is avoided. Pest populations may adapt to and become unaffected by cultural controls that cause direct pest mortality as the example with corn rootworms in the Midwestern U.S. illustrates (Levine and Oloumi-Sadeghi, 1991). Those that do not provide needed habitat or that increase biological control are more sustainable and would be ecologically preferred. Not incorporating green organic matter and planting soybean with no-till in the Midwest to prevent seed-corn maggot damage is an example of a cultural control that operates by not providing needed habitat for a pest (Hammond and Stinner, 1987). Herzog and Funderburk (1985) provide numerous examples of plant resistance and cultural controls that increase biological control.

A large amount of knowledge and expertise is needed to develop and implement effective IPM strategies for arthropod pests. This complexity sometimes leads to an overreliance on approaches that are overly simplistic and easy to use without adequate consideration of fundamental ecological information. This is especially a problem for severe and difficult-to-control pests. Previous scientists writing on sustainable agriculture have focused on this problem. Gliessman (1990) discusses this subject and reviews the related scientific literature involving sustainable agriculture. The recommended remedy is employment of pest management strategies over the biogeographical region that improve biological control by an understanding of interactions between host plants, arthropod herbivores, and their natural enemies.

Considerable public funds are spent developing IPM programs to understand pest ecology, crop response to pest injury, and the economics of crop production. Both public and private funds are spent developing and evaluating efficacy and safety of management tactics. There also are costs associated with training practitioners and implementing IPM programs. The economic and ecologic benefits of IPM are widely recognized in the U.S. and Europe and are gaining recognition in Third World countries (Goodell, 1984). However, the high costs to government and industry and the amount of expertise needed to develop and implement IPM programs result in frequent breakdowns and abuses. The sustainability of agricultural production systems is a major consideration of IPM programs, but greater emphasis must be placed on eliminating abuses that lead to negative effects on long-term economic sustainability of agricultural systems, eventual loss of efficacy of control tactics, or unnecessary depletion of natural resources.

As discussed, any type of management tactic can be used inappropriately in an IPM program, resulting in undesirable economic or ecologic consequences. However, most emphasis is placed on abuses with pesticides. It should be noted that there is no compelling scientific reason to believe that pesticides, if used according to IPM theory, are incompatible with sustainable agricultural systems. If pesticides were unavailable, crop losses from arthropods, even with the addition of other means of management would be great (e.g., Pimentel, 1986). Consequently, elimination of appropriate pesticide usage would increase land area needed to produce food, leading to destruction of natural habitat and potentially depleting soil, water, and energy resources. The economics and economic sustainability of agricultural production systems also would be affected (Zilberman et al., 1991).

CONCLUSIONS AND RECOMMENDATIONS

IPM theory is based on scientific principles of ecology and economics. Long-term profitability of crop production systems, sustained efficacy of control tactics, and conservation of natural resources are important considerations. In our view, the concept of IPM is completely compatible with the goals of sustainable agriculture. Indeed, we think IPM is essential for truly sustainable agricultural systems. IPM programs have been developed and implemented for numerous crops in many production situations with desirable consequences. Special attention in this chapter was given to arthropods inhabiting soybean. It should be noted that successful IPM programs have been developed for soybean produced in numerous geographical areas in both temperate and tropical climates. Although soybean IPM may represent a more mature system than exists for many commodities (Wintersteen and Higley, 1992), it illustrates what is possible with sufficient research efforts.

The therapeutic approach for managing pests in IPM programs has been very successful in most situations. Reduced inputs of pesticides have resulted, with highly desirable economic and ecological benefits. This has led to a view by many that IPM equates to the therapeutic approach. Preventive control tactics, however, are economically and ecologically desirable in many pest situations and we urge enhanced development and use of preventive control tactics in the future. We agree with ecologists that greater emphasis needs to be placed on approaches that improve biological control through an understanding of the interactions between host plants, arthropod herbivores, and their natural enemies. Preventive control tactics are especially important for effective management of difficult-to-control and severe pests. Development of resistance to managemant tactics by these pests continues to be a major problem, and preventing resistance development is a major issue for sustainable agriculture.

Sometimes only nominal information on pest ecology, crop response to pest injury, and the economics of crop production are needed to develop effective, sustainable IPM strategies. In other cases, developing and implementing long-term management solutions that do not deplete natural resources are very difficult. The costs associated with developing, implementing, and continually refining IPM programs are large in all situations. Additional costs to develop and evaluate efficacy and safety of management tactics are great. Government and industry have large financial investments in agriculture, and sustainability in all its aspects obviously is in everyone's best interest. Laws and regulatory actions should promote development and implementation of effective, sustainable pest management strategies. Government also has a responsibility to establish priorities and provide necessary funding to achieve the same.

Public health and environmental safety in relation to agriculture already are recognized as important government responsibilities. However, overreliance on and unnecessary use of individual control tactics also requires government regulation. In some cases, pesticide resistance management strategies need to be considered when developing pesticide labels. Requiring application of a pesticide based on documented need or prescription may also be necessary to eliminate unnecessary applications. Prescription may be an appropriate alternative to cancellation of a pesticide in some situations. Legal mandates also need to be developed to prevent overreliance on pest-resistant crop varieties.

In all instances, scientists developing and implementing management programs should be involved directly with policy decisions. Although there are clearly problems associated with the practice of IPM, it is both compatible and essential to sustainable agriculture and provides an appropriate approach for providing economically and ecologically sound solutions for reducing the deleterious effects of pests. A greater emphasis on interdisciplinary research is needed to rectify some of the problems in the practice of IPM. Above all, there must be a greater understanding and consideration of fundamental ecological and economic information.

REFERENCES

All, J. N. 1980. Consistency of lesser cornstalk borer control with Lorsban in various corn cropping systems. *Down to Earth. 36*(2):33–36.

Altieri, M. A. 1981. Weeds may augment biological control of insects. *Calif. Agric.* 35:22–24.

Altieri, M. A., A. van Schoonhoven, and J. Doll. 1977. The ecological role of weeds in insect pest management systems: a review illustrated with bean (*Phaseolus vulgaris* L.) cropping systems. *PANS* 24:185–206.

Altieri, M. A. and W. H. Whitcomb. 1979. The potential use of weeds in the manipulation of beneficial insects. *HortScience.* 14:12–18.

Bardner, R. and K. E. Fletcher. 1974. Insect infestations and their effects on the growth and yield of field crops: a review. *Bull. Entomol. Res.* 64:141–160.

Beall, G. A., C. M. Bruhn, A. L. Craigmill, and C. K. Winter. 1991. Pesticides and your food: how safe is "safe"? *Calif. Agric.* 45(4):4–11.

Beeman, R. W. 1982. Recent advances in mode of action of insecticides. *Annu. Rev. Entomol.* 27:253–258.

Boote, K. J. 1981. Concepts for modeling crop response to pest damage. *ASAE Pap. 81-4007.* American Society of Agricultural Engineers, St. Joseph, MI. 24 pp.

Brandenburg, R. L. and G. G. Kennedy. 1982. Intercrop relationship and spider mite dispersal in a corn/peanut agroecosystem. *Entomol. Exp. Appl.* 32:269-271.

Brandenburg, R. L. and D. A. Herbert, Jr. 1991. Effect of timing on prophylactic treatments for southern corn rootworm, *Diabiotica undecimpunctata howardi* Barber (Coleoptera: Chrysomelidae), in peanut. *J. Econ. Entomol.* 84:1894-1898.

Broersma, D. B., R. L. Bernard, and W. H. Luckman. 1972. Some effects of soybean pubescence on populations of the potato leafhopper. *J. Econ. Entomol.* 65:78-82.

Buntin, G. D. and L. P. Pedigo. 1983. Seasonality of green cloverworm (Lepidoptera: Noctuidae) adults and an expanded hypothesis of population dynamics in Iowa. *Environ. Entomol.* 12:1551-1558.

Burleigh, J. G. 1972. Population dynamics and biotic controls of the soybean looper in Louisiana. *Environ. Entomol.* 1:290-294.

Cheshire, J. M., Jr., J. E. Funderburk, D. J. Zimet, T. P. Mack, and M. E. Gilreath. 1989. Economic injury levels and binomial sampling program for lesser cornstalk borer (Lepidoptera: Pyralidae) in seedling grain sorghum. *J. Econ. Entomol.* 82:270-274.

Coaker, T. H. 1980. Insect pest management in *Brassica* crops by inter-cropping. *I.O.B.C. W.P.R.S. Bull.* 3:117-125.

Crossley, D. A., Jr., G. J. House, R. M. Snider, R. J. Snider, and B. R. Stinner. 1987. The positive interactions in agroecosystems. In: G. J. House and B. R. Stinner (Eds.) *Arthropods in Conservation Tillage Systems.* Misc. Publ., Entomology Society of America, Lanham, MD. pp. 73-81.

Denno, R. F. and M. S. McClure. 1983a. Variability: A key to understanding plant-herbivore interactions. In: R. F. Denno and M. S. McClure (Eds.) *Variable Plants and Herbivores in Natural and Managed Systems.* Academic Press, New York, pp. 1-12.

Denno, R. F. and M. S. McClure. 1983b. *Variable Plants and Herbivores in Natural and Managed Systems.* Academic Press, New York.

Fenemore, P. G. 1982. *Plant Pests and Their Control.* Butterworths, Wellington, N. Z.

Funderburk, J. E., D. L. Wright, and I. D. Teare. 1988. Preplant tillage effects on population dynamics of soybean insect predators. *Crop Sci.* 28:973-977.

Funderburk, J. E., I. D. Teare, and F. M. Rhoads. 1991. Population dynamics of soybean insect pests *vs.* soil nutrient levels. *Crop Sci.* 31:1629-1633.

Gallun, R. L. 1977. Genetic basis of Hessian fly epidemics. *Ann. N.Y. Acad. Sci.* 286:255-274.

Gallun, R. L. and G. S. Khush. 1980. Genetic factors affecting expression and stability of resistance. In: F. G. Maxwell and P. R. Jennings (Eds.) *Breeding Plants Resistant to Insects.* Wiley, New York. pp. 64-85.

Geier, P. W., L. R. Clark, D. J. Anderson, and H. A. Nix (Eds.). 1973. Insects: studies in population management. *Mem. Ecol. Soc. Aust.* 1:1-295.

Georghiou, G. P. and R. Mellon. 1983. Pesticide resistance in time and space. In: G. P. Georghiou and T. Saito (Eds.) *Pest Resistance to Pesticides.* Plenum Press, New York. pp. 1-46.

Gilbert, N. 1989. Biometrical interpretation. *Making Sense of Statistics in Biology*. Oxford University Press, New York.

Gliessman, S. R. 1990. Agroecology: Researching the ecological basis for sustainable agriculture. In: S. R. Gliessman (Ed.) Agroecology. *Researching the Ecological Basis for Sustainable Agriculture*. Springer-Verlag, New York. pp. 3-10.

Goodell, G. 1984. Challenges to international pest management research and extension in the Third World: do we really want IPM to work? *Bull. Entomol. Soc. Am.* 30(3):18-26.

Gould, F. 1986. Simulation models for predicting durability of insect-resistant germplasm: a deterministic diploid, two-locus model. *Environ. Entomol.* 15:1-10.

Gould, F. 1988. Evolutionary biology and genetically engineered crops. *BioScience.* 38:26-33.

Hammond, R. B. and L. P. Pedigo. 1982. Determination of yield-loss relationships for two soybean defoliators by using simulated insect-defoliation techniques. *J. Econ. Entomol.* 75:102-107.

Hammond, R. B. and B. R. Stinner. 1987. Seedcorn maggots (Diptera: Anthomyiidae) and slugs in conservation tillage systems in Ohio. *J. Econ. Entomol.* 80:680-684.

Hatchett, J. H., T. J. Martin, and R. W. Livers. 1981. Expression and inheritance of resistance to Hessian fly in synthetic hexaploid wheat derived from *Triticum tauschii* (Coss) Schmal. *Crop Sci.* 21:731-734.

Herzog, D. C. 1979. Variety selection and planting date manipulation as cultural controls for the velvetbean caterpillar, *Anticarsia gemmatalis* (Hubner). In: F. T. Corbin (Ed.) *World Soybean Research Conference II: Abstracts*. North Carolina State University, Raleigh. p. 23.

Herzog, D. C. and J. E. Funderburk. 1985. Plant resistance and cultural practice interactions with biological control. In: M. A. Hoy and D. C. Herzog (Eds.) *Biological Control in Agricultural IPM Systems*. Academic Press, Orlando, FL. pp. 67-88.

Herzog, D. C. and J. E. Funderburk. 1986. Ecological bases for habitat management and cultural control. In: M. Kogan (Ed.) *Ecological Theory and Integrated Pest Management Practice*. Wiley, New York. pp. 217-250.

Herzog, D. C. and J. W. Todd. 1980. Sampling velvetbean caterpillar in soybean. In: M. Kogan and D. C. Herzog (Eds.) *Sampling Methods in Soybean Entomology*. Springer-Verlag, New York, pp. 107-140.

Higley, L. G. and L. P. Pedigo. 1991. Soybean yield responses and intraspecific competition from simulated seedcorn maggot injury. *Agron. J.* 83:135-139.

Higley, L. G. and W. K. Wintersteen. 1992. A novel approach to environmental risk assessment of pesticides. *Am. Entomol.* 38:34-39.

Hokkanen, H. M. T. 1991. Trap cropping in pest management. *Annu. Rev. Entomol.* 36:119-138.

Huffaker, C. B. and R. F. Smith. 1980. Rationale, organization, and development of a national integrated pest management project. In: C. B. Huffaker (Ed.) *New Technology of Pest Control*. Wiley, New York. pp. 1-24.

Janzen, D. H. 1968. Host plants as islands in evolutionary and contemporary time. *Am. Natur.* 102:592-595.

Jensen, R. L., L. D. Newsom, and J. Gibbens. 1974. The soybean looper: effects of adult nutrition on oviposition, mating frequency, and longevity. *J. Econ. Entomol.* 67:467–470.

Kogan, M. and D. C. Herzog (Eds.). 1980. *Sampling Methods in Soybean Entomology.* Springer-Verlag, New York.

Kogan, M. and S. G. Turnipseed. 1987. Ecology and management of soybean arthropods. *Annu. Rev. Entomol.* 32:507–538.

Knipling, E. F. 1979. The Basic Principles of Insect Suppression and Management. USDA-SEA Agric. Handbook 512. U.S. Government Printing Office, Washington, DC.

Leonard, B. R., D. J. Boethel, A. N. Sparks, Jr., M. B. Layton, J. S. Mink, A. M. Pavloff, E. Burris, and J. B. Graves. 1990. Variations in response of soybean looper (Lepidoptera: Noctuidae) to selected insecticides in Louisiana. *J. Econ. Entomol.* 83:27–34.

Letourneau, D. K. 1990. Two examples of natural enemy augmentation: a consequence of crop diversification. In: S. R. Gliessman (Ed.) *Agroecology. Researching the Ecological Basis for Sustainable Agriculture.* Springer-Verlag, New York. pp. 11–29.

Levine, E. and H. Oloumi-Sadeghi. 1991. Management of Diabroticite rootworms in corn. *Annu. Rev. Entomol.* 36:229–256.

Newsom, L. D. and D. C. Herzog. 1977. Trap crops for control of soybean pests. *La. Agric.* 20:14–15.

Ordish, G. and D. Dufour. 1969. Economic bases for protection against plant diseases. *Annu. Rev. Phytopathol.* 7:31–50.

Ottens, R. J. and J. W. Todd. 1979. Effects of host plant on fecundity, longevity, and oviposition rate of a whitefringed beetle. *Ann. Entomol. Soc. Am.* 72:873–839.

Painter, R. H. 1951. *Insect Resistance in Crop Plants.* Macmillan, New York.

Panizzi, A. R. and F. Slansky, Jr. 1985. Review of phytophagous pentatomids (Hemiptera: Pentatomidae) associated with soybean in the Americas. *Fla. Entomol.* 68:184–214.

Pedigo, L. P., R. A. Higgins, R. B. Hammond, and E. J. Bechinski. 1981. Soybean pest management. In: D. Pimentel (Ed.) *Handbook of Pest Management in Agriculture*, 1st ed., Vol. 3. CRC Press, Boca Raton, FL.

Pedigo, L. P., S. H. Hutchins, and L. G. Higley. 1986. Economic injury levels in theory and practice. *Annu. Rev. Entomol.* 31:341–368.

Pedigo, L. P., L. G. Higley, and P. M. Davis. 1989. Concepts and advances in economic thresholds for soybean entomology. In: A. J. Pascale (Ed.). *World Soybean Research Conference IV. Proceedings.* Realización, Buenos Aires, Argentina. pp. 1487–1493.

Pedigo, L. P. and L. G. Higley. 1992. A new perspective of the economic injury level concept and environmental quality. *Am. Entomol.* 38:12–21.

Pimentel, D. 1986. Agroecology and economics. In: M. Kogan (Ed.) *Ecological Theory and Integrated Pest Management Practice.* Wiley, New York. pp. 299–319.

Poston, F. L., L. P. Pedigo, and S. M. Welch. 1983. Economic injury levels: reality and practicality. *Bull. Entomol. Soc. Am.* 29:49–53.

Rabb, R. L. and F. E. Guthrie (Eds.). 1970. *Concepts of Pest Management.* North Carolina State University, Raleigh.

Roush, D. T. and J. A. McKenzie. 1987. Ecological genetics of insecticide and acaricide resistance. *Annu. Rev. Entomol.* 32:361-380.

Shelton, M. D. and C. R. Edwards. 1983. Effects of weeds on the diversity and abundance of insects in soybeans. *Environ. Entomol.* 12:296-298.

Southwood, T. R. E. 1978. *Ecological Methods with Special Reference to the Study of Insect Populations*, 2nd ed. Chapman and Hall, New York.

Sprenkel, R. K., W. M. Brooks, J. W. Van Duyn, and L. L. Dietz. 1979. The effects of three cultural variables on the incidence of *Nomuraea rileyi*, phytophagous Lepidoptera, and their predators in soybean. *Environ. Entomol.* 5:205-209.

Stansly, P. A. and P. P. de Sevilla. 1990. Pesticide use in Ecuador. *J. Agric. Entomol.* 7:203-215.

Stansly, P. A. and G. J. Orellana. 1990. Field manipulation of *Nomuraea rileyi* (Moniliales: Moniliaceae): Effects on soybean defoliators in coastal Ecuador. *J. Econ. Entomol.* 83:2193-2195.

Stern, V. M., R. F. Smith, R. Van Den Bosch, and K. S. Hagen. 1959. The integrated control concept. *Hilgardia* 22:81-101.

Stinner, R. E., J. Regniere, and K. Wilson. 1982. Differential effects of agroecosystem structure on dynamics of three soybean herbivores. *Environ. Entomol.* 11:538-543.

Stone, J. D. and L. P. Pedigo. 1972. Development and economic-injury level of the green cloverworm on soybean in Iowa. *J. Econ. Entomol.* 65:197-201.

Sullivan, M. J. 1985. Resistance to insect defoliators. In: R. Shibles (Ed.). *World Soybean Research Conference III. Proceedings.* Westview Press, Boulder, CO. pp. 400-405.

Tabashnik, B. E., N. L. Cushing, N. Finson, and M. W. Johnson. 1990. Field development of resistance to *Bacillus thuringiensis* in diamondback moth (Lepidoptera: Plutellidae). *J. Econ. Entomol.* 83:1671-1676.

Tammes, P. M. L. 1961. Studies of yield losses. II. Injury as a limiting factor of yield. *Tijdschr. Plantenziekten* 67:257-263.

Teigen, J. B. and J. J. Vorst. 1975. Soybean response to stand reduction and defoliation. *Agron. J.* 67:813-816.

Turnipseed, S. G. 1977. Influence of trichome variations on populations of small phytophagous insects in soybeans. *Environ. Entomol.* 6:815-817.

van Lenteren, J. C. and J. Woets. 1988. Biological and integrated pest control in greenhouses. *Annu. Rev. Entomol.* 33:239-269.

Welter, S. C. 1989. Arthropod impact on plant gas exchange. In: E. A. Bernays (Ed.). *Insect-Plant Interactions.* CRC Press, Boca Raton, FL.

Whitcomb, W. H., H. A. Denmark, A. P. Bhatkor, and G. L. Greene. 1972. Preliminary studies on the ants of Florida soybean fields. *Fla. Entomol.* 55:130-142.

Wintersteen, W. K. and L. G. Higley. 1992. Advancing IPM systems in corn and soybeans. In: A. R. Leslie (Ed.) *Successful Implementation of Integrated Pest Management for Agricultural Crops.* Lewis Publishers, Chelsea, MI.

Wuensche, A. L. 1976. Relative Abundance of Seven Pest Species and Three Predaceous Genera in Three Soybean Ecosystems in Louisiana. M.S. thesis, Louisiana State University, Baton Rouge.

Zilberman, D., A. Schmitz, G. Casterline, E. Lichtenberg, and J. B. Siebert. 1991. The economics of pesticide use and regulation. *Science.* 253:518–522.

8

Economics of Sustainable Agriculture

David C. White, John B. Braden, and Robert H. Hornbaker

TABLE OF CONTENTS

INTRODUCTION

In the last half of the 20th century, the United States has seen a 38% drop in the share of consumer spending going for food. Over the same period, there have been dramatic increases in agricultural yields, and fewer and fewer farmers are feeding nearly 65% more people in this country and still producing food and fiber for export.[1] In these respects, the success of U.S. agriculture is unparalleled. However, in recent years, many have come to view the success with growing ambivalence.

The technologies underlying the success place great demands on rural environments and natural resources. Soil resources are being lost in many areas far faster than they are naturally regenerated.[2] Streams, rivers, and lakes, once clear, are brown with the eroded soils.[3] Drinking water supplies and aquatic habitats are being threatened by toxic chemicals and excessive levels of nutrients.[2] In many farming areas, complex terrestrial ecosystems have been replaced by large expanses of relatively sterile cropland.[4]

The technologies used in agriculture are vitally dependent on petroleum and natural gas which are nonrenewable resources.[5] In areas where irrigation is important, there is increased competition for the water and often increased cost of obtaining it.[1] And, in spite of the ever-increasing quantities of foodstuffs produced by American farmers, as well as the long shelf-lives and attractive appearance of conventionally produced and processed foods, consumer confidence in the safety and healthfulness of the food supply is fragile.[6] Incidents of suspected chemical contamination of grapes and apples illustrate how suddenly and significantly consumer confidence in food can plummet.

The system that has produced plentiful, seemingly ever cheaper food has also changed the workforce, the capital/labor mix, and the land ownership patterns of agriculture—profoundly altering rural society.[7] And, the system has come to depend heavily on government assistance. The cost and return trends shown in Figure 1 reveal the degree of that dependence. Since World War II, production expenditures have increased relative to market revenues, reducing farming profits. At the same time, government payments (essentially shown in Figure 1 by the difference between gross income and market income) have accounted for a growing share of net farm revenues. In a time of mounting concern about government indebtedness, the sustainability of a food production system reliant on government largess may be especially questionable.[6]

Concerns over environmental impacts, resource depletion, safety of conventionally produced food, the vigor of rural society, and the expense of government farm programs all underlie calls for new approaches to agriculture.[6,8] "Sustainable agriculture" is an umbrella term encompassing the kinds of new approaches to farming being called for; farming in ways that protect the environment, conserve natural resources, reduce the use of potentially toxic chemicals, and increase financial independence.

Sustainable agriculture has certainly come to mean different things to different people, but weaving through the many meanings of the term is a common thread of mistrust of the economic reasoning that produced our conventional agriculture. The suspect economic reasoning has urged the adoption of an industrial production model for agriculture: specialization, labor-saving innovation, and intensified use of purchased inputs. For more than a generation, farmers have heard that synthetic fertilizers and pesticides are vital to their operations. The advice has come from

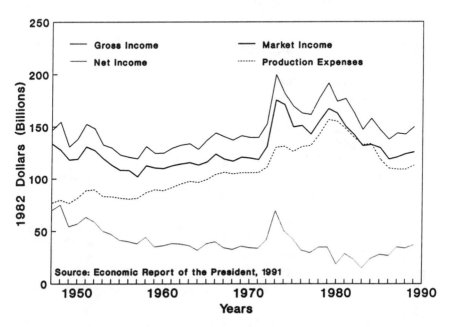

FIGURE 1. Farm income and expenses, 1947 to 1989.

reputable sources—extension advisers, farm managers, and bankers—with convincing economic arguments supporting their advice.

A critical aspect of the current interest in sustainable agriculture is a search for an economic framework that is more complete in its accounting of the trade offs implied by farming and agricultural policies. A framework is sought that somehow acknowledges, elevates in importance, and responds to previously unpriced resource, environmental, and safety impacts of farming while safeguarding the financial future of farmers. An important aspect of the framework is a recognition that agriculture is inseparable from the greater society and cannot be remedied in isolation.

This chapter is devoted to the economic dimensions of agricultural sustainability—from describing a broad conceptual framework within which the problems can be assessed, to discussing the record of efforts to make marginal improvements. It begins with a discussion of how economic theory relates to the general subject of sustainability. The next section presents a framework for understanding sustainability as a moving target and progress toward it as a dynamic, incremental process. Then the literature on applied economic studies of "alternative" production practices is reviewed, including some assessments of the effects of public policies on farming decisions and economics. The chapter is summarized and some conclusions are drawn about the role of economics in the ongoing debate over sustainability.

DEFINING SUSTAINABLE AGRICULTURE

A widely used definition for sustainable agriculture is one adopted in 1988 by the American Society of Agronomy.[9]

> A sustainable agriculture is one that, over the long term, (1) enhances environmental quality and the resource base on which agriculture depends, (2) provides for basic human food and fiber needs, (3) is economically viable, and (4) enhances the quality of life for farmers and society as a whole.

This is a valuable guiding principle, but important questions remain to be answered before its multidimensional and intertemporal ideals can be translated into prescriptive decisions. For example, how long is long term? And, while priorities will be evident in the details of any implementation, the definition makes no comment about the relative importance of its four elements.

Economics deals with trade off assessment and priority setting, and the economics literature provides a number of concepts and theories that will be useful in addressing the kinds of trade offs required to achieve a sustainable agriculture.

Scope of Concern

Agriculture is an integral part of a larger social and natural world. Like everyone else in that larger world, farmers and agricultural policy makers have concerns that range in scope over both time and space. Concerns can vary in the time dimension from immediate to far beyond our own or even our children's lifetimes. Spatially, concerns can range from one's household, neighborhood, or business to the state of the entire globe. Most people focus predominantly on concerns that are immediate and proximate. The broader the perspective and the longer the timeframe, the more abstract the concern and the less likely it is to affect an individual's behavior.[10] Sustainability is a concern that, by its long-term and inclusive nature, requires attention to issues beyond everyday concerns. Even if all individuals had sustainability of their own enterprises as a goal, it does not follow that society as a whole would necessarily be sustained. There are sure to be incompatibilities between enterprises, including competition for resources. As a corollary, the failure of an individual enterprise may not necessarily mean a loss to society of the productivity of the underlying resources. From a broad social perspective, it is the permanence of the productivity of resources — rather than the permanence of tenure of individual owners, managers, or even technologies — that is most pertinent to sustainability. From this perspective, survival of an individual enterprise may not be important or desirable. Of course, social concern for the fate of individual enterprises is not so clear-cut, because enterprises typically provide multiple benefits, such as employment, self-sufficiency, and dignity, as well as economic product. However, whatever the social benefits of particular enterprises, a sustainable agricultural sector may look very much different from the agricultural sector of today.

An aggregate, social perspective shifts the focus of sustainability from the ups and downs of individual enterprises or specific product lines to the underlying strengths and weaknesses of the industry as a whole. In this view, there is no question of agriculture's economic survival in some form. It is ensured by the need to feed and clothe a growing world population and the fact that livelihoods can be made filling those needs. The relevant questions are: Can those future social needs be fully met? At what sacrifice? And, with what redistribution of wealth? Ultimately, questions about sustainability are global and very long term, and they change constantly in response to such dynamic factors as technological advances, new knowledge and skills of managers, and changes in social institutions, as well as changes in the quality of the natural resource base.

Growth vs. Sustainability

The notion of sustainability is tested most forcefully when confronted with the notion of *economic growth*. Economic growth is believed to lead to a rise in general prosperity—if it is equitably distributed through the population at large. Economic growth becomes an imperative for society because it must be maintained at a pace more rapid than population increase if prosperity is to increase.

While economic growth seems straightforward in concept, its characteristics are widely debated, and its causes are poorly understood. Without knowledge of precisely how growth occurs and which economic sectors or resources may be most limiting, there is reason to hesitate before attempting to exert control over it.

Economic growth is usually defined as an increase in the gross national product (GNP). The conventional measure of GNP emphasizes reproducible goods and services—the products and services represented in the commercial pages of a phone book—with adjustments made for wear and tear of the capital stock. However, no account is taken of wear and tear on the environment, depletion of the natural resource stock, or changes in the quality of human capital. As a result, this conventional definition of economic growth really captures only a portion of what comprises "quality of life" or prosperity. Further, the definition fails to include factors that may vitally affect the potential for growth in the future.

The limitations of defining prosperity and measuring growth have been vividly illustrated by Repetto et al.[11] They determined that nearly half of the officially reported increase in GNP in Indonesia in recent years has been accounted for by reductions in that nation's stock of petroleum, forest, and soil resources. The depletion of soil and petroleum is irreversible, and the depletion of forests will take decades to reverse. Thus, much of Indonesia's recent growth may have come at the expense of its future opportunities to grow.

Agriculture's most direct contributions to the improvement in the quality of life are employment for the farmers and nutrition for society as a whole. However, agriculture competes for natural resources with other activities that also contribute to the quality of life. Land devoted to agriculture cannot be used for housing or recreation. Water drawn for irrigation is unavailable for recreation, industry, and sanitation. Synthetic chemicals applied to crops sometimes degrade ecosystems and contaminate water supplies and can present a health danger to farmers and consumers. Questions of sustainability suggest that a better accounting of all aspects of quality of life is needed, and gains in agricultural output must be balanced against undesirable side effects and against both contemporary and future competing demands for the resources being used.

In its need to balance immediate growth in agricultural output against the potential for future output, society faces uncertainties about the future and a lack of information about available options. There are no easy answers. More and more evidence indicates that the dependence on nonrenewable resources and the environmental degradation associated with today's agricultural technology will constrain future agricultural production; but there is also real fear that abandoning the current technology will simply accelerate the onset of the kind of general shortfall in agricultural production being forecast. On the other hand, the history of technological breakthroughs in agriculture is encouraging. Continued advances from research, such as in the area of biotechnology, hold at least the promise of reducing dependence on the natural resources used to produce food. However, reliance on the timely development of benign technologies, in order to provide uninterrupted growth in agricultural output and improvement in the general quality of life, may be dangerously wishful thinking.

Economic growth and sustainability are not mutually exclusive, but there certainly is conflict between them. The nature and severity of the conflict varies spatially and temporally. The spatial variation reflects global differences in natural resource endowments, in social attitudes about their use, and in pressures of human population and standard of living. Over time the conflicts vary with growth in both population and standard of living, as well as with the direction and rate of technology development. Historically, markets have played a crucial role in resolving the conflicts and enabling economic progress.

The Role and Adequacy of Markets

It is intuitive that resource depletion, in agriculture or in other sectors of society, will necessarily limit future economic growth, and there certainly is evidence to support that intuition. However, history argues against broad-brush pessimism on this point. Questions of the adequacy of natural resources have fascinated and frustrated economists since the Reverend Thomas Malthus' dismal 18th century forecasts. Malthus foresaw humanity deprived, depraved, and malnourished because its appetites would inevitably overtax the capacity of the available farmland to produce food.[12] While it may be that only the timing of his predictions was wrong, his predictions have yet to be realized on any kind of general scale, two centuries later.

Since before Malthus' time, markets have played central roles in determining what, how much, when, and where goods and services will be available. They mediate between current economic decisions and their future consequences. Markets create incentives to search out economic growth opportunities, and market-based competition exerts discipline against wasting costly resources. These are powerful tools for generating economic growth, but perhaps less powerful in assuring sustainability.

In an ideal economic world, all planning for the future could be conducted through markets. Prices would guide investment, savings, and resource conservation toward future needs. Realistically, however, the world is far from perfect and market guidance to the future is far from complete. Markets alone provide no incentives to minimize environmental impacts unless actual costs are attached. As a result, environmental goods and services are abused, either through ignorance or through pursuit of short-sighted or narrow objectives — indeed, market pressures to minimize monetary costs may encourage environmental abuses. Markets are also inadequate in properly rationing natural resources for which ownership claims are difficult to enforce: e.g., forests, fisheries, and the atmosphere. Absence of secure property rights for natural resources can cause markets to skew economic growth, often to the disadvantage of those resources and the quality of the environment.

On balance, markets probably promote sustainability more than they hurt, but they alone are insufficient. Other forms of guidance are also required. Reconciling the "Jekyll and Hyde" qualities of markets is a major challenge of government and the legal system. With suitable government and legal guidance, markets can assist in addressing resource conservation and environmental protection concerns.

In terms of changing the direction of society (or agriculture), markets are perhaps better understood as institutions for pursuing social goals more than for deciding what those goals ought to be. Where society is currently heading comes, in part, from the routine market interactions. Judging the current direction to be unsustainable is, in effect, judging the markets to be in need of fundamental adjustments that are not expected to occur spontaneously. The problem of deciding what changes to make remains. Economists have made efforts to clarify the issues involved in "sustainable" economic development and have suggested alternative sets of priorities and rationale for addressing them.

Paradigms for Developing Finite Natural Resources

There are a number of economics texts that are centrally concerned with the concept of sustainability.[13-16] In one of those texts, Pearce and Turner[16] frame the issues by asking: ". . . how should we treat natural environments in order that they can play their part in sustaining the economy as a source of improved standard of living?" They then provide five responses that represent different perspectives on the relationships between economic prosperity, technology, and natural resources.

One response flows from what Pearce and Turner call the "sustainability paradigm."[16] This paradigm holds that economic growth can endure only if it is accompanied by improving and increasing natural capital. Thus, sustaining economic prosperity in the face of population growth requires investment in, not depreciation of, natural resources and the environment — planting more trees, enhancing soil quality, protecting water supplies, and so forth.

The second response is what Pearce and Turner call the "trade off paradigm."[16] This idea is that drawing down the stock of natural and environmental resources makes possible a build up in the stock of reproducible (human-made) capital. The new mix of capital stocks will subsequently make the economy less dependent on natural resources. A corollary is that conserving the natural resource stocks slows the accumulation of reproducible stocks and thereby prevents the economic growth that would be possible through more efficient use of natural resources. The trade off perspective recognizes some depletion of natural resources as necessary for sustained prosperity. In fact, resource depletion may even indicate transformation to a less resource-intensive economy. In this light, the depletion of Indonesia's natural resources, described earlier,[11] might be argued to be a necessary part of a sustainable growth path.

A third response proposed by Pearce and Turner might be called the "substitution paradigm."[16] In this view, humans are able to produce machines, processes, or institutions that accomplish much of what natural resources accomplish in sustaining economic activity. Furthermore, the paradigm forecasts continued expansion of the capacity to emulate or replace natural systems, or to separate humans from them. From this perspective, it is the flow of services provided by assets that is important to economic growth, not whether the assets are natural or artificial. For example, pest control can be achieved either through management of natural predation or through chemical means. As long as the needed services can be obtained most cheaply from natural resources, it makes sense to invest in those resources. However, when synthetic sources are cheaper, enhancing the natural assets no longer makes sense.

The fourth response might be called the "uncertainty paradigm."[16] Its roots are in economic perspectives on decision making under ignorance. If delaying a decision can provide better information on the future value of an undeveloped resource, then a rational decision maker should (at the margin) resist making irreversible commitments before the better information becomes available.[17] According to this paradigm, the potential to make better decisions in the future mediates against irreversible degradation of the resource base. This view does not lead to definite conclusions about the degree of conservative bias, only that there should be such a bias.

The uncertainty paradigm is strengthened by, but does not depend on, aversion to risk. Risk aversion is a willingness to sacrifice to achieve certainty rather than to bear risk (for example, to purchase insurance for which the costs exceed the expected payoff). Pearce and Turner[16] contend that an appropriate rule for safeguarding future economic growth is to avoid further net degradation in the natural capital stock. They recognize that the rule must accommodate changes in the mix of natural capital as nonrenewable resources are used up. They justify their position on the grounds that: (1) the current stock will at least give future generations as many economic options as the current generation has; (2) even if reproducible capital can substitute for natural capital, the poorest countries, which also face the most population growth pressure, simply cannot afford the reproducible capital that would be required; and (3) given our great uncertainty about future needs for natural capital and the potential for artificial substitutes, maintenance of the natural resource base is a wiser course than irreversible degradation.

The uncertainty argument has motivated other scholars to warn against irreversible degradation of natural systems, but with different prescriptions about where the line should be drawn. For example, Ciriacy-Wantrup[18] and Bishop[19] argue that irreversible destruction of breeding stocks and habitat should be avoided so that renewable resource populations can be revitalized if and when their value and importance is established. They call this approach the "safe minimum standard of conservation." Unlike the proposal of Pearce and Turner, the safe minimum standard is consistent with reductions in the natural capital stock as long as the reductions do not destroy the potential to reinvest in and renew the resources. This is roughly the philosophy behind the legal protection accorded endangered species. It does not, however, provide guidance on depletion of nonrenewable resources.

Finally, a fifth paradigm arises from a philosophical perspective, not an economic one. Call this the "intergenerational equity" paradigm. The central contention is that principles of justice mediate against a diminution of the choices available to future citizens.[20,21] Thus, current citizens are obliged to pass on a resource base of natural and artificial assets that will not preclude options for the future. As a moral principle, this notion of intergenerational equity is probably widely accepted. The problem comes in determining which changes in the resource base are irreversible and cannot be offset by other changes. That difficulty leads back to the four paradigms that rest on different views of economic possibilities—views differing in such things as: time preference, optimism about technology, aversion to risk, and value judgements (including assigning value to the intrinsic worth of elements of the natural world).

These paradigms are examples of efforts to formulate decision rules for a society faced with choices involving the use of resources. The differences between the paradigms offer useful insights into the complex and value-laden nature of the decisions to be faced. In the meantime, economic decisions are made every day and cannot await the kind of social consensus these paradigms suggest.

Renewable Resource Management

In addition to the broad considerations of the role of natural resources in economic growth, economic theories also deal with resource use issues from a more concrete perspective of microeconomics, or enterprise management. In resource management theory, the concept of sustainability is probably best known as a decision rule for harvesting a renewable resource, such as timber or fish. The idea is to harvest only the net growth, leaving a level of breeding stock unchanged from year to year to continue producing the desired growth. In this light, "maximum sustainable yield" (MSY) is the highest rate of harvest that can be attained indefinitely.

The MSY rule is a purely biophysical rule that fails to incorporate social trade offs and values.[22] For example, cost of harvest and consumer demand can interact at any point in time to create pressure for harvesting either more or less than exactly what is sustainable.

A third factor that the MSY rule does not consider is the value of time—based on the idea that because today's dollar could be made to grow (compounded) at the available interest rate, interest rates reflect the premium society places on time. Accordingly, dollars to be received in the future must be discounted, at the same interest rate, to be made comparable to today's dollars. Because of this value of time, a quantity of resource is worth more harvested sooner rather than later, and although exceptions can be imagined, a simple MSY policy generally will not maximize value as seen from the present. Compared to an MSY stocking level and harvest schedule, the present value of the resource could actually be increased by using more of the growing stock immediately and making smaller perpetual harvests thereafter. At real (inflation adjusted) interest rates of 3 to 5% that are typical in today's economy, shifting to such a maximum economic yield (MEY) rule can have dramatic effects on stocking rates and optimum yields.

An MEY rule is driven by social values more than by realities of the natural world, and discounting has been criticized for giving unwarranted priority to the preferences of current decision makers.[23] There is longstanding disagreement about the proper level of the discount rate.[24,25] However, discounting should not be dismissed lightly. Borrowing and investing are fundamental to our economy and individuals cannot ignore the monetary value of time imposed on them by interest costs. From the individual's perspective discounting is essential to a clear comparison of alternative long-term investments. If the concept of sustainability includes the necessity of economic growth, and economic growth depends on the allocation of all scarce resources (natural and artificial) to their most productive uses, the winnowing of alternatives by individual investors may actually promote the sustainability of overall development.

It is less clear whether discounting is appropriate when considering the collective choices made by society as a whole. From a societal viewpoint, it is difficult to agree on an appropriate discount rate, it is virtually impossible to justify projecting the use of any particular discount rate many years into the future, and it may be inconsistent with intergenerational equity.

These concepts from natural resource economics provide valuable but limited insights into the sustainability issues of agriculture. For example, the MSY model does illustrate the concept of sustainability, but the natural growth model that is central to it is inappropriate for representing agricultural production in two respects. First, the multiyear character of the model is crucial to many of the insights it provides (e.g., the distinction between MSY and MEY is largely due to the multiyear influence of discounting). In contrast, most agricultural production is in annual or seasonal growth cycles. Second, the human role in the natural growth model is essentially limited to when and how much to harvest. Agricultural technologies exert much more profound control over the entire growth process.

In many ways, agriculture is better represented by an industrial production model. The natural elements in agriculture are soil and climate, which have limited agricultural potential without significant human intervention. This human intervention greatly increases the complexity of the process whose sustainability is in question. The complexity includes the introduction of other inputs into the system—inputs that must, themselves, then be tested for sustainability. Many of these inputs are not renewable.

Nonrenewable Resource Management

A nonrenewable resource is one that exists in a finite, exhaustible supply. It is a resource that, if it regenerates at all, is at a rate so slow as to be irrelevant within human timeframes. For example, some deep aquifers may receive recharge but only tens or hundreds of years after the excess rainfall infiltrates the soil surface and percolates beyond the reach of plant roots. Because the supply of such a resource is limited, its use necessarily implies a loss of future options. Optimal depletion theory is an economic framework for understanding the trade offs and for appropriately allocating the resource over time.[16]

Optimal depletion theory derives from two essential concepts. The first is that the use of a quantity of a nonrenewable resource carries an opportunity cost. Opportunity cost is a term, widely used in discussions about resource use and management decisions, to recognize that there is some value to a path not taken—that choosing one thing implicitly imposes a cost related to other options not selected. In optimal depletion theory, the opportunity cost is equal to the value of the same quantity of the nonrenewable resource at some future date.

Economically efficient consumption of an ordinary (i.e., manufactured) good occurs when its price equals its production cost. In contrast, efficient consumption of a nonrenewable resource occurs when the price paid equals the production (extraction) cost plus the opportunity cost. Recognizing the existence of opportunity costs means that nonrenewable resources cost more and therefore should be consumed less than if they were reproducible.[26]

The marginal influence on consumption just described has no explicit time dimension. The second concept needed to complete the optimal depletion theory relates to how opportunity costs behave over time. For efficient allocation of the resource through time, the opportunity cost must exert a constant influence. As in the discussion of renewable resources, the value of time must be accounted for. This means that for the influence of the opportunity cost to remain constant over time, the opportunity cost itself must actually change over time at a rate equal to the appropriate discount rate.[26]

The problem, of course, is for society to know and use the appropriate opportunity cost and discount rate. In a practical sense, opportunity cost can be calculated in the present by subtracting extraction cost from the price paid by the consumer, both of which are known. This suggests that markets are, in effect, imputing future resource values for society as a whole. These imputed values are known by a number of different terms such as: royalty, rent, net price, and marginal profit.[26] Doubt certainly remains about whether these market residuals are adequate to allocate limited resources properly between present and future consumption.

A number of other factors also complicate the theory of optimal depletion. If we assume that the reserves that are least costly to extract are used first, production costs (including exploration) will rise with time. With an appropriate opportunity cost, the price of the resource should rise, reducing consumption, and effectively rationing the increasingly scarce resource. On the other hand, new discoveries can act as a shock to the optimal depletion schedule, perhaps not only changing the cost of production, but also implying that the opportunity cost needs to be recomputed. A third circumstance is when there is another resource or technology that can provide an alternative to the nonrenewable one in question, i.e., a backstop technology. Under these conditions, the cost of the backstop effectively limits opportunity cost to the difference between the cost of the backstop minus the production cost of the nonrenewable resource. When the opportunity cost of the nonrenewable resource reaches that point, use of the resource is curtailed because the backstop becomes the more cost effective.[16]

Given the high levels of nonrenewable resources used in conventional agriculture (e.g., fossil fuels, fossil water, and mineral fertilizers), understanding the issues involved in their sustainability is crucial to dealing with the sustainability of agriculture itself. As with the broad paradigms of sustainable development, economics provides useful understanding of the issues involved in using these nonrenewable resources. However, the decisions made by society are ultimately a reflection of a set of fundamental values explicitly or implicitly embraced — not the result of theoretical paradigms.

In spite of the fact that available economic theories fail to establish the value sets ultimately needed to fully prescribe a "sustainable agriculture", economists recognize that the ideas of sustainability and permanence have real and useful meanings. Concepts like sustainability and permanence may spring from outside of economics, but they can certainly influence the calculus of economic decision making, providing important standards against which observed conditions and trends can be judged. This usefulness is not dependent on having full social consensus or a detailed goal. Change is inherent in agriculture, in society, and in the natural environment. Given such a dynamic decision environment, sustainability is perhaps best viewed as an essential but fluid goal that continually retreats into the future ahead of us.

TOWARD SUSTAINABILITY

Seeing sustainability as a general direction in which to head, rather than a fully defined objective, farmers and policy makers can make reasonable improvements without worrying whether a fully prescribed system of sustainable agriculture is, or ever can be, universally agreed to. Within this framework of gradual evolution, three overlapping levels of progress have been suggested. In a gradient from "low sustainability" to "high sustainability", progress can be characterized as going from *efficiency* changes (first and easiest), to *substitution* changes, then finally to fundamental *redesign* changes.[27] Having this framework in mind will be helpful to understanding and evaluating the applied, farm-level research that will be discussed later.

Even gradual progress toward sustainability requires, at least in the short-run, an increase in management input. As the farm operators change production practices, the efficiency, substitution, and/or redesign changes will require more management attention to ensure that the alterations produce their intended results.

Efficiency

Efficiency changes progress toward sustainability by reducing inputs for the same output, increasing output for the same input and/or reducing wastes. Examples of efficiency changes include banding rather than broadcasting fertilizers and spot rather than whole-field applications of pesticides.[27] Efforts to increase production efficiency are consistent with standard management techniques — with reduced environmental impacts and resource demands as coincidental benefits. The advantages that efficiency improvements provide toward sustainability are largely accepted, as far as they go. Their shortcoming is that maximizing efficiency can be a dead-end path in relation to sustainability, if the system being made more efficient is basically unsustainable.

Substitution

Substitution changes can increase sustainability by replacing limited or environmentally troublesome inputs with alternatives that are less limited or troublesome. Substitution changes might include replacing herbicides with mechanical weed control, or using organic rather than synthetic fertilizers. These kinds of changes can produce improvements, but, as with efficiency changes, having too narrow a perspective when electing to make such a change can result in its being misdirected or short-sighted. For example, a given technology may require less energy but increase other environmental impacts. If the lower energy cost of the new technology makes it cheaper to use, it could become even more widespread than what it is replacing, resulting in a net *increase* in environmental impact.[27]

Production economics emphasizes the importance of changing the mix of inputs in response to new technologies and shifts in relative prices. Economic analysis of input substitution (for example, assessing the trade offs between higher fuel and fertilizer prices in the future and present capital costs of switching to fuel- and fertilizer-saving technologies) is currently widespread in agricultural economics research. Using these economic techniques to focus directly on sustainability may be computationally more complex, but the economic issues are basically the same.

Redesign

Compared to substitution, redesign changes are more fundamental and involve whole production systems. In terms of agriculture, the emphasis would be on production systems, perhaps farms, as a whole — unique in time and space. The systems approach offers an opportunity to simultaneously pursue complex goals such as ecological and economic vitality, self-sufficiency, and diversity. An example of a redesign change might be to integrate, or reintegrate, livestock into a grain farm to develop a more ecologically balanced and self-sufficient operation.[27] Redesign changes would likely also incorporate efficiency and substitution changes. For example, changing from a monocultural grain production system to a multiyear crop rotation system can involve numerous substitution and efficiency effects. It becomes very difficult to characterize, measure, and value the sustainability of such very different systems.

The categories of efficiency, substitution, and redesign provide a context within which the nature and magnitude of changes undertaken in agriculture can be described. Whatever the nature and extent of a proposed change, agricultural decision makers are still faced with the choice of whether or not to make the change. Benefit-cost analysis provides a systematic process for assessing such a decision.

Benefit-Cost Analysis

Benefit-cost analysis refers to a basic set of "common principles and analytical techniques" that can be used to "appraise decision problems in the light of objectives chosen by the decision maker".[28] The process will not guarantee any particular financial outcome. It need not even emphasize financial considerations. That is up to the decision maker. It will serve to provide discipline and a systematic approach to assessing the pros and cons of an impending decision. Used objectively, the approach can help evaluate trade offs by at least exposing those costs and benefits that the decision maker can measure (monetary or otherwise).

Benefit-cost analysis can provide a structure for summarizing the information obtained through processes as simple and subjective as personal estimates or as sophisticated as computer budgeting and optimization techniques. The approach has been proven useful at all levels of decision making from the individual farmer to the government policy maker.

In the introduction, we mentioned the need for an economic framework that does a more complete accounting of trade offs. Benefit-cost analysis provides a basic structure for such a framework. The problem, of course, is not in the simple arithmetic of summing the pluses and minuses. The real difficulty is in obtaining a broad consensus of what should be counted and what values should be assigned.

The reality is that "sustainable agriculture" is a concept that may be so value laden and dynamic that it is futile to await a social consensus on its final definition. Nevertheless, the words themselves suggest a general goal, and it is not only possible but important to consider opportunities to take steps in that general direction. It is also important to measure the resulting progress. There is a growing body of economic studies of alternative farming systems, many of which have been developed specifically in response to concerns for the sustainability of farming.

APPLIED ECONOMIC RESEARCH OF ALTERNATIVE FARMING SYSTEMS

The constant task for farmers, if they are to remain on the land, is to manage the available physical, financial, and labor resources to produce marketable products at a profit.[29] As long as this "profit or perish" imperative remains at the ground level of agriculture, the resource, environment, and rural society aspects of sustainability must be assessed in terms of their impacts on the productivity and profits of farming. Can farmers make management changes to improve in one or more of these dimensions of sustainability? Will proposed changes at least maintain current levels of output? Will they provide stable or increased incomes? If society insists on some remedial change to correct a particular resource or environmental concern, how can farmers provide it with minimal loss of income? The benefits associated with a more "sustainable" agriculture will be realized on a broad scale only if the systems providing those benefits are profitable for the farmer. Of course, the necessary profit can include government subsidies and/or higher prices paid by consumers as a result of regulations that ensure that particular, costly farming practices are followed.

The possibility that productivity and profits must be sacrificed to achieve other aspects of a sustainable agriculture sets the stage for the applied economic research in this area. The general approach that is predominant in the literature is to identify one or more alternative farming systems that are, in one way or another, an improvement over more conventional systems. The analysis then attempts to quantify and interpret the trade offs involved in adopting such a system. Other research approaches (e.g., surveys and econometric analysis of aggregate data) have been used effectively to investigate attitudes and general behavior patterns.

It is argued that questions of agricultural sustainability are questions less about the individual (natural and technological) parts of a production system as about how those parts interact to form a whole. A major difficulty faced by researchers is that an agricultural production system is so complex that it can be considered both in total and in detail only for a specific place and time. The site-specific results that are produced by studies of this sort are not necessarily reproducible elsewhere. It is not surprising that the reported studies have been too few and too limited in scope to permit general conclusions about the profitability of "sustainable" systems compared to "conventional" systems.[6,8] Nevertheless, there are lessons to be learned from the literature.

This section will discuss the reported research, looking at it from three different perspectives. First, the tools that economists bring to the study of agricultural alternatives will be discussed by considering the analytical techniques that have been employed. The next section describes some of the sustainability issues that have gotten attention in the literature and what the research has to say about them. Finally, the limitations of the research will be examined and some general observations drawn.

Research Techniques

The central element common to most of the research is a production budget, which accounts for the costs, outputs, and income produced by a particular farming system. The reported studies differ widely on how the budgets are developed, what they include, and how they are interpreted.

The simplest use of budget analysis is a case study. This approach has been used a number of times in the sustainable agriculture literature[30,31] with the second half of the National Research Council study, *Alternative Agriculture*,[8] undoubtedly providing the most prominent example. The case study technique uses site-specific data obtained from the subjects being studied, and analysis is basically limited to an accounting of pertinent information about each case, including but not limited to economic data. Obviously, the outcome of an assessment of this type is determined by the cases selected for study.

In an effort to support more general interpretations, a representative budget approach is often employed. This is probably the most prevalent technique found in the literature. In some studies, the representative budgets are based on average values obtained by sampling groups of similar farms.[32,33] Some researchers have attempted to increase the breadth of their potential audience by basing the budgets on data representative of an entire region, such as small-grain, row-crop farming systems of the Northern Plains.[34] Other studies take a more hypothetical approach and construct budgets from a number of data sources including individual farms (to establish a consistent set of resources), research plots (to establish general agronomic responses), and market data (to obtain general price information).[30,35] The representative budget is more hypothetical than the case study approach and allows considerable flexibility. However, the researcher retains responsibility for the choice of what to represent,[36] and testing the sensitivity of the analysis to assumptions that were made and the data that were used is critical to any kind of general application of the results.

Sensitivity analysis has been used to suggest that the characteristically more diverse production systems of sustainable agriculture produce less variability in economic outcomes and are, therefore, less risky.[36] A stochastic version of the basic budget technique – using statistical distributions of experimental plot yields and historical prices – has been used to address this question directly by introducing risk explicitly into the analysis.[37]

The analyses just referred to provide basic information for making comparisons between alternative farming systems. Many of the resource, environmental, and social concerns associated with sustainable agriculture remain outside a traditional budgeting process, and considering them requires the decision maker to extend the analyses beyond normal, monetary bounds. The complexity of the questions involved often means that comparative analyses are inconclusive. Some research efforts have used optimization techniques to better focus the analysis on the trade offs. For example, linear programming can provide estimates of the economic value of various resources (land, water, labor, etc.) being traded off when choosing one system over another. The technique allows the economics of alternative management practices to be assessed under explicitly imposed resource, economic, or legal constraints.[38]

The techniques discussed to this point have not had an explicit time dimension beyond 1 year – obviously incomplete for addressing the "long-term" aspects of sustainability. Simulation is often used to assess alternative farming systems in respect to long-term impacts such as soil erosion impacts[39] or difficulties faced during transition from a conventional farming system to a low-input system.[40]

Budgeting, optimization and simulation are all techniques typically used for microeconomic analysis, focusing on enterprise- or business-level decision making. In order to draw general conclusions from them about agriculture at a regional or sectoral level, policy makers must assume that the analyses are representative of a general population (which may or may not be valid and is not easily tested). In contrast, econometric analysis of aggregate economic data offers an approach to comparing the choices available to farmers that is, essentially by definition, representative. For example, time series data from the USDA Farm Costs and Returns Survey have been used to compare the profitability of farms broadly classified by the crops they produce and their expenditures on chemicals.[41]

Another area of interest concerning sustainable agriculture, particularly at the policy-making level, has to do with the attitudes and experiences of agricultural producers. This interest occasions a very different type of research: attitude surveys. Surveys of farmers have explored why they have or have not adopted alternative farming systems,[42] tested relationships between attitudes and behavior relative to resource and environmental issues,[43] and assessed the experiences of farmers who have attempted to adopt alternative systems.[32]

Research Findings

Reduction in Chemical Use

Undoubtedly the most emotion-laden issue associated with sustainable agriculture is the reduction or elimination of agricultural chemicals. A number of studies have attempted to assess the economic potential for individual farms to reduce chemical use. The results have been mixed.

There is evidence that profits can be maintained (perhaps increased) while chemical use is reduced, ostensibly through the adoption of more efficient production systems. For example, the analysis of the USDA Farm Costs and Returns Survey data has shown that farms with the corn-soybean mix with livestock and with low or medium chemical expenditures per acre have relatively higher returns on assets and returns per acre.[41] A similar aggregate study in Illinois showed that farmers spending the most for purchased inputs per acre harvested more bushels, but earned less profit per acre, compared to farmers spending less on inputs.[44]

In addition to the trends in aggregate data, individual cases of successful abandonment of chemicals have been studied. For example, one farm, where no herbicides, lime, or fertilizers have been used since 1971, consistently produced crop yields above county averages.[30] Further, studies using the hypothetical or representative farm approach have also suggested potential gains. In one comparison study, all farms had similar economic returns, with those farms using chemicals having higher production levels, but also offsetting higher energy consumption and higher production costs.[45] However, these results have not been universal. Some simulations based on yield data from research plots have shown farming without synthetic chemicals to equal or exceed conventional farming in economic performance, but other similar models have shown an economic advantage for the conventional systems.[36] In studies of farming systems representative of Nebraska[37] and the Northern Plains,[34] profitability was compared between systems employing different rotations, tillage practices, and chemical treatments. While reduced-chemical systems had lower costs, there seemed to be little relation between the level of chemical use and average profitability over a number of years.

Reduced Soil Loss and Tillage

While maintenance of a productive soil is central to the idea of a sustainable agriculture, analysis of the economics of soil erosion has been either a study of policy implications or of extremely subtle and very long-term effects. Mueller studied the long-term soil impacts and profitability of alternative production systems on a 400-acre farm representative of corn farms in Wisconsin.[39] In the short term, costs associated with conservation tillage (i.e., any tillage system that maintains at least 30% of the soil surface covered from harvest until protection is provided by the growing leaves of the next crop) exceeded those of conventional tillage systems. However, in the long term, costs were lower for conservation tillage systems, suggesting that these systems could offer farmers an economical solution to soil erosion problems.

Domanico, Madden, and Partenheimer examined how a profit-maximizing farmer would react to different soil erosion restrictions.[38] Under the hypothetical circumstances that they analyzed, they found that no-till was always the most profitable tillage practice. If erosion was unconstrained, no-till was followed by the conventional tillage and then by an organic system. When soil erosion was constrained, the organic became more profitable than the conventional system.

Reduction in Energy Use

Energy use is another prominent concern associated with the idea of sustainable agriculture, and just like chemicals, the economic viability of reducing energy use is subject to much study and debate.

It is generally claimed that low-input farming systems require less energy. However, in analyses where energy reduction has been documented, there is almost always a clear trade off. One study of wheat producers in Pennsylvania and New York showed the conventional producers averaged higher energy inputs and almost one third higher yields of wheat than did the low-input producers. The low-input producers had lower cash expenses, but if their greater labor input was valued at a wage available within the community, their total costs were higher.[33]

Viability under Changing Conditions

The robustness of an alternative farming system implies an ability to maintain economic viability in the face of adverse conditions. An unknown future can present two quite different kinds of adversities.

One type of hazard arises from future variations in such factors as weather and markets. The risk introduced into particular farming systems by these factors can be estimated statistically and the expected performances of the systems can be modeled and compared. Some researchers have suggested an advantage for sustainable systems, arguing that they would be less susceptible to natural and economic risk because of their greater diversity.[36] One study that attempted to quantify riskiness found that rotations did have an advantage over continuous cropping systems. However, in the same study, reducing chemical use demonstrated no such advantage.[37]

The second type of unknown hazard is one that is completely unpredictable. This type includes such things as new regulatory constraints, the loss of a market due to changing consumer tastes, or dramatically changing prices for scarce inputs. When this type of hazard occurs, it essentially redefines the environment in which farming decisions must be made. While the risk associated with this kind of hazard can not be estimated probabilistically, the impact of such a hazard can be simulated. For example, the study by Domanico et al.[37] used linear programming to impose soil erosion limits and demonstrate the relative economic susceptibility of conventional, no-till and organic systems.[37] In their study, changing erosion constraints did cause the relative performance of the systems to change.

In another study, the economic performances of alternative systems were compared with and without government subsidies. Goldstein and Young[35] demonstrated an example of how the economic advantage of conventional high input, high yield systems can be dependent on subsidized prices. In the circumstances that they were analyzing, when the subsidies were removed, the more diverse, lower input system became the more profitable.

Long-Term and Transition Effects

The profits of a farm in the process of shifting to an alternative management system are determined by a combination of factors. In a paper addressing transition to organic farming, Dabbert and Madden[40] organize these factors into five categories: rotation adjustment (introduction of more, or less, profitable crops), biological transition (establishment of a new relationship within natural processes and a new role for the farmer), price (changes in commodity quality can raise or lower prices), learning (the farmer's new production and marketing roles must be learned), and a perennial effect (after some period of adjustment there is some permanent residual change). They report farmers' experiences during a transition period vary from severely reduced to slightly increased *yields*, with an even greater range in the associated *profits*.

Based on a number of studies, the long-term yield effects of adopting low-input production systems do not appear to be dramatic — whether increased or decreased. Goldstein and Young[35] report yields stabilizing 5 to 10% below those from conventional systems, and Helmers, Langemeier and Atwood[37] found yields to be lower but less variable. On the other hand, there are other case studies of alternative systems that have reported yields actually higher than the local county averages.[6]

Attitudes, Characteristics, and Experiences

Numerous surveys have been made to gain insights into the adoption of sustainable agriculture and to understand the barriers to adoption that exist. Findings indicate obstacles such as lack of easy access to reliable information on alternative production systems, problems with obtaining information about special markets, and credit discrimination.[42] Surveys of practitioners have also provided evidence of the nature and level of success possible under a particular system. Lockeretz and Madden[32] contacted organic farmers to get information on their operations, their financial conditions, and ideas on how to succeed. The majority of the respondents were operating complex crop and livestock systems, and their generally solid successes were encouraging. However, their reliance on special marketing channels for their products indicated that critical elements of their success might not transfer to mainstream agriculture.

A survey was also used to explore the relationship between the structure of the farming sector and environmental impacts. Heffernan and Green[43] compared large-scale and small-scale farmers' concern for and care of the environment. Large farms were found to have lower estimated soil loss than small farms. However, this relationship was found to be largely due to the land in the larger farm having less potential for erosion.

Insights and Open Questions

The cited literature reflects only a relatively few alternative farming systems, both in terms of the "conventional" systems examined and the "non-conventional" system being compared. The result is that to date only a fraction of new strategies that might be employed under the vastly different circumstances across the country have been examined. Especially in the early literature, the analyses were typically of an all-or-nothing decision between "conventional" agriculture and an "organic" system which employed virtually no purchased chemical inputs. The value of these studies is limited in drawing general conclusions or formulating a general strategy for reducing purchased inputs, because they provide little information about the more likely systems which are gradual shifts from conventional systems.[6]

Other criticisms of these early research efforts have included concerns about using small sample sizes, making comparisons between sets of farms that were poorly matched, and a lack of multiyear data. These problems make it difficult to clearly assign reasons for the observed differences in costs and/or incomes. While more recent studies have improved on these earlier shortcomings, the studies remain largely anecdotal. Each tends to address a different combination of crops, rotations, production practices, or climatic region. They are also very individual in their assumptions about the influences of market prices, government agricultural policies, and even crop yields. And, as a group, the studies fail to fully consider the impact of risk arising from such influences.[6]

The research also shows, however, that the economic impacts are very dependent on circumstances. The bulk of the research has examined cases where profits were being made, but many of these success stories rely on lower input costs and higher commodity prices. In these examples the farmers have found market niches where, by producing crops without chemicals, they have been able to tap a source of higher prices for "organic" products.

There is a persistent taxonomy problem in discussions of sustainable agriculture, especially in the reporting of site-specific applied research. The lack of common categories for classifying tillage practices, pest control regimes, crop rotations, etc. makes it difficult to summarize results and draw general conclusions. For example, a study of Northern Plains production systems (with benefits toward sustainability) showed that the profit of one "alternative" system was 30% higher and another 48% lower than the baseline "conventional" system.[34] And, just as alternative or sustainable agriculture is a variable concept, so is "conventional" agriculture. In Nebraska, profits under reduced-input rotational systems were found to be lower than intensive "conventional" systems using rotations, but higher than other "conventional" systems using continuous row crops.[37]

In spite of the limitations of the research, there is evidence of profit potential in alternative farming strategies. The most promising potential for increasing profit seems to be in lowering production costs, primarily through the use of fewer agricultural chemicals. While these alternative strategies require greater management and labor to control pests, the farmer may be financially better off either because the increases in the cost of replacement inputs are not enough to erase the savings in chemicals, or because the farmer is better able to retain the increased labor and management costs, through the use of "unpaid" family effort.

AGRICULTURAL SECTOR

Implications of Farm-Level Findings

In general, the farm-level research just discussed explored the feasibility of alternative systems for operating a farm, or an enterprise within a farm. However, the results from some of those studies do offer insights into influences to and from beyond the farm gate.

An obvious connection between social policy and farmer behavior in the United States is federal agricultural programs. For example, Goldstein and Young[35] compared the profitability of low-input legume-based rotation with a more intense and erosive cropping system typical of farming in the highly erodible Palouse region of Washington. They found that the conventional system required more than twice the expenditures for inputs, but produced a cash crop every year and was more profitable with the government commodity program that was in place at the time. However, when the federal program was removed, the more environmentally benign alternative system became the more profitable.

While government involvements can be counterproductive to the economic viability of more sustainable systems, it is not clear that farmers operating in a free market could make the necessary changes on their own. Since lower input cost is a nearly universal characteristic of alternative systems, whether these systems also achieve higher profits will also depend on the prices received for the mix of crops grown. Since alternative systems tend to emphasize crop diversity and rotations, being able to market the crops that are produced is critical. In general, the crops grown for their environmental and soil fertility benefits have a lower market value than conventional crops like corn and soybeans.[6] And, if this is true now, a massive shift toward more diverse rotations would be expected to exacerbate the situation, e.g., by flooding the smaller hay and small grain markets, and underproducing for the corn and soybean markets. These macroeconomic effects are important, for they will determine whether sustainable systems gain widespread adoption.

Regional or Sectoral Studies

The literature includes a number of studies that are aggregate in nature and address the characteristics, historical changes, and likely future responses of farming regions, or U.S. agriculture as a sector.

Lowrance and Groffman[46] studied the conversion from high- to low-input agricultural systems at the regional level. Historical data on changing inputs and changing land uses were used to project possible effects of conversion on productivity and environmental quality on areas in Michigan and Georgia. The historical trends in nitrogen inputs and in the areal extent of crop production showed a strong positive relationship in both regions and suggested that a reversal of historical trends would have environmental benefits. Their results also indicated important regional differences, with the benefits of conversion to low-input systems much more pronounced in Michigan than Georgia.

A study by Dvoskin and Heady[47] used an interregional linear programming model to determine the response of U.S. agricultural production to future economic conditions, with an emphasis on energy availability and prices. Their results showed two of the most important responses to alternative scenarios of limited energy to be reductions in irrigation and in nitrogen fertilizer use. These responses have significant implications for both the overall productivity and the regional distribution of U.S. agriculture.

A recent study by Knutson et al.[48] addressed the national implications of eliminating various classes of chemicals from U.S. agriculture. Using a large-scale aggregate economic model, they examined a number of scenarios that were various combinations of bans on herbicides, insecticides (and fungicides), and inorganic fertilizers. Their model showed that broad-scale banning of chemical use would cause both undesirable and desirable changes in the existing patterns of environmental impact and resource use, as well as a number of significant economic and social adjustments.

Aggregate studies of this kind are certainly limited by both the data and the analytical tools available, relative to the magnitude and complexity of the problem being tackled. For example, the Knutson study has been criticized for not examining more likely scenarios of partial bans and/or gradual reductions in chemical use. The problem that the researchers face is how to design meaningful policy options such as partial chemical bans for the tremendous diversity of agriculture across the U.S. The vast array of crops being produced, the local differences in the production environments, the range of chemicals available for each crop, available alternatives to the chemicals, and other variables make designing a systematic and objective reduction policy very problematic.

Nevertheless, large-scale aggregate modeling does provide a way of assessing broad agriculture policies that might be proposed by the political process. Such efforts are bound to be controversial because of their inherent analytical limitations and the complexity and high stakes involved in proposing significant changes to status quo agriculture.

SUMMARY AND CONCLUSIONS

Our introductory comments contrasted the tremendous increase in agricultural productivity in the United States since World War II with some areas of environmental and social stress that have been linked to it. The areas of concern include: environmental degradation, depletion of critical natural resources, fragile consumer confidence in food safety, impoverishment of rural society, growing dependence on government intervention, and a general mistrust of economic reasoning that is seen as underlying the development of today's agriculture. The contrast between the concerns and the unprecedented productivity set the stage for the inquiries and discussion of the rest of the chapter.

We then presented a conceptual discussion, establishing a working definition of sustainable agriculture and exploring a wide range of economic theory related to the general concept of sustainability. Important conflicts were described between sustainability and other objectives that motivate human decisions, and alternative decision frameworks were presented that may improve our ability to recognize and reconcile those competing objectives.

We then argued that even though there are contradictions, conflicts, and constant change inherent in the details of a sustainable agriculture, the general principle of sustainability is worthy and there are changes in that direction that make sense. A framework was described for marking progress toward sustainability, and benefit-cost analysis was presented as a systematic procedure that can extend the horizons of decision making to better assess the sustainability of outcomes.

The discussion then shifted to a body of applied research reflecting the kind of pragmatic effort to improve on the status quo just described. The studies varied considerably in the research approaches taken, the geographic settings, and the types of agriculture being investigated. Even though only a small fraction of possible alternative farming systems have been studied, the studies suggest that there is at least a potential for profit from systems that are more environmentally benign and more self-sufficient, both in terms of resource inputs and in terms of dependence on government programs. The degree of self-sufficiency and the level of profit were both found to be very dependent on the circumstances in which the alternative systems were studied. Other studies were discussed that have examined the aggregate impacts of proposed shifts in agricultural technology. These studies showed varying effects on total food supplies, levels of government payments to farmers, food costs to consumers, income to farmers, and so on.

Neither the theoretical nor the applied research presented in this chapter conclusively demonstrates an economic advantage or disadvantage of increasing sustainability, and yet, the stresses that were cited earlier as becoming evident in the environment and in society suggest that our current path may well not be sustainable. How do we decide what changes should be made? What role should economics play in those decisions?

Agriculture is a basic human activity performed at the frontier where society interacts with nature. There are other activities at this frontier (e.g., mining and logging), but none is more critical to society or more broadly invasive of nature than agriculture; and, whether or not we humans are yet ready to foreswear our efforts to dominate nature, it is becoming increasingly apparent that it is in our self-interest to monitor this frontier for undesirable or unexpected changes. Undesirable changes, like the depletion of water and fossil fuel resources, warn of real limits to our dominance. Unexpected changes, like global warming and the ubiquitous presence of manufactured chemicals, raise doubts about the wisdom of some of our past choices. However, the demand for the products of agriculture is increasing, and sudden radical changes in how that demand is met seem both unlikely and unwise, at least in the short term.

In important ways, agricultural products are quintessential economic commodities — they are in continual demand, many require processing, they are perishable, and their supplies are unevenly distributed over both space and time. These characteristics mean that there is an inevitable role for middlemen to buy, store, process, transport, and sell agricultural products, and in the process arbitrate between producers and consumers who are often widely separated geographically. This chain of transactions between producers and consumers, and the institutions (e.g., laws and markets) that have developed to support such transactions, are the economy.

Economics is frequently identified as the culprit behind short-sighted or misdirected choices. That reaction is like shooting the messenger. A market economy does not assign values for society as much as it reflects the values of society. If it is in the interest of society for agriculture, or any other human endeavor, to move in a new direction, a different set of values needs to be brought into play. The values can be expressed within the economy through changes in what society is willing to pay for various goods and services (including premiums or discounts reflecting preferences about production technology). The values can also be imposed on the economy through laws, regulations, and evolving ethics that effectively constrain production and/or consumption decisions, with the market economy adjusting accordingly.

Economic decision making is a process of balancing costs and benefits. Questions about sustainability generally relate to perceived shortcomings in three areas of the process:

1. A failure to recognize, understand, and consider all effects of our actions (externalities).
2. A failure to assign proper values to all outcomes (pricing).
3. A failure to place appropriate value on time and future generations (discounting).

Balancing costs and benefits is relevant to all levels of decision making, from national and international policy making to the choices of individual producers and consumers. And, making improvements in the areas of shortcoming should mark progress toward increased sustainability.

The earlier discussion of alternative paradigms for sustainable development illustrates different viewpoints on judging technological progress, valuing nature, and providing for the future. Though helpful to understanding the value-laden nature of sustainability, these paradigms are, themselves, insufficient. Each requires more information about relative values before their reasoning is complete enough to guide decisions. The concept of "sustainable agriculture" is a similar attempt to frame humanity's ongoing self-interest. The expectations expressed in the principle are so inclusive that further priorities (or values) are needed before a goal is fully defined.

In a very real sense, continuing technological advances, increasing awareness of environmental implications, and growing demand for food combine to make sustainability a moving target. And, for agriculture to be both as productive and environmentally benign as possible on an infinitely varied landscape, it must be tailored to fit each unique site. This makes sustainability infinitely varied in application.

Because of the need to fit agriculture to the site, society must ultimately rely on individual farmers to know their land and to interact with nature as agents of the rest of present and future humanity. Nevertheless, we are all in this together. Society cannot afford a cavalier attitude about its food supply, ignoring linkages between the ethics and lifestyle of the general population and those of farmers, and the impacts we all have on our natural resources.

The kinds of decisions that society will have to make to sustain itself will be increasingly complex and far-reaching. Since not all social behavior is economic and not all values have monetary prices, economics alone will be insufficient for making these decisions. Nevertheless, because of the fundamental role economic behavior does play in society, economics will be an essential ingredient. Further, frameworks developed and used by economists can be broadened and refocused to provide useful organization to the decision making needed at all levels of society.

ACKNOWLEDGMENTS

This chapter has been made possible in part through Project No. ILLU-05-0331 of the Illinois Water Resource Center and Project No. G-15601 of the Agricultural Experiment Station, College of Agriculture, University of Illinois. Portions of this chapter are based on Economics of Sustainable Agriculture: A Literature Review, a Department of Agricultural Economics Staff paper by D. C. White, R. D. Ott, and R. H. Hornbaker.

REFERENCES

1. Economic Report of the President, U.S. Government Printing Office, Washington, D.C., 1991, Table B-96.
2. The Second RCA Appraisal. Soil, Water, and Related Resources on Nonfederal Land in the United States. Analysis of Conditions and Trends. Soil Conservation Service, U.S. Department of Agriculture, Washington, D.C., 1987, 50.
3. Clark, E., J. Haverkamp, and W. Chapman, *Eroding Soils: The Off-Farm Impacts. Conservation Foundation*, Washington, D.C., 1985.
4. The Crisis of Wildlife Habitat in Illinois Today, Illinois Wildlife Commission Rep. 1984–85, Illinois Department of Conservation, Springfield, IL, 1985, 9.
5. Gever J., R. Kaufmann, D. Skole, and C. Vörösmarty, *Beyond Oil: The Threat to Food and Fuel in the Coming Decades*, Complex Systems Research Center, University of New Hampshire, A Project of Carrying Capacity, Inc., Ballinger Publishing, Cambridge, MA, 1986.
6. GAO (U.S. Government Accounting Office), Alternative Agriculture: Incentives and Opinions, GAO/PEMD-90-12, U.S. Government Printing Office, Washington, D.C. February, 1990.
7. U.S. Congress, Technology, Public Policy, and the Changing Structure of Agriculture: Volume 2. Background Papers, Part D: Rural Communities, OTA-F-285, Washington, D.C., 1986.
8. National Research Council, *Alternative Agriculture*, National Academy Press, Washington, D.C., 1989, 10–13; 65–85.
9. American Society of Agronomy, 1988 Annual Convention, Anaheim, California, quoted in *Am. J. Altern. Agric.*, 3(4): 181, 1988.
10. Meadows, D. H., D. L. Meadows, J. Randers, and W. W. Behrens, *The Limits to Growth*, Universe Books, New York, 1972, introduction.
11. Reppetto, R., W. McGrath, M. Wells, C. Beer, and F. Rossini, *Wasting Assets: Natural Resources in the National Income Accounts*, World Resources Institute, Washington, D.C., 1989. 27–60.
12. Malthus, T., *An Essay on Population*, 6th ed., Ward, Lock and Co., London, England, 1826.
13. Kneese, A. V., R. U. Ayres, and R. C. d'Arge, *Economics and the Environment: A Materials Balance Approach*, Resources for the Future, Washington, D.C., 1970.
14. Page, T., *Conservation and Economic Efficiency: An Approach to Materials Policy*, John Hopkins University Press, Baltimore, 1977.
15. Pearce, D. W., E. Barbier, and A. Markandya, *Sustainable Development: Economics and Environment in the Third World*, Edward Elger, London, 1989.
16. Pearce, D. W. and R. K. Turner, *Economics of Natural Resources and the Environment*, Johns Hopkins University Press, Baltimore, 1990.
17. Arrow, K. J. and A. C. Fisher, Environmental preservation, uncertainty, and irreversibility, *Q. J. Econ.*, 88(2):312–319, 1974.
18. Ciriacy-Wantrup, S. V., *Resource Conservation*. University of California Press, Berkeley, CA, 1952.

19. Bishop, R. C., Endangered species and uncertainty: the economics of the safe minimum standard, *Am. J. Agric. Econ.*, 60(1):1017, 1978.
20. Norton, B. G., Environmental ethics and the rights of future generations, *Environ. Ethics*, 4(4):319–330, 1982.
21. Norton, B. G., Intergenerational equity and environmental decisions: a model using Rawls' veil of ignorance, *Ecol. Econ.*, 1:137–159, 1989.
22. Conrad, J. M. and C. W. Clark, *Natural Resource Economics: Notes and Problems.* Cambridge University Press, Cambridge, England, 1987, 70–77.
23. Hyde, W. F., *Timber Supply, Land Allocation, and Economic Efficiency*, Johns Hopkins University Press, Baltimore, 1980, appendix B.
24. Baumol, W. J., On the social rate of discount, *Am. Econ. Rev.*, 58(4):788–802, 1968.
25. Lind, R. C., Ed., *Discounting for Time and Risk in Energy Policy*, Resources for the Future, Washington, D.C., 1982.
26. Fisher, A. C., *Resource and Environmental Economics.* Cambridge University Press, Cambridge, England, 1981.
27. MacRae, R. J., S. B. Hill, J. Henning, and A. J. Bently, Policies, programs, and regulations to support the transition to sustainable agriculture in Canada, *Am. J. Altern. Agric.*, 5(2):76–92, 1990.
28. Sugden, R. and A. Williams, *The Principles of Practical Cost-Benefit Analysis.* Oxford University Press, Oxford, England, 1978.
29. Milligan, R. A. and B. F. Stanton, What do farm managers do? In: *Farm Management: How to Achieve Your Farm Management Goals*, USDA Yearbook of Agriculture, U.S. Government Printing Office, Washington, D.C., 1989, 2–6.
30. Madden, J. P. and P. O'Connell, *Early Results of the LISA Program*, Agricultural Libraries Information Notes, USDA/National Agricultural Library, 15(6/7):1–10, 1989.
31. Adams, R. M., Global climate change and agriculture: an economic perspective, *Am. J. Agric. Econ.*, 71(5):1272–1279, 1989.
32. Lockeretz, W. and P. Madden, Midwestern organic farming: a ten-year follow-up, *Am L. Altern. Agric.*, 2(2):57–63, 1987.
33. Berardi, G., Organic and conventional wheat production: examination of energy and economics. *Agro-Eco Syst.*, 4, 367–376, 1978.
34. Dobbs, T. L., M. G. Leddy, and J. D. Smolik. Factors influencing the economic potential for alternative farming systems: case analyses in South Dakota, *Am. J. Altern. Agric.*, 3(1):26–34, 1988.
35. Goldstein, W. A. and D. L. Young, An agronomic and economic comparison of a conventional and a low-input cropping system in the Palouse, *Am. J. Altern. Agric.*, 2(2):51–56, 1987.
36. Cacek, T. and L. Langer, The economic implications of organic farming, *Am. J. Altern. Agric.*, 1(1):25–29, 1986.
37. Helmers, G. A., M. R. Langemeier, and J. Atwood, An economic analysis of alternative cropping systems for east-central Nebraska, *Am. J. Altern. Agric.*, 1(4):153–158, 1986.
38. Domanico, J. L., P. Madden, and E. J. Partenheimer, Income effects of limiting soil erosion under organic, conventional, and no-till systems in eastern Pennsylvania, *Am. J. Altern. Agric.*, 1(2):75–82, 1986.

39. Mueller, D. H., R. M. Klemme, and T. C. Daniel. Short- and long-term cost comparisons of conventional and conservation tillage systems in corn production, *Am. J. of Soil Water Conserv.*, 40(6):466–470, 1985.
40. Dabbert, S. and P. Madden, The transition to organic agriculture: a multi-year simulation model of a Pennsylvania farm, *Am. J. Altern. Agric.*, 1(3):99–107, 1986.
41. Nehring, R., Lower input use can boost returns, *Agric. Outlook*, USDA/ERS, Rockville, MD, 28–29, June 1989.
42. Blobaum, R., Barriers to conversion to organic farming practices in the midwestern United States. In: *Environmentally Sound Agriculture*, W. Lockeretz (Ed.). Praeger Publishers, New York, 1983, chap. 16.
43. Heffernan, W. D. and G. P. Green, Farm size and soil loss: prospects for a sustainable agriculture, *Rural Soc.*, 51(1):31–42, 1986.
44. Hornbaker, R. H., Economic evaluation of low input efficient farms in Illinois, presented at the Illinois Rural Leaders' Forum: Sustainability of Agriculture, Springfield, IL, April 6, 1989.
45. Lockeretz, W., R. Klepper, B. Commoner, M. Gertler, S. Fast, D. O'Leary, and R. Blobaum, Economic and energy comparison of crop production on organic and conventional corn belt farms. In: *Agriculture and Energy*, W. Lockeretz (Ed.). Academic Press, New York, 1977, 85–101.
46. Lowrance, R. and P. M. Groffman. Impacts of low and high input agriculture on landscape structure and function, *Am. J. Altern. Agric.*, 2(4):175–183, 1987.
47. Dvoskin, D. and E. O. Heady, Economic and environmental impacts of the energy crisis on agricultural production. In: *Agriculture and Energy*, W. Lockeretz (Ed.). Academic Press, New York, 1977, 1–17.
48. Knutson, R. D., C. R. Taylor, J. B. Penson, and E. G. Smith, *Economic Impacts of Reduced Chemical Use*, Knutson and Associates, College Station, TX, 1990, chap. 5.

9

Making Sustainable Agriculture the New Conventional Agriculture: Social Change and Sustainability

Steve Padgitt and Peggy Petrzelka

TABLE OF CONTENTS

INTRODUCTION

Several of the previous chapters in this volume review technological developments which hold promise for altering agricultural production practices in ways that, if adopted, would make agriculture more environmentally benign. The preceding chapter focuses on economic issues which must be addressed in the ongoing deliberation of making sustainable agriculture the new conventional agriculture. While technology and profitability remain necessary prerequisites for adoption of new practices, a host of other cultural and social factors also intervene and affect decision making. In this chapter, we extend the discussion to include ideological and social considerations that will inhibit and/or facilitate change to more sustainable systems of agriculture.

Today, a series of technological, environmental, economic, ideological, and social factors is converging. They constitute a much different milieu for agricultural change than has existed in recent years, perhaps even several decades. New questions are being asked about the direction and future of agriculture with new and diverse parties involved in the questioning.

In this chapter, we briefly present a short review of agricultural systems to illustrate three points. One is the continuous nature of change and lack of sustainability of past systems of agriculture; the second is the inherent exploitive nature of past systems of agriculture on the environment; and the third is the complexity of systems in a finite world with an ever-increasing population to feed.

It is necessary to examine the change which is occurring using both macro and micro sociological perspectives. Viewing agricultural systems using a macro approach enables us to look at technology in the larger context of how it is related to social and ideological systems. Failing to acknowledge the relationships among the systems has led to adverse impacts, both on the environment and on rural communities. Examination of these systems leads us to three world views (or paradigmatic themes) of agriculture—subsistence, commercial, and sustainable. This macro approach provides a broad context from which to interpret current efforts to push agriculture in new directions.

While acknowledging that external trends and institutional structure define major parameters, the micro perspective also recognizes that change on the land in our private enterprise system is a result of individual producer decisions. In the considerations at the micro level, we discuss prerequisites needed to nurture such trends as well as approaches to accelerate making sustainable agriculture the new conventional agriculture.

AGRICULTURE IN HISTORICAL PERSPECTIVE

Most of human history is a chronicle of agricultural pursuits. Over time and even today, the bulk of the world's inhabitants have been agriculturists. Only in the past few centuries has food been produced in sufficient quantities and with enough reliability so that significant proportions of the population were free to seek other endeavors.

Across the planet, major systems of agriculture have emerged, prospered, and then vanished. Although change and transition are inevitably necessary and desirable characteristics of any sustainable system, transitions need to be gradual and not unduly disruptive. When survival has required totally new technologies, or when degradation of land prompted abandonment of once productive territory, then a particular system of agriculture did not serve its people very well.

As long as populations were relatively small and territory readily plentiful, nomadic and transitory systems were accommodated. However, the potential for adverse environmental consequences burgeoned as societies evolved from being small bands of hunters and gatherers, began using sharp sticks to stir the soil to cultivate plants, and when they selectively bred animals for livestock and grazed them on fragile semiarid lands.

There are ample lessons of where once productive and seemingly sustaining systems have collapsed. The one-time rich Tigris-Euphrates River valley is one example. The originally fertile sub-Sahara region of Africa that fed the Roman Empire is another. Here in the United States, prosperous Indian farmers of the Mesa Verda area of southwestern Colorado deserted their prosperous settlements when productive but fragile lands succumbed to erosion and depletion.

These kinds of catastrophic situations do not appear imminent with conventional agriculture. However, more and more, scientific knowledge is confirming the sensitive and delicate balance of ecological systems, and the extent to which human systems, including agriculture, have depleted and degraded resources and created serious ecological imbalances. Agriculture is inherently disruptive of natural processes.

Today, we face a number of ecological and environmental dilemmas. Growth of the world's population and the consumption patterns of certain societies, namely, our own, are pressuring the finiteness of known resources as never before. Degradation of land and water resources and consumption of limited reserves of fossil fuel energy are becoming well documented.

Not surprisingly, modern American agriculture is identified as a culprit. For example, in areas of the Midwest where much of the world's prime agricultural land is found, in little longer than a century, half of the topsoil has been lost to wind and water erosion. Additionally, the highly mechanized, chemically dependent, and extensive monocropping system of agriculture that characterizes much of U.S. farming has raised other concerns about the land. These include concerns over long-term productivity of the land, the impacts of erosion, increased compaction, loss of organic matter, water transport characteristics, an altered soil ecosystem, etc.

Both surface and ground water quality are being degraded. For some time, soil sediment, mostly from excessive tillage practices, has reduced water quality and restricted uses of lakes, rivers, and streams. Residues from agriculture fertilizers and pesticides have caused nutrient overload and ecological imbalances of surface water. More recently, the public's concern for the safety of drinking water supplies has been further expanded as chemical residues have increasingly been documented in ground water.

Agriculture is also a large user of energy. This includes major inputs into the manufacture and operation of modern machinery as well as agriculture's dependence upon petrochemically based fertilizers and pesticides. Additionally, when foodstuffs are produced long distances from processing and consumption sites, transportation becomes a significant energy consideration. Although agriculture interests have promoted the production of certain commodities, (i.e., corn for ethanol), as an alternative to fossil fuels for energy, this is not as efficient an alternative as sometimes assumed.

While agriculture is not the only culprit contributing to the demise of the natural environment, evidence reveals it plays a major role in disrupting the existing ecological and environmental systems. A change in the dominant conventional methods of agriculture is necessary to begin reversing this disruption.

THREE COMPONENTS OF AGRICULTURE

Interpreting persistence and change in society is the heart of sociology. Beginning with the industrial revolution, early sociologists offered interpretations and meanings to trends that were perplexing to many persons. No individual theory, and perhaps even the discipline's collectivity of social theories, provides full explanation. Nevertheless, most give some insight into the processes as well as the diversity from which change arises and is interpreted—by persons experiencing it as well as by more detached observers attempting to explain it. Early approaches were often descriptive and explanatory rather than prescriptions for altering the course of events. For some analysts, change was seen as an evolutionary process with ever-increasing specialization and complexity, and individuals and groups had little power to alter the course of events. Interpretations of consequences varied. Some were basically optimistic and saw the trends as progress,[1] while others were more pessimistic, viewing change as cyclical and disruptive.[2]

Likewise, today's changes in agriculture, including the interest in more sustainable systems, are viewed with great diversity. In part, this is because making sustainable agriculture the new conventional agriculture involves, simultaneously, subtle change and a major transformation. Although certain technologies will be refinements in, consistent with, and easily adopted into existing production systems, others will not. Further, for the broader notion of sustainable agriculture to prevail, there will need to be—for many producers as well as related support systems—fundamental alterations not only in the technological sense, but also in ideological and social contexts. Ultimately, sustainable agriculture is as much a cultural approach to farming as it is new techniques of production.

Therefore, it is instructive to view agriculture not only as existing in the natural environment, but also as comprising three dynamic and interrelated systems—the technological, the social or organizational, and the ideological. By technological, we refer to the methods and materials used to produce agricultural commodities as well as methods by which supportive knowledge is generated. The social component refers to organizational arrangements and structures, along with accompanying norms or rules guiding the behavior of people. The ideological component is the ideas, values, beliefs and philosophies providing a general framework for defining and viewing one's world. Because change in each of these components influences and is influenced by the other two, ultimately the three need to be considered as a whole. Initially, however, the components will be singled out in order to examine them more closely.

The technological element often receives high priority when describing the systems of agriculture. A dominant technology or set of technologies tends to be identified and used to label a whole era.[3] For example, hoe culture, rudimentary or advanced plow production, mechanical agriculture, petrochemical age farming, or the promised biotech era illustrate how dominant technologies have been used to characterize a whole farming system.

In the 20th century, especially, technological changes in agriculture have been pervasive and revolutionary. Sometimes technology appears disproportionately dominant, and the importance of the social and ideological elements is lost. To focus only on the technology, however, would be an error. Although development and adoption of innovative technologies at times appear to take a path of their own, this would not occur if viable countervailing social and ideological systems had been present. More likely, however, has been the situation when ideological and social systems have been supportive of the technological. Clearly, this has been the case in the United States. In his classical study of American values (i.e., socially shared ideas about what is good, right, important, and desirable) Williams identified several themes highly supportive of technology, including such guideposts as efficiency and practicality, progress, science and rationality, achievement and success.[4] These values ultimately become expressed in compatible kinds of organizational arrangements as well as social norms influencing behavior of citizens.

Conceivably, it is possible for another set of values to emerge which would be less amenable to technological development. Or, there may be values that place a strong emphasis on certain kinds of social structures and, thereby, inhibit technological development. For example, Amish agriculture places high importance on family and less on individuality as values. Similarly, it tends to reject certain kinds of technological developments, but accepts others. If such a belief system was predominant in U.S. society, a quite different set of priorities would likely have emerged for American agriculture over the past several generations, and, consequently, a different set of technologies and social arrangements.

In addition to highlighting a dominant technology to describe a system of agriculture, it is also appropriate to focus on distinctive social structures. In so doing, such classifications as family farms or corporate agriculture, free or controlled market economies, people-land relationships (i.e., owner-operator, tenant, sharecropper), may emerge as dominant labels. Certain arrangements become valued more than others, and some may be more internally consistent with given technologies than others. For example, a system of family farms has been a value at least since the Hamilton-Jefferson dialogues. However, if we look at the kinds of technologies being developed, emerging production systems, and the record of families living on the land, there are obvious incongruities. Although U.S. agriculture is still identified as a system of family farms, the decline in the number of families in farming over the past half century is dramatic. Increasingly a bi-modal distribution of agriculture production characterizes U.S. production.

The sociologist Ogburn held that the normal pattern for social change was first innovation in technology and then disparity between ideological and organizational elements.[5] When incongruous kinds of conditions evolved, whether across or within the three elements, strain was placed on a system and this set in motion a natural drive toward greater system equilibrium. Anthropologists and rural sociologists have documented numerous examples where events originally viewed as relatively insignificant technological substitutions, i.e., a steel ax for a stone ax[6] or wet rice cultivation for wild rice gathering,[7] have resulted in major ramifications for social and ideological systems. Often, there has been a tendency to consider mostly the primary direct impacts of adoption of innovations. Secondary, indirect consequences generally have not been adequately anticipated, nor have they been equitably distributed across segments of the population.

There is a long-standing controversy over whether the ideological system, (ideas, values, and beliefs) can direct change in the social and technological systems, or whether instead the social and technological experiences determine the content of one's beliefs and values. In his classical work, *The Protestant Ethic and the Spirit of Capitalism*, Weber argued the rapid changes occurring with industrialization were brought about in part because a specific type of belief and value system emerged.[8] Today's debate on sustainability and people-environment relationships has prompted similar discussions. A number of writers and scholars are suggesting that a qualitatively different belief and value system is developing and that it could result in potentially as significant a transformation as Weber described. The hypothesis is directed at shifts in the larger U.S. societal context and, perhaps, the entire Western World.[9] If these observations have merit, then sustainable agriculture is unquestionably intertwined in this larger movement.

DIALECTICAL ANALYSIS: A PERSPECTIVE FROM WHICH TO DESCRIBE AGRICULTURAL CHANGE

One tool early sociologists used to describe change was dialectical analysis.[10] This approach posits the notion that eras become characterized by a dominant, fairly stable, and reasonably coherent set of beliefs, values, social institutions, and technologies (thesis). Over time, however, for a variety of reasons, these fail to meet the needs of at least some individuals and segments of society. As a result, an opposing position (sometimes articulated as polar opposites) emerges (antithesis). Eventually, a new sense of reality and social arrangements dominate (synthesis). Supposedly, this new set is of a higher form and is a mediated product of earlier positions.

A dialectical framework can be used to contrast and illustrate conventional vs. a more environmentally attuned form of agriculture, whether the new form is called organic, regenerative, alternative, or something else. The recent popularity of the term "sustainable agriculture" may already mean some mediated understanding is starting to emerge, i.e., synthesis.

In our discussion of ideological bases of agriculture, we use three rather than two contrasting systems. The three systems of agriculture used to exemplify this are subsistence, commercial, and sustainable. Expanding the typology to three systems will help to illustrate continuity and change across two major transformations and thus reduce the tendency to conceptualize only along a dialectic framework. All three systems are associated with certain technologies, world views, and social expectations. After introduction of the typology, several of the defining dimensions of the three systems will be examined.

Subsistence Agriculture

Subsistence agriculture is a term that has been applied to earlier eras and to developing countries even today. Rogers has argued that the hand-to-mouth life characterizing this level of production results in a subculture unto itself.[11] As such it has self-perpetuating tendencies. In subsistence agriculture, managing for the needs of the day diminishes energies and resources that otherwise could be available for longer-term concerns. The prevalence of and consequences from uncontrolled disease, unreliable food supplies, unstable political systems, limited technologies, and ravages of the environment preoccupy people's lives. This creates difficulties for individuals to sufficiently detach themselves from their situations and begin to operate under a different vision or toward a different system of agriculture. Although individuals may have special circumstances or talents to change, the masses are tied to narrow parameters of change.

Commercial Agriculture

The label "commercial agriculture" has been widely used to depict contemporary U.S. agriculture. Given the propensity for purchased inputs, marketed outputs, and money as a medium of exchange, this is an appropriate label and contrasts sharply with subsistence agriculture. Accepted principles of commercial agriculture include virtues of expansion, growth, and dominance over nature. Investment, debt, and a sense of satisfaction with current production practices reduce agility and motivation to change to more sustainable systems. Commercial agriculture carries with it technologies, world views, and social expectations that inhibit the freedoms individual farmers have to radically change their modes of operation.

Sustainable Agriculture

Defining sustainable agriculture is elusive, and probably will remain so. There are several reasons. One reason is the inevitable ambiguity that exists in the direction of any social movement. Usually the more distinctly that positions are articulated (e.g., organic agriculture) the less likely they are to become the dominant or conventional system. Another reason is the emphasis that may eventually be given to different dimensions of sustainability that have become a part of the debate, such as the three that Douglass identifies: sustainability as food sufficiency, as stewardship, or as community.[12]

Figure 1 presents the three systems of agriculture. As is the case with many typologies, this serves to illustrate rather than precisely define. In some instances, the categories are speculative. This is especially the case

Defining Dimension	Subsistence	Commercial	Sustainable
Social Identity	Family	Self	Community
World of Reality	Past	Present	Future
Major interpersonal processes	Conflict	Competition	Cooperation
Nature of change	Uncontrolled & uncontrollable	Controlled	Planned & anticipated
Relationship to nature	Vulnerable to	Control over	Harmony with
Interpersonal relations	Mutual distrust	Individual rights	Community needs
Natural resources	Finite & consume	Develop & consume	Finite conserve & preserve
Motivational drive	Safety & security	Self achievement	Community accomplish
Role of State	Undeveloped, unstable, meeting needs of those in power	Coordinate protect rights laissez faire	Regulate
Knowledge base	Tradition	Science & technology	Science Technology mediated/ indigenous
Technological development	Borrowed or serendipitious	Supported, faith as solution to problems	Controlled for collective good

FIGURE 1.   Three systems of agriculture compared.

for sustainable agriculture. The delineated systems, however, represent qualitatively different ideologies. Any given system will comply only to some degree of approximation.

Social Identity

Although the family will be a major social institution and establish social identity for persons in all three systems of agriculture, in subsistence systems, allegiance to the family is a stronger social norm.[13] Personal goals are often subordinated in order to fulfill family obligations. In commercial agricultural systems, family ties are likely to be less restrictive, particularly intergenerational obligations. In a relative sense, this frees persons to pursue their self interests. Sustainable agriculture places greater priority on collective interests; thus, community concerns will have greater importance in defining one's social identity and be more important than in either of the other two systems.

World of Reality

In comparison to other systems, those in subsistence agriculture are more likely to look to the past to define reality and are not as apt to assess opportunities to affect future situations. This arises, in part, from their dependence upon tradition as a basis for knowing and understanding the world around them and also for tradition to provide necessary structures for survival. In commercially based systems where self interest is accepted as a guidepost, there is more attention to setting goals and pursuing accomplishments, albeit the time span is relatively short, i.e., usually within one's own life. Natural resources are viewed as being relatively infinite from the commercial agriculture perspective; therefore, the opportunity to develop is high and the need to conserve is low. This contrasts to the sustainable viewpoint, where limits to the planet's resources are defined as real. Adherence to sustainability implies conservation of resources for the benefit of present and future populations, not just accumulated wealth of individual entrepreneurs.

Relationship to Nature

In subsistence agriculture, people are vulnerable to the forces of nature and have limited control over them, whereas the stance in commercial agriculture is manipulation and control over nature. Sustainable agriculture is a blend of the two and is labeled harmony with nature.[14] These views impact how risk is managed. With little control over one's destiny, risk is something the subsistence agriculturist avoids. For the commercial agriculturist, risk is accepted and calculated for the windfall advantages that can accrue from taking risks. In sustainable systems, risk is not so much the challenge of opportunity and probabilistic assessments that characterize commercial agriculture, but more the notion that through knowledge and understanding risk is managed and minimized for the benefit of all.

Technological Development

Technological development in subsistence systems tends to be characterized as the result of external diffusion and serendipity. Limited resources and a tendency to perceive limited good in the world are contributing factors. Although commercial agriculture may fail to acknowledge the importance of heritage and external diffusion, there is a strong emphasis upon and faith in experimentation and new technologies to resolve both anticipated and unanticipated problems. From the sustainable perspective, technology is a tool, but its development needs to be monitored and directed to serve a collective good.

Role of the State

In subsistence systems, political entities are frequently unstable and tend to serve those in power as much as the populace. Governments in commercial systems are stable and emphasize rights of individuals. Sustainable systems give greater attention to the collective good, and when deemed appropriate have less reservation about controlling the behavior of individuals through regulatory measures.

Modernization theory has arisen and has been used to describe transformation processes from subsistence to commercial systems, although the labels are more often "traditional" to "modern" societies.[15] A major change in values, including those that facilitate motivation for a "better" life, is emphasized, as well as the need for changes in social institutions and organizations that support achievement of new goals. An increase in literacy is among the intervention strategies utilized to initiate a self-sustaining growth for the transformation, presumably because more literate populations will invoke "rational decision making" and avoid tradition as the dominant factor to justify any particular belief or action.

The transition from commercial to sustainable systems, likewise, involves major changes in values and adjustments in institutional and organizational arrangements. Increases in literacy facilitated the switching from subsistence to commercial systems, particularly the transition from an emphasis upon familial interests to self-interest. Not only were skill levels increased, but values were changed, particularly the priority on self-interest. Similarly, different values and new skills will contribute to a transition from commercial to sustainable systems. Today, these are not altogether clear, but many of the dimensions illustrated in Figure 1 will be important. A major challenge in the transition to sustainable systems is mediating self-interest and collective good. It may be that the role of the state will become more prominent in this mediation, though currently, this is not a popular notion.

SUSTAINABLE AGRICULTURE AS INDIVIDUAL PRODUCER CHANGE

In this section we discuss change from the perspective that, ultimately, making sustainable agriculture the new conventional agriculture will happen as individual farmers begin to change their cultural and management practices in ways that are less threatening to the environment. Several of these practices were discussed in earlier chapters. The challenge is to make these practices attractive, viable options for farmers to adopt. In the social scientist's world, this is behavioral change; and the same general principles for change apply, whether they are related to individuals altering a personal habit such as diet or farmers changing their pest management practices. Unless change is being externally coerced, at least four prerequisites must be present for voluntary behavioral change to occur. These are: (1) an awareness of a problem; (2) knowledge of alternatives; (3) motivation for change; and (4) resources to implement changes. The discussion of these prerequisites will focus on the individual farmer level, including current experiences, and then turn to various actions needed to raise the potential for adopting sustainable practices.

FOUR PREREQUISITES FOR BEHAVIORAL CHANGE

Awareness of a Problem

Failure to be aware of or recognize a problem in one's own circumstances can be a major obstacle to voluntary change. As a rule, people have greater difficulties in acknowledging shortcomings of their own actions than they do seeing faulty behavior of fellow persons or groups. When applied to farming, this principle suggests farmers may be aware of certain sustainability and stewardship problems when characterizing agriculture in general, but do not perceive or are not aware of problems associated with their own operations. Indeed, this has been documented repeatedly in regard to two sustainability issues in agriculture, soil conservation and water pollution resulting from farm chemical use.

Although large pluralities of farmers are likely to express concern and agree that erosion problems exist in their locality and elsewhere, far fewer admit that such is the case on their own farms. The differences in perceptions are often dramatic. For example, in an especially erosive watershed, Bultena et al. recently found a majority of farmers had observed erosion to the extent that land values decreased, but most (90%) reported having little or no erosion themselves.[16] The farther away the soil erosion dilemma is from one's own farm, the more observable the problem is to the farmer.

A similar pattern has emerged in perceptions about problems associated with farming's dependence upon agricultural chemicals and their effects on water quality. Like the general public, farmers also tend to feel agricultural production is too dependent upon commercial fertilizers and chemical herbicides, but concern is generalized and not acknowledged as applicable for their own operations. In Iowa, where 95% or more of corn producers use commercial nitrogen fertilizer and chemical herbicides, Lasley found three fourths of the farmers agreeing that agriculture was too dependent upon these inputs.[17] Yet only half as many identified their own operations as having any level of chemical intensiveness. Even fewer, often in the same single digit proportions as for soil erosion, acknowledge serious problems of water quality on their farms.

The discrepancy between farmer's generalized awareness of degrading environmental/ecological conditions and their failure to perceive or acknowledge problems on their own farms is probably a combination of inaccurate appraisal (by farmers and by those working on agricultural issues), legitimate innocence, confusion in interpreting contradictory information, and a process of denial to reduce cognitive dissonance or mental stress when assessing the actual conditions. In the case of soil erosion, where the problem is long-standing and well documented, the denial hypothesis appears credible in many instances. Unfortunately, overcoming denial is a more difficult barrier than mere innocence.

In comparison to soil erosion, failure to acknowledge a ground water pollution problem on one's farm is more understandable, and at present may not be as fraught with denial. The issue is relatively new. Documentation is missing in many areas. When chemical residues are found, the source is not always obvious or traceable. In the case of pesticides, residues are often in trace amounts, detected at a few parts per billion and below health standards. Although farmers express concern about health effects of low level contamination in drinking water, scientific evidence of consequences is generally lacking or inconclusive. As a result, those familiar with the use of pesticide products may tend to underestimate the potential of their adverse effects.[18]

Knowledge of Alternatives

In a conventional approach to assessing knowledge about farming alternatives, a list of specific practice options would be developed and depth of understanding of these options would be determined. This approach, however, would fail to capture some key aspects of sustainable agriculture, primarily because more than reducing sustainable agriculture to a list of practices, it must be viewed as a holistic integration. Elements essential within this integration include existing scientific and indigenous knowledge, new ideas, and detailed information about on-farm characteristics. Knowledge of internal resources and production patterns are typically the on-farm characteristics which are the most basic and beneficial, but often absent or under used in management.

Unfortunately, farmer's knowledge of these characteristics appear limited by two factors. First, there is a tendency to follow past conditioning (i.e., the commercial agriculture framework) and ask about or seek the latest "technological fix" rather than concentrating efforts on refining existing knowledge and technologies. Second, many farmers do not have, or often falsely assume they do have, the kinds of on-farm records necessary to make refinements for sustainable agriculture. In many operations, detailed enterprise records are not kept or effectively utilized to adjust or fine-tune practices.

To a considerable degree, sustainable agriculture alternatives involve a transition from substituting capital for labor to substituting management for capital. In the recent past, management has increasingly become a part of purchased product inputs. This "safe rather than sorry" stance happens when fertilizer rates beyond recommended levels are applied, when insecticide applications are used even though infestation risks are low, or when herbicides are applied as though weed problems were uniform across all fields. In a fully capitalized large-scale operation, or among farmers less motivated by management detail, such practices may be conscious choices, but they may carry profit and environmental sacrifices. In the absence of sound on-farm records these practices may appear to be logical and appropriate decisions. However, with adequate information, it is revealed that such practices, while often the common solutions, are not always the prudent choices.[19]

Knowledge of production practices that are more sustainable is also limited by the lack of up-to-date and accurate information on alternative agriculture. In a study of 168 farmers belonging to the sustainable agricultural organization Practical Farmers of Iowa (PFI) major needs were more information and more educational programs about sustainable agriculture. Presently, these farmers rely primarily on their personal experiences for information about sustainable practices.[20] They see that a critical need exists for generating new knowledge on how to farm wisely with nature.

Motivation for Change

Lockeretz notes, sustainability is ". . . a synthesis of ideas originating from various sources, out of various motivations."[21] Studies conducted on farmers practicing sustainable agriculture examine, among other factors, the farmers' motivations for doing so. In the Iowa PFI study, the primary reasons for using sustainable practices were to ". . . improve personal and family health, to farm in a way that is better for the soil and environment, and to pass on productive land to future generations." The researchers note that ". . . increased profits were less important than factors relating to health, the environment and future productivity".[22] Taylor et al., in their study of 32 South Dakota farmers who practice sustainable farming, found that the primary reasons for doing so include: ". . . to be a good steward of the soil, to reduce pollution of ground and surface water, to raise a residue-free, high quality product; and to reduce possible harmful effects of farm chemicals on the health of farmers and their families".[23] Secondary reasons noted as motivating farmers to change from conventional to sustainable farming include problems with conventional farming, ideological misgivings about conventional farming, contact with an individual suggesting alternatives, and farming more in harmony with nature.

A related study surveyed both conventional and sustainable farmers. In examining the relationship of farmers' conservation attitudes to actual behavior, Lynne and Rola found that not only economic variables, but also stronger attitudes towards conservation action ". . . result in a higher probability that the farmer will take action to conserve soil".[24] These findings are consistent with other studies conducted on motivations for practicing alternative agriculture, and reveal that farming practices may be adopted for a number of reasons which are not all solely economic in nature.

Certainly, farmers become motivated if survivability is threatened. Survivability of farmers, especially in terms of economic conditions, but also environmental and other factors noted above, received much attention throughout the 1980s. The attention focused in part on farmers who had made sound management decisions within the commercial agriculture framework, but were caught in a host of changing relationships within agriculture. As suggested earlier in Figure 1, emerging studies lend support to the notion farmers who are motivated to adopt sustainable practices are committed to the ideological and social aspects of the sustainable agriculture movement. However, it has also been suggested that those who have recently adopted sustainable practices tend to be primarily motivated by personal economic goals, in an attempt to try another course through which to recover from the 1980s farm crisis.

Thus, it appears several factors are converging at the micro level to motivate farmers' change. These include the defining dimensions presented in Figure 1, self-interest, economic motivations, and influences on decision making. Knowledge and comprehension of these factors is crucial in understanding what underlies farmers' motivations.

Resources for Change

Information as a resource comes in many forms; farmers' own experiences, indigenous knowledge, literature, research, neighbors, field agents, field days, dealers, etc. With this large influx of information, farmers are soon faced with an information overload, attempting to sort out large amounts of (often contradictory) information. Therefore, while information is an important resource for the sustainable farmer, the critical resource for the farmer is the ability to synthesize this information, and make the best management choices possible for his/her farming operation.

Sustainable agriculture requires the farmer to be a well-informed manager, possessing management and analytical skills. If reliance on synthetic fertilizers and pesticides is reduced, then the farmer must more fully understand the complex interrelationships among crops, weeds, insects, diseases, and soil fertility in order to overcome risks of crop damage and encourage the factors that maintain competitive yields.[25] This cannot be accomplished without reliable on-farm records. However, record keeping may be what farmers like to do least.

At a general level, farmers may be knowledgeable of many practices promoted in the quest for sustainable agriculture. For example, such basic elements as setting realistic yield goals, giving nutrient credits for manure or legume crops, using rotations as pest control, scouting before pesticide application, reducing tillage operations, etc. are not new ideas for farm operators. In fact, most farmers already claim they incorporate many of these practices. Often, they believe either their current practices are at an optimum, or that alternatives are not viable for their particular operations.

However, when specific actions are profiled, many farmers do have opportunities to refine a number of their practices to improve profit and protect the environment.[26] Among a population of farmers selected as cooperators in a targeted integrated crop management project in Iowa, only about one in six kept field-based records.[27] Yield goals were often based on inappropriate criteria and fertilization rates were increased above those levels. Use of banding rather than broadcasting herbicides was rare, even though cultivation was nearly universal. These examples reveal that in-depth knowledge by the farmer on their plots could have led to refinement in a number of their farm practices, a refinement which would improve both the farmers' profitability and the environment.[28]

If farmers are not knowledgeable of or do not adhere to best management practices within the commercial framework, they are unlikely to modify or integrate these practices into a more sustainable framework. Indeed, one of the problems that has plagued various projects geared to more environmentally sound yet profitable systems has been the extent to which many farmers do not have reliable records from which to assess alternative practices.

Acquiring and applying these new resources takes time, and farmers' time is scarce. Over the past generation, farmers have accepted the principles of efficiency of specialization and economy of scale. As a result, many operations have been expanded to the point that certain seasons, especially planting and harvesting, are periods of frenzied activity. Increasingly, farmers hold jobs off the farm. These farmers may have very little time for farming and obtaining in-depth knowledge of the farm operation. Devoting more time to the farm operation becomes an opportunity cost measured by lost off-farm income. More on-farm work also means less time available for leisure, socializing, and other activities valued by the farmer.[29]

FACILITATING BEHAVIORAL CHANGE

The preceding discussion reveals that, overall, farmers are limited in their awareness of farm specific environmental problems, knowledge of sustainable alternatives, motivations to practice sustainable methods, and resources available for changing to a sustainable system. Our discussion now turns to examining some possibilities for increasing these prerequisites to behavioral change.

Our first prerequisite for voluntary change asserts that individuals must be aware of and accept situations as problematic for themselves before they will seek alternatives. This is based on the premise that once aware, people are prepared to seek alternatives and then act in their best self-interest. Research reveals that, while farmers are generally aware of environmental and other sustainability problems, not all are mindful of or believe the specific problem pertains to them. More efforts are needed to increase farmers' awareness of environmental problems on their land.

Raising awareness of problems at a general level may be advanced through promotion of and education about certain concepts and practices. Mass media methods are most effective in creating awareness at a general level, often by giving prominence to case examples or extraordinary events. This is a beneficial and apparently successful first step, but the media are relatively ineffective in providing sufficient depth to assist farmers in determining problems in specific instances, especially if the determination is somewhat technical. Even specialized media channels, such as the farm press, lack this kind of targeted message. General messages are effective, however, in prompting individuals to ask further questions.

A more specific way to increase awareness is experiential learning. Experiential learning environments not only heighten awareness of a dilemma, but also provide a means to solve the problem. Examples of this type of learning which relate to sustainable agriculture include applied workshops on a relevant topic, such as water quality and on-farm research.

Workshops have been successful in drawing rural families into a dialogue and prompting them to take action. For example, discussions about water quality issues can be supplemented with demonstrations and then instructions for proper water collection methods for testing. After collection of the samples, the participants are taught to analyze their water and given instructions on what to do based on the results. By examining for themselves the quality of their drinking water, awareness of a potential problem may be increased for farmers and their neighbors.

When most farmers become knowledgeable about making practices work on their own farms they talk to other farmers and inspect first hand the "evidence" for its applicability to their farming operations. Likewise, on-farm research involving test plots and demonstrating detailed record keeping may also provide discussion among the farmers and their peer network(s). An experiential learning environment such as this could supply hard evidence to the farmer and neighboring farmers concerning the extent of soil erosion or other conservation problems existing in their area. A key element is for farmers to be actively engaged in dialogue accompanying activities.

When the awareness of ecological/environmental problems has been raised and questions are being asked, it is essential that the kind of assistance needed be available to provide credible answers, or that educational opportunities exist so that farmers can become sufficiently informed to make determinations for themselves. A support structure to assist with this is essential if farmers are to find workable solutions once the quest for sustainable alternatives has begun. This support structure may involve (and is not limited to) individuals (e.g., other farmers practicing sustainable agriculture, dealers), organizations (e.g., alternative agricultural groups, cooperatives), and institutions (e.g., extension services, community colleges, alternative agricultural foundations). These support structures can be effective not only in providing assistance, but also in raising awareness of sustainability issues.

The PFI study reported earlier revealed a desire among sustainable agriculture practitioners for information and educational programs. A first step to furthering knowledge about sustainable agriculture alternatives is to alter the way farmers currently view their operations. Since Ryan and Gross first studied the adoption of hybrid seed corn in 1943, farmers have been conditioned to seek knowledge about and adopt commercially based innovations;[30] the time curve from introduction of an innovation to its adoption by a majority of farmers has steadily decreased as farmers endorsed such guiding principles of commercial agriculture as specialization and efficiencies of scale, maximizing yields, and substituting labor for capital.

As a result, the kinds of innovations that captured farmers' interests and were deemed worthy of their search for more knowledge were ones furthering these trends. Typically, the alternatives adopted replaced on-farm resources with purchased ones, or substituted one purchased input for another. Aggressive farmers actively sought new knowledge about and then adopted latest technologies in order to gain early windfall advantages. To avoid being thrown off a treadmill, others quickly adopted.[31] Unfortunately, the competitive advantage of many changes was soon lost, and to be sustainable within an economic competitiveness context, a new round of innovations and adoption became necessary.

A different approach to viewing farm operations is necessary when considering sustainable agriculture. Rather than looking for the latest approach being handed down by Land Grant colleges and universities or marketed by agribusinesses, farmers need to become aware that sustainable practices must be uniquely defined for each farming operation, given the goals, skills, and financial and natural resources available for that specific operation. Agricultural sustainability cannot be achieved through a predefined set of management practices or a recipe for success. The educational programs on sustainable agriculture may need to take a systems approach, where the farmer is the researcher; experimenting, refining, and testing alternative technologies and cropping mixes.[32] Currently, knowledge of sustainable alternatives involves as much a commitment to seek and experiment as it does practicing various alternative methods.

Underlying this commitment to seek and experiment are the farmers' motivations for doing so. To understand what can be done to increase farmers' motivations and adoption of sustainable practices, it is essential to know and understand what is involved in the farmers' decision-making process. Recently, the Office of Technology Assessment, U.S. Congress identified 12 factors influencing agricultural decision making.[33] These are summarized in Figure 2.

Identifying and removing the impediments/barriers to adoption of sustainable agricultural practices is a logical step to increasing farmers' motivations. An increase in motivations at the individual level may

1. Farmers are a heterogenous group with unequal abilities and unequal access to information and resources for decision making;

2. Farmers' decisions are based on their fundamental reasons for farming; their objectives may not be clearly defined or articulated;

3. Economic factors exert important, but not sole, influences on farmer decision making;

4. Farmers typically make production decisions within short time frames, which discourages investments in resource production measures;

5. Farmers make changes slowly;

6. A farmer's innovation decision process consists of several sequential stages;

7. Farmers adopt "preventive innovations" more slowly than incremental innovations;

8. Individual farm characteristics appear to explain only a small portion of conservation adoption behavior; institutional factors (e.g. farm programs, credit availability) probably are highly influential;

9. Studies on adoption of farm practices have rarely examined the physical setting of adoption decisions or the extent of resource degradation as it relates to adoption of remedial farm practices;

10. Farmers tend to underestimate the severity of soil and quality problems on their own farms;

11. Farmers are most likely to adopt technologies with certain characteristics;

        a) relative advantage over other technologies
        b) compatible with current management objectives
        c) easy to implement
        d) capable of observed or demonstrated
        e) capable of being adopted on an incremental or partial basis;

12. Decentralized information exchange among farmers promotes a wider range of innovations than do more centralized diffusion channels.

FIGURE 2.   Factors influencing farmers' decision making. (From Beneath the Bottom Line, Office of Technology Assessment/Congress of the United States, Washington, D.C., 1990.)

require addressing factors at the macro level. For example, statement 8 from Figure 2 suggests that institutional factors are probably highly influential when explaining farmers' adoption of conservation behaviors. One obstacle to adoption of sustainable agricultural practices, suggested by those who work on rural and agricultural issues, is agricultural policies that advocate production efficiency and encourage specialization and extensive use of chemicals.[34]

Once impediments/barriers have been reduced and alternatives identified, two additional resources are necessary: (1) management skills and (2) time. It appears at present that in many operations, both may be problematic.

In Iowa (and we assume elsewhere), skeptical views held by many farmers towards sustainable farming practices are a definite obstacle to adoption of such practices. However, Bultena suggests,

. . . diffusion could be accelerated by placing increased priority on outreach programs for reaching conventional farmers who, while generally sympathetic to the agronomic, environmental, and social goals of the alternative agriculture movement, may feel trapped in conventional agriculture, that is, persons who, while open to change, lack the management skills, needed information, social support, and financial resources to successfully engage in a new form of agricultural production.[35]

Consequently, while the motivations may be there for some farmers, limitations of various resources are acting as barriers to adoption of sustainable agricultural practices. Some farmers are threatened with the idea they have to be more "management smart".[36] Individuals, organizations, and institutions committed to sustainable agricultural issues need to realize these farmers may be lacking in the necessary management and analytical skills to successfully farm in a more sustainable manner and that the kinds of assistance that were effective within the commercial agriculture context may be less appropriate in the new context. Educational opportunities must be provided so farmers can be informed, trained, and nurtured while developing the necessary skills. These opportunities need to be in an environment where there is continuous dialogue between the farmer and the individual(s) providing assistance; where farmers are actively involved in their on-farm decision making, and where they are made aware of what their farming practices are and what steps are needed in order for practices to be more environmentally benign, yet maintain acceptable production and profit levels.

Who will provide assistance to farmers lacking the skills and resources necessary for a transition to sustainable agriculture is still unknown. Some farmers may need to buy more management input from cooperatives or other private sector organizations. Record-keeping may become a purchased input. Dealers may also be more involved in farming decisions in the future. Dealers' expertise on local conditions would be a critical asset for the implementation of sustainable practices on specific sites.[37]

Extension specialists may also provide this assistance, through nurturing of skills and working alongside the farmer. During World War II and the early post-war period, extension farm management specialists implemented programs of long-term whole farm planning. These Farm and Home Development (F & HD) programs were widely recognized as being very successful in assisting farm families in developing long-range plans and implementing them into successful operations. While the program was criticized by some as being too expensive and eventually duplicating private sector services, the personnel who were involved in helping deliver the program to the individual farm families and the families themselves felt that it was one of the most successful programs that had ever been carried out.[38]

Extension specialists may need to initiate and pilot parallel kinds of programs. For example, there is interest in such services as pest scouting, enterprise records, etc., but not always at levels that are commercially profitable. If pilot programs demonstrate these to be advantageous, then interest will grow and a new service industry might be born as pilot programs are converted to private entrepreneurship. This role would include examining economic and other impact changes in an operation may bring to the farming system and acquiring access to the specific expertise that is needed in various parts of the systems planning. County staff may need to become more specialized in some areas of expertise, depending on the type of farming that is or may become predominant in their local area. This will be seen as expensive in the short run, but in the long run, it could make a significant contribution to a more profitable and environmentally sound system.[39] Behavioral change will not occur simply through a lecture. People are more likely to change their behavior if they are shown specific examples of what to do and are given a chance to practice their new skills, so that they build confidence and belief in their ability to change.[40]

CONCLUSION

The dialogue on the four prerequisites for behavioral change proposes and brings awareness to issues more than it offers definitive steps to instigate change. This results from several factors. One is the complex nature of sustainability and our evolving understanding of it. A second factor limiting our conclusions is that, to a considerable degree, sustainability is context and situation specific, i.e., what management practices qualify as sustainable in one context could be regarded as unacceptably exploitive in another. Third, because agricultural producers are not a homogeneous group, diversity rather than consensus may be the rule, and one must always be cautious of over generalizing positions and preferences within the farm population. Nevertheless, for sustainable agriculture to become a dominant form, its adherents need to be expanded so that sustaining practices account for a larger component of agricultural production. For now, commercial agriculture is the dominant system. It has tremendous momentum, and the extent to which sustainable agriculture ideology and practices will intervene to place agriculture on a different course is yet unclear.

While technology and profitability are necessary prerequisites for adoption of different farming practices, social and ideological considerations also have a vital part to play in understanding adoption behavior.

In examining the change which is occurring, both a macro and a micro sociological perspective need to be considered. Examination of agriculture through the three world views (subsistence, commercial, and sustainable) provides us with a broad historical context in which to interpret current events and efforts to push conventional agriculture in a new direction. Examination at the micro level recognizes that ultimately, effective change needs to occur at the individual level, and reveals the many factors involved in individual change processes.

Social change is often a slow and frustrating process. As noted previously, sustainable agriculture is difficult to define. There are no "cookbook recipes" advising those working on sustainable agricultural issues. However, it is essential that those working in the area of sustainable agriculture understand where this transformation in American agriculture is coming from, and what is involved in the farmers' decision making processes. This is a challenge to those working in making sustainable agriculture the new conventional agriculture. As with farmers, awareness of the issue is the first step in the process.

REFERENCES

1. Comte, A., *The Positive Philosophy*, H. Martinueu, Trans. and Ed., Bell, London, 1915.
2. Spengler, O., *The Decline of the West*, Knopf, New York, 1962 (originally published in 1918).
3. Smith, T.L. and Zopf, P.E., Jr., *Principles of Inductive Rural Sociology*, F.A. Davis Company, Philadelphia, 1970.
4. Williams, R.M., *American Society: A Sociological Interpretation*, Prentice-Hall, Englewood Cliffs, NJ, 1970.
5. Ogburn, W.F., *Social Change*, Viking, New York, 1950.
6. Sharp, L., Steel axes for stone age Australians, *Hum. Organ.*, 11, 17, 1952.
7. Rogers, E.M., *Communication and Development: Critical Perspectives*, Sage Publications, Beverly Hills, CA, 1976.
8. Weber, M., *The Protestant Ethic and the Spirit of Capitalism*, Scribner, New York, 1958.
9. Milbrath, L.W., *Envisioning a Sustainable Society*, State University of New York Press, Albany, 1989.
10. Weiss, F.G., Ed., *Hegel, the Essential Writings*, Harper and Row, New York, 1974.
11. Rogers, E.M., *Communication and Development: Critical Perspectives*, Sage Publications, Beverly Hills, 1976.
12. Douglass, G.K., *Agricultural Sustainability in a Changing World Order*, Westview Press, 1984.
13. Rogers, E.M., *Communication and Development: Critical Perspectives*, Sage Publications, Beverly Hills, CA, 1976.

14. Beus, C. and Dunlap, R., Conventional versus alternative agriculture: the paradigmatic roots of the debate, *Rural Sociol.*, 55, 590, 1990.
15. Lerner, D., *The Passing of Traditional Society*, Free Press, New York, 1967.
16. Bultena, G.L., Hoiberg, E.O., Korsching, P.F., Padgitt, S.C., and Malia, J.E., Siltation of the Red Rock Reservoir: Farmers' perspectives on causes and solutions, Sociology Rep. No. 161, Iowa State University, Ames, 1990.
17. Lasley, P., Kettner, K., Duffy, M., and Edelman, M., Iowa farm and rural life poll, Rep. Pm-1369, Iowa State University Extension, Ames, 1989.
18. Douglas, M., *Risk Acceptability According to the Social Sciences*, Russell Sage Foundation, New York, 1985.
19. Smidt, T.B., Connelly, K.A., Thoreson, D.R., and Voss, R.D., Integrated Crop Management Project, Integrated Farm Management Demonstration Program, Rep. Pm-1417, Iowa State University Extension, Ames, 1990.
20. Korsching, P.F. and Malia, J.E., Institutional support for practicing sustainable agriculture, *Am. J. Altern. Agric.*, 6, 17–22, 1991.
21. Lockeretz, W., Open questions in sustainable agriculture, *Am. J. Altern. Agric.*, 3, 174, 1988.
22. Malia, J.E. and Korshcing, P.F., Practicing sustainable agriculture in Iowa, presented at Rural Sociological Society meetings, Seattle, August 5 to 8, 1989.
23. Taylor, D.C., Dobbs, T.L., and Smolik, J.D., Sustainable agriculture in South Dakota, Res. Rep. 89-1, South Dakota State University Economics Department, Brookings, 1989.
24. Lynne, G.D. and Rola, L.R., Improving attitude-behavior prediction models with economic variables: farmers action toward soil conservation, *J. Soc. Psychol.*, 128, 12, 1987.
25. Crosson, P., What is alternative agriculture?, *Am. J. Altern. Agric.*, 4, 28, 1989.
26. Padgitt, S.C., Monitoring audience response to demonstration projects, Rep. IFM 8, Iowa State University Extension, Ames, 1990.
27. Padgitt, S.C. and Boraas, T., Model farms demonstration project, in Preliminary Baseline Findings and Data Book, Iowa State University Extension, Ames, 1991.
28. Smidt, T.A., Connelly, K.A., Thoreson, D.R. and Voss, R.D., Integrated crop management project, Integrated Farm Management Demonstration Program, Rep. Pm-1417, Iowa State University Extension, Ames, 1991.
29. Crosson, P., What is alternative agriculture?, *Am. J. Altern. Agric.*, 4, 28, 1989.
30. Ryan, B. and Gross, N.C., The diffusing of hybrid seed corn in two Iowa communities, *Rural Sociol.*, 8, 15, 1943.
31. Cochrane, W.W., *The Development of American Agriculture*, University of Minnesota Press, Minneapolis, 1979.
32. Flora, C.B., Research priorities for a sustainable agriculture, in *Leopold Center for Sustainable Agriculture Conf. Proc.*, Leopold Center for Sustainable Agriculture, Ames, 1991, 5.
33. Office of Technology Assessment and Congress of the United States, *Beneath the Bottom Line*, U.S. Government Printing Office, Washington, D.C., 1990.

34. Duffy, M. and Chase, C., Impacts of the 1985 food security act on crop rotations and fertilizer use, Staff Pap. 213, Iowa State University Department of Economics, Ames, 1989.

35. Bultena, G.L., A comparison of conventional and sustainable farmers in Iowa, in *Leopold Center for Sustainable Agriculture Conf. Proc.*, Leopold Center for Sustainable Agriculture, Ames, 1991, 51.

36. Mullins, S.B., What farmers expect—and need—from sustainable agriculture, in *Leopold Center for Sustainable Agriculture Conf. Proc.*, Leopold Center for Sustainable Agriculture, Ames, 1991, 37.

37. Keeney, D., The role of input dealers in sustainable agriculture, *Leopold Lett*, 3, 3, 1991.

38. Jennings, V.M., The role of Extension in implementing integrated management system components of the 1990 food, agriculture, conservation and trade (FACT) act, presented at the Wheat Industry Resource Committee meeting, Vail, CO, July 8, 1991.

39. Jennings, V.M., The role of Extension in implementing integrated management system components of the 1990 food, agriculture, conservation and trade (FACT) act, presented at the Wheat Industry Resource Committee meeting, Vail, CO, July 8, 1991.

40. Bandura, A., *Social Foundations of Thought and Action: A Social Cognitive Theory*, Prentice-Hall, Englewood Cliffs, NJ, 1986.

10

Challenges for the 21st Century

Jerry L. Hatfield and Dennis R. Keeney

TABLE OF CONTENTS

INTRODUCTION

Change is a given component of human existence. Paradoxically, change is also greatly feared by societies, particularly their public and private institutions. While some have pictured sustainable agriculture as a move to the past, the preceding chapters illustrate well that it is a forward-looking concept that emphasizes and embraces change.

Concepts of sustainability in vogue today will change as technology develops, as societies evolve in their relationships to nature, as economies adjust to sustainable development, and as individual lifestyles adjust to more efficient utilization of the natural resources available.

Throughout the preceding chapters, invited authors have examined several components of sustainable agriculture systems. The authors have shown, or at least attempted to show, a scientific rationale for each of the components of what we categorize as sustainable agriculture. Harwood (1990) outlined the history of sustainable agriculture and concludes with the thought that "we should cherish the diversity of thought and experience that provides the 'raw material' for evolution of a new paradigm." These chapters are a portion of that "raw material" we hope will begin to form the basis for changing agriculture. Our goal in this chapter is to link some of the current issues into a framework where research and technology transfer can provide an impact.

It is hard for us to visualize farming practices of the 21st century because of the rapid changes of the past 50 years. In this time, agriculture in the developed world has become dependent on fossil fuels for power and for petrochemicals such as pesticides (see Chapter 1). A prime example is the corn yield increase in the Midwest after the introduction of hybrid corn. Part of the increase can be attributed to the genetic improvement while another portion of the trend is due to technological advances such as improved planting and harvesting equipment, fertility, pest control, and irrigation. These changes occur relatively slowly and are seldom noticeable from year to year. These changes have largely been brought on by a combination of changing technology and markets. They have resulted in major shifts in the role of the agrarian society from one of steward of land resources to one of large scale farming and in some cases industrialized agricultural systems, with a resulting decline in the number of farms and farmers (Smith, 1992). Now society is calling on agricultural science to help improve sustainability of our agriculture.

Many recent reports, including those released by the National Research Council on *Alternative Agriculture* and *Sustainable Agriculture Research and Education in the Field*, have increased our awareness of the changing values and issues in agriculture (see also papers by Lockeretz, 1988; Busby, 1990; Weil, 1990; Francis and Youngberg, 1990, Bues and Dunlap, 1990; Keeney, 1989, 1990; Wilken, 1991). These articles and reports illustrate that alternatives to conventional farming systems currently exist, largely through the ingenuity of individual farmers. While the biological and physical scientists can study these systems and perhaps improve on them, adoption by the farming community usually is slow to nonexistent. Social scientists and economists must be involved as these systems are researched, and farmers, who must be the users of these systems, must be involved in all phases of their development and research.

ENVIRONMENTAL ISSUES

Several issues are driving the concern for a more sustainable agriculture. Environmental concerns undoubtedly top the list, but often environmental issues and sustainable agriculture are considered as synonyms, whereas in reality they often diverge and sometimes operate at cross-purposes.

The environmental issues of most concern include water quality, soil quality, soil erosion, wildlife preservation or enhancement, species diversity, air quality (primarily odors), and visual quality (scenic values). Of these issues, water quality and soil erosion appear to generate the greatest emphasis. They are interlinked, and controlling one source of environmental damage may enhance degradation of another environmental component. For example, the move to no-till to reduce soil erosion might require more agrichemicals with current methods, enhancing the potential for ground and surface water quality degradation. There are a number of reports that verify the existence of agricultural chemicals in water supplies (CAST, 1985; Hallberg, 1989a,b; Fedkiw, 1991; Holden and Graham, 1992). Concern over water quality directly involves human health risks; soil erosion is far less glamorous and in general is a phenomenon that occurs without much public observation. Thus it is often not the environmental issue that garners strong public attention. Indeed, Crosson (1991) questions the commonly held theory that soil loss is a threat to agricultural sustainability in North America. However, it is also recognized that agriculture is the remaining unregulated nonpoint pollutant source and that sediment is the major source of this contamination.

These concerns have prompted a detailed examination of the role of agricultural practices on the environment. Over the past decade, countless articles and many divergent opinions have been expressed regarding agriculture's dilemma of growing food while turning a profit and protecting the environment (CAST, 1985). Government intervention with commodity and conservation programs has produced a confusing collage of incentives, rules, and regulations that also influence, often negatively, the move to more sustainable and environmentally responsible farming systems. It is no wonder that the farm producer who wants to change is often unable to do so. In many instances, the appropriate systems research has not been done, or the information has not been tailored to meet farmer needs, or environmentally responsible systems are not profitable (Bromley, 1991; Rosmiller, 1992; Smith, 1992).

There is considerable debate over the role of education and voluntary change as opposed to regulation as the major agents of change. Either approach requires information. Indeed, a sustainable agriculture can be classified as an information-rich rather than a strictly technology-based form of farming (Francis et al., 1990; Lockeretz, 1991; Keeney, 1990; Doll and Francis, 1992). There is a perception that farmers are unwilling to change practices to remove any potential for environmental contamination. However, experience indicates that if the risk of change is minimized through incentives, insurance, and other mechanisms, change can be brought about (Nowak, 1992). Most of the management practices that are sustainable are already cost-effective. The risk involves getting them incorporated into the system and developing the management skills to use them efficiently.

Atmospheric quality is an emerging issue as we more fully recognize the role of the atmosphere as a medium for chemical transport. A related concern, odors, has long been a bane to concentrated livestock systems and is a problem that is extremely difficult to solve since it involves both technical and social issues that are hard to quantify.

The other issues listed earlier, particularly those involving wildlife and landscapes, are also critical and necessary to a sustainable agriculture. They are even more difficult to assess and thus to address on a research basis. However, there is a general consensus that the public regards agriculturists as stewards of the land and its associated resources, a view well expounded by Leopold (1949). These issues must be addressed on a systems basis, usually at the landscape or watershed scale.

Water Quality

In the early 1980s, the public became aware that agriculture through use of farm chemicals might be affecting the quality of surface and ground water. This concern was recognized by the scientific community and has led to many research and education activities. However, it is well recognized that sediments eroded from agricultural soils and associated chemicals have negatively affected surface water quality and cause great off-site costs due to siltation of water courses and lakes and impoundments. Hillel (1991) illustrates the impact of agricultural practices throughout the modern history of man. He concludes that if we are to improve and sustain our production, we must change our attitudes about the soil and the interactions of farming with the soil.

Erosion has been the primary surface water quality problem associated with farming and has also had a major impact on productivity. Pierce (1991) reviewed the research on soil erosion effects on productivity and concluded that there is not a well-defined relationship between erosion and productivity for most soils. He stated that the problems with defining these relationships were due to a lack of consistent measurement techniques with which to quantify soil erosion losses. Visual observations of sediment in streams and rivers are easy to make and often cause the most public reaction. However, the public concern has been heightened with analyses showing detectable levels of agrichemicals in the water supply. As an example, during the spring of 1990 there were 14 days in which the nitrate nitrogen levels exceeded the 10 ppm drinking water standard in Des Moines, Iowa and, in 1991, the number of days was 15. The immediate correlation which was drawn related application of commercial fertilizers and, in particular, anhydrous ammonia in the fall to these increases. Recent studies by Keeney and DeLuca (1993) have shown that over the past 45 years of observations on the Des Moines River, the nitrate concentrations have not substantially increased in spite of a shift from mostly soil, legume, and manure sources to fertilizer nitrogen. These results illustrate that we live in a complex world and that our soil management and tillage practices change the environment and that there are not simple cause and effect relationships between farming practices and water quality. Observations of nitrate nitrogen in Walnut Creek watershed near Ames, Iowa show that there is a large temporal variation within a season and differences among seasons due to weather and possibly management practices. These interactions are difficult to sort out without long detailed records and well-designed studies. Unfortunately, we do not often have this level of detail in our records, and the observations are not at a scale which can be directly related to farming practices except in a general sense. The results by Keeney and DeLuca (1993) suggest that farming in central Iowa was not designed for high levels of nitrogen efficiency in the 1940s nor is it at that level now. Clearly, system level changes are needed (Keeney, 1992).

Water quality issues are easier to identify than they are to solve. An example of this is the Presidential Initiative to Enhance Water Quality, as described by Onstad et al. (1991). This program has a goal to evaluate the impact of current farming practices on water quality and to develop and evaluate new practices which would enhance water quality. Studies are being conducted in Iowa, Minnesota, Missouri, Nebraska, and Ohio as part of this program to test the current practices and to determine what potential effect changes would make in water quality. This project involves the scientific expertise from seven federal agencies and an even larger number of state and local agencies within each state. The movement of agrichemicals from the surface to the aquifer is a complex process and requires the integration of several disciplines and spatial scales if the results are to be applicable outside of the study area.

The Iowa State Well Water Survey showed that agrichemicals and nitrate nitrogen were present in the domestic wells from shallow ground water wells throughout Iowa (Hallberg et al., 1990; Kross et al., 1990). Detectable amounts of coliform bacteria found in wells suggest that the wells are directly coupled with human activity and may not be an unbiased representation of the quality of the aquifer. Well construction and farmstead management may be directly related to these observations. Techniques to date the age of water within the aquifer and to determine the pathways of movement to the aquifer will be required before the relationship between farming practices and water quality can be identified.

To the present, ground water quality has been a greater public concern than surface water quality. However, recently observed concentrations of agrichemicals in the Mississippi and Missouri rivers as well as many observations of high levels of contaminants in surface impoundments used for drinking have raised concerns about the linkages between agriculture and surface water quality (Goolsby et al., 1991). Agrichemicals which are attached to the surface of soil particles are moved with sediment in runoff and often exhibit high concentrations in surface water after a runoff event. Farming practices which increase infiltration and decrease runoff do not exhibit concentrations above the drinking water standard because water runoff is minimized. Hatfield (unpublished data, 1992) found that atrazine concentrations were above the drinking water standard in surface water only after runoff events within agricultural watersheds. The surface water concentrations of atrazine were dependent upon the length of time from the application to the runoff event. Concentrations in tile drainage from fields were below the 3 ppb level throughout the year. These results suggest that there are several different levels of complexity in determining the relationship between tillage and chemical practices and water quality. The linkage between surface runoff, chemical movement, and surface and ground water quality must be better understood before we can develop farming systems which are environmentally benign.

Herbicide and nitrate levels in the ground and surface water dominate the water quality issue in the Midwest U.S. grain region. In order to protect water resources, research and education must directly address both weed control and nitrogen management. New low rate herbicides, soil erosion control with residue management that might require more herbicide and nitrogen fertilizer, advances in nitrogen availability testing for corn, recognition of nitrogen contributions from soil, and legume residues and manures must be coupled in systems approaches that ultimately reduce water contamination.

The changes in the soil surface biological and physical environments will create new challenges for weed, insect, and disease management. At present, we do not fully comprehend these impacts on soil biotic factors. Water quality goals must be set, recognizing the changes which may occur as a result of changes in farming practices. Often, the effects of these changes are not readily detectable over the short term, thus requiring long-term studies.

Regnier and Janke (1990) proposed several alternatives for weed control which reduce the need for agricultural chemicals. To date, these methods have not been adopted widely. There are many social, technical, and economic reasons for the lack of widespread adoption of their techniques for weed control. But it is likely that these techniques will be part of regulations and incentives over the next decade. We must initiate on-farm systems studies to develop alternative weed control approaches in row-crop farming, taking into account the need for residue management and soil erosion control.

Reducing herbicide use will intuitively reduce herbicide contamination of ground water. However, lowering nitrogen inputs sufficiently to bring nitrate concentrations to background levels (often assumed to be 2 ppm of nitrate N or less) (Fedkiw, 1991) has always proved elusive (Keeney, 1986). Many nitrogen transformations affect mobility and availability of this nutrient. Both mobility and availability are interrelated with the water status of the field and the watershed, as well as with the residence time and pathways of water in the watershed. An interesting example is the comparison between the findings of Hallberg and co-workers in the Big Spring watershed (Hallberg, 1989b) and the Des Moines River watershed, both largely agricultural watersheds. The Big Spring watershed is smaller, composed of thin soils over limestone, with numerous sinkholes and channels leading to fairly rapid response time of water in the outlet, an artesian spring. Nitrate levels now appear to be decreasing in response to a concerted educational program reducing nitrogen use and accounting for residual nitrogen in the soils. A comparison of historical nitrate levels in the Des Moines River watershed showed little change over nearly 50 years. This watershed has poorly drained soils, is largely tile-drained, and has highly productive high organic matter soils that release considerable nitrogen. The conclusion reached in the latter study was that the solution to lowering nitrate levels in the lower Des Moines River must go far beyond just source reduction, and we must look at the management practices within the watershed to increase the sinks while lowering the sources of nitrogen. Any planned educational effort for this watershed will require a broader approach than that used in Big Springs.

Cover crops may help retain nitrogen in the biocycle and therefore reduce source loading (Meisinger et al., 1991). Cover crops also offer the advantage of protecting soil from erosion. Cover crops can influence the nitrate leaching and ground water quality through modifications in the water budget, affecting soil nitrate concentrations, and through synchronizing competition during the water recharge season (Meisinger et al., 1991). Overall, the results show that nitrate concentrations in the leachate below the root zone range from 20 to 80% of the concentrations of the leachate under no cover crop (Meisinger et al., 1991). From reviewing literature, they found that grasses and brassicas were two to three times more effective in reducing nitrate leaching than were legumes. They also suggested that there were some knowledge gaps which should be addressed, and these included an improved prediction of nitrate concentration in the soil profile. This requires an improved understanding of the N cycle and the mineralization process in particular. There needs to be additional information on the fate of N recovered by the cover crop to ensure that cover crop decomposition and mineralization are synchronized with the N requirements of the subsequent crop (Meisinger et al., 1991).

Additional research also is needed to develop cover crop systems for climates with short growing seasons and/or low water availability. Cover crops that also fix nitrogen (some of the low growing clovers such as berseem) or have other economic values would be welcomed. Planting of cover crops without realizing an economic gain explains why this practice is not adopted currently. Alternately, incentives will be needed to increase cover crop use.

Atmospheric Quality

The quality of the atmosphere is usually an issue only when there is a marked degradation of the environment or when a health hazard exists. It is common to associate atmospheric quality problems with areas of high population and industrial density. Generally, we do not perceive that there is a problem with air quality associated with agricultural activities except for the isolated incident of blowing dust, pesticide drift, and the odors from manure handling and application. However, the atmosphere has long been recognized as a major transport medium for volatile organics, especially pesticides (Atlas and Giam, 1988), and Iowa monitoring studies have shown high levels of herbicides in Iowa rainfall in the spring (Nations and Hallberg, 1992). The issue of global climate change, including the role of soils as a source/sink for greenhouse chemicals such as methane, carbon dioxide, and nitrous oxide is bringing increasing urgency to atmospheric studies. Atmospheric quality has emerged as a major issue and will continue to increase in importance over the next century. We must include this issue in research design and evaluation.

The recent report by Nations and Hallberg (1992) showed that agrichemicals were present in the precipitation over Iowa. They showed that over a 2-year period, the detectable concentrations ranged from 0.1 to 40 μg l^{-1} with most observations below 1 μg l^{-1}. Atrazine was found to be present in the largest concentrations and the concentration patterns could be related to application times. Their study raises some very important questions which need to be addressed regarding the transport of

chemicals from the soil surface to the atmosphere. The process of volatilization losses from the soil surface is not clearly understood and will have
to be evaluated as part of the efforts to link farming practices to environmental quality. It is not clear whether the chemicals which are lost are
carried as gas due to volatilization or are bound to small dust particles
which escape when a field is cultivated. Pesticide drift and nontarget
crop damage are evidence of movement in the atmosphere and will need
to be addressed through new technology for applications. These answers
will be necessary before we can completely close the mass balance for
agrichemicals and nitrate. Nitrate concentrations in the rainfall can contribute from 20 to 30 kg ha^{-1} to the soil; however, this amount is rarely
added to the nitrogen balance for the crop. The half-life for ammonia in
the atmosphere is low, and studies have shown that ammonia from feedlots can add significantly to the nitrogen load of water bodies.

Odors from manure storage, handling, and application to the soil are
probably the most direct association which is made between farming
practices and atmospheric quality. It is also the area in which knowledge
is least available. Odors are subjective, and quantification of them is
extremely difficult and expensive. Systems for utilization of manures
must direct concern toward methods for reducing odors during storage,
handling, and application. Manure is a resource in terms of nutrient
content and returns of organic matter to the soil. Unfortunately, costs of
handling and application at crop utilization rates over large areas away
from the source often exceed those of commercial fertilizer. Until the
mind-set that regards manure as a "waste" rather than a resource is
overcome, it is hard to make gains on the wise use of this resource. There
is a renewed effort in evaluating the potential role of manures, particularly with the disposal problems from the large animal feeding operations
and the linkage between the manure storage and handling to water quality. Manures are in theory a valuable resource for sustainable agricultural
systems in that they ensure internal recycling of nutrients and improve
soil quality through application of organic matter. However, there are
more questions than available technology at the present time. The development of affordable accurate application equipment alone would be a
great advance. Other needed research areas include rapid methods to
determine the nitrogen content of manures and ways to mesh manure
application with residue management systems.

Use of Agricultural Chemicals

The evolution of agricultural practices from the 1960s to the present
time has incorporated the use of agrichemicals (also known as pesticides)
for weed, insect, and disease control. During this period, the land use

patterns have also changed to more monocultures and limited crop rotations which has increased the reliance on the use of chemicals. The case studies described within *Alternative Agriculture* show examples in which individuals replaced agrichemicals with other methods of control and crop rotations and increased management. Cook (1991) states that in the Pacific Northwest, crop rotations in the wheat areas may not be a viable solution and that an understanding of the complete soil and plant system will be necessary before making any changes. His hypothesis is that the current methods have focused on the physical and chemical processes, and before we can make any advances in changing the practices, an understanding of the biological system will be required.

The fate of agrichemicals in the environment is poorly understood. To develop alternative practices will require the interaction of agronomists, entomologists, pathologists, hydrologists, agroclimatologists, geneticists, sociologists, and economists to fully develop an understanding of the changes which are occurring within the farming system. Our research on agrichemicals has been limited to the disciplinary problem which the chemical was devised to control, that is, diseases and fungicides, insects and insecticides, weeds and herbicides. Future efforts with respect to environmental quality will have to be an integrated and interdisciplinary approach. An example of a successful program for the Pacific Northwest is the STEEP (Solutions to Environmental and Economic Problems) project (Papendick, 1991). This program was designed to incorporate the expertise of several disciplines and to also link the research with an educational program. Importantly, it incorporated growers into the project design and evaluation. Some of the impacts of this program have been seen in the area of tillage and plant management, plant design, erosion and runoff prediction, pest management, socioeconomics of erosion control, and soil erosion-productivity relationships (Papendick, 1991). This project can be considered as an example of the scope that projects will have to embrace if changes are to be made in chemical use and management practices.

Analysis of the effects of chemicals and pest control will require an improved understanding of the economic relationships among the pest populations and crop productivity (Zimdahl, 1991). In the case studies in *Alternative Agriculture*, the farmers had developed these relationships for their own farms in order to guide their decision-making process in terms of cultivation and insect control. Any change in the farming practice will have to be closely evaluated from a thorough economic analysis which includes all of the production costs and may have to be extended to include the off-site costs, for example, sediment deposition or pesticide drift.

ECONOMICS OF SUSTAINABLE AGRICULTURE

The changes which occurred from the 1960s to the present time were driven by farm policy and conservation measures as much as any other factors. The 1985 and 1990 Farm Bills may be regarded as landmarks in changing the face of agriculture. These acts promoted a change in the management of the land through the adoption of conservation control measures on all highly erodible land. Earlier Farm Bills in the 1970s which promoted open trade encouraged more intensive and extensive use of the land. Acceptance of the recommendations was not taken at face value, and there has been much controversy about the different ways in which the land surface could be managed to achieve the goals put forth in the 1985 and 1990 Farm Bills. Total resource management has received increased recognition. The need to better understand the environmental impacts of different practices and how valuable these practices may be to agriculture is increasingly recognized (Gray, 1991). Researchers, extension specialists, and farm managers are pressed for answers that are not available. Farmers, however, have continued to change, and the latest survey from the Conservation Technology Information Center (CTIC) shows a rapid increase in the number of acres which are planted under some type of residue management (Becherer, 1991). Becherer (1991) projected from the CTIC surveys that by 1995 the no-till acreage will double from 14 to 28 million acres, ridge-till acreages will quadruple from 2.7 to 11 million acres, and mulch tillage will be practiced on 100 million acres compared to the 55 million currently farmed with this technique. Part of this change has been due to increased dictate for erosion control. Residue management systems are largely a win-win situation. They may require smaller tractors, less time, and are adaptable to large acreages. However, evidence to date indicates we have not learned to run these systems without the use of at least as much, and often more, agrichemicals than used for conventional systems. Also, these systems discourage rotations and lessen the options for use of animal manures. Clearly there are questions to be asked regarding the trade offs of conservation tillage, particularly no-till, and other objectives of farming.

Lal et al. (1990) summarized some of the recent comparisons of conventional tillage and reduced tillage research and concluded that "conservation tillage can be made an integral part of sustainable agricultural systems through practically oriented, multidisciplinary research." Technological advances in planting equipment to ensure stand establishment and weed control with postemergence chemicals have promoted the adoption of some of these practices. However, these changes have raised some questions in terms of the most effective nutrient management and pest management strategies when there is a change in the residue management. To fully achieve the goal of a reduced input will require better management strategies which will provide an improved decision-making complex. Economists will have to become an integral part of the multi-disciplinary research teams in order to have an economic structure from which decisions can be made and comparisons can be completed.

More complete economic analyses are needed for conventional and alternate farming systems (Ikerd, 1990). Virtually no attention is paid to off-site effects of farming practices (Hitzhusen, 1991). Siltation, pollution of water supplies, and loss of wetlands and wildlife are examples. The issues of farm size and declining rural populations need more economic consideration. Recent studies at the Leopold Center (Duffy and Melvin, unpublished) have shown that medium-sized (400 acres) farms can be more profitable than large farms (greater than 1000 acres) if labor-intensive animal enterprises are included. This flies in the face of the "bigger is better" philosophy that has dominated row crop agriculture since the 1960s.

SOCIOLOGICAL ADAPTATION OF TECHNOLOGY AND TECHNOLOGY TRANSFER

The initial reaction to sustainable agriculture research was that all inputs were to be replaced and that farming was going to revert to a system with all mechanical weed control, limited fertilizer inputs, and ultimately, reduced yields. It is now clear that the goals of sustainable agriculture are the same as those of sustainable development, namely, to meet the needs of the present without compromising the ability of future generations to meet their needs (Thomassin et al., 1991). The approach is to develop systems which embrace the concepts of total resource management while providing the flexibility to evolve and change to incorporate both our current understanding and technology. It is naive to think that we would revert to prop-driven airplanes and two-lane highways for our travel; likewise, it is naive to think that agriculturalists will suddenly ignore all of the available technology in their decision-making process. Evidence indicates that this technology will be an important part of sustainable systems of the future. The challenge is to construct the technology and provide the technology transfer so that agriculture can be sustainable.

A larger underlying theme involves the best way to communicate the research findings to the farmers. The case studies cited in *Alternative Agriculture* reveal that the farmers were involved in on-farm research and the National Research Council report on *Sustainable Agriculture Research and Education in the Field* shows the value of on-farm research. This approach has been supported by numerous authorities writing on the issue of research for a sustainable agriculture (Anderson and

Lockeretz, 1991, 1992; Benbrook, 1991; Franzluebbers and Francis, 1991; Papendick, 1991; and Schaller, 1991).

For example, Shennan et al. (1991) described the use of on-farm comparisons for tomato production in California and concluded that on-farm research provided valuable insights into the variety of management options and farm level responses that were not possible with research plots. Papendick (1991) described the effective incorporation of the farmers into the objectives of the STEEP program and the active role of the wheat growers in the Pacific Northwest in securing support for the project. The structure of the Leopold Center for Sustainable Agriculture at Iowa State University, Ames, Iowa provides for the incorporation of farmers into each of its interdisciplinary issue teams in order to provide immediate feedback of the issues being addressed and the effectiveness of the research approach.

On-farm comparisons are necessary to complete our understanding of the adaptation of technology. Sustainable agriculture is not a discipline but a complex integration of several disciplines, and farmer involvement is necessary to provide the input of the decision maker. Shennan et al. (1991) were able to complete a statistical analysis of the differences between the conventional and organic tomato farms through sampling and surveys. They were able to identify factors which could be considered attributes of each system and the common factors of each system. These analyses provide a quantitative comparison of farming practices which removes the bias and perception of the system response. On-farm research may become more commonplace to achieve two goals: comparison of the response of a management practice at the field scale, and incorporation of farmer involvement in both design and technology transfer (Smolik and Dobbs, 1991).

The National Soil Tilth Laboratory has been involved in the comparison of two farming systems in Central Iowa. One of the farms is that operated by Thompson and described in *Alternative Agriculture*, and the other is an adjoining conventional farm. The comparisons have centered on the differences in soil structure, soil biology, and infiltration processes. These results are described in several reports (Berry and Karlen, 1993; and Karlen and Colvin, 1992). The experiences gained from these comparisons provide insights into how the basic responses can be compared at the field scale in order to provide a more quantitative understanding of the differences among practices. The water quality projects described by Onstad et al. (1991) also use field level comparisons of farming systems to assess the impact of farming practices on water quality. The Iowa portion of this project, which is conducted at the Walnut Creek watershed near Ames and at the Deep Loess Research Station near Treynor, involves field scale studies and sampling within fields. These studies are placed in a context of the impact of the current practice and a change in either chemical, fertilizer, or tillage management at the field scale. Studies of this type will require use of more comparative statistics, with emphases on both spatial and temporal analyses. There is a need to develop methods which can accommodate a field and farm level comparison with the precision and reliability required for transfer to the user.

Nowak (1992), in evaluating why farmers did or did not adopt residue management strategies, concluded that the limitations were because farmers were either unable or unwilling to adopt new technologies. The reasons behind these two constraints reveal much about our current technology. The reasons for the farmers' attitudes of being unable to adopt technology were lack of information, high costs of obtaining information, too much complexity in the system, too expensive, excessive labor requirements, too short a planning horizon, availability and accessibility of supporting resources considered to be limited, lack of adequate managerial skills, and lack of control over the adoption decision. He stated that the main reasons that farmers were not willing to adopt residue management practices were that the information conflicted with existing information or was inconsistent; the information lacked relevance; there were conflicts between current production goals and the new technology; ignorance about the technology either by the farmer or the promoter of the technology; practice is not suited to the physical setting; practice increases risk of negative outcomes; and belief in traditional practices (Nowak, 1992). Each of these reasons reveals that we have much to do in the research and information transfer field to not only make our research relevant but to also incorporate the farmer into the evaluation process. This indicates that researchers in agriculture can not be isolated from reality. Ways must be found for researchers to work with farmers and to be rewarded for the extra effort this takes relative to bench scale or plot scale activities.

Technology transfer of sustainable agriculture information has forced a reexamination of the current methods by which information is moved from the researcher to the user. There is an increasing awareness that our current methods may not be adequate and do not allow for effective transfer. Extension specialists should be an integral part of the research team and should be considered as investigators, rather than consumers of research information. Again, the STEEP program and the water quality projects have integrated extension and education efforts as part of the infrastructure. There are also local groups which act both as their own information-gathering and information-exchange networks. Examples of these are the local farmers' groups which have as a mission to evaluate and share information on different practices. The Practical Farmers of Iowa conduct on-farm research, usually in cooperation with Iowa State researchers, but also share the information among farmers through newsletters, on-farm demonstration plots, and research reports. Sustainable agriculture research has caused a reexamination of the infrastructure of information sharing. There are many aspects which are positive in this change, and these have been an increased involvement of the farmer in research discussion and an awareness of the value of on-farm research. Involvement of the farmer will also provide for an assessment of the risk management aspects associated with changes in either technology or management. The integrative nature of sustainable agriculture and the focus on resource management require that a holistic approach be considered in research management and planning. This concept is reinforced by the ongoing studies supported by the Northwest Area Foundation in St. Paul, Minnesota. They have begun to survey farmers' attitudes and change in practices and have concluded that there is an intricate fabric which blends the farming practices with the social and economic layers in rural America. Their studies are in the beginning stage and will provide a valuable source of information about the rate of change of farming practices and the research gaps (Northwest Area Foundation, 1992).

QUALITY OF FOOD

Quality of the produce will become an increasingly important issue for the 21st century. The increasing awareness of diet and nutrition will change the public's definition of quality food. There are limited examples comparing farming system effects on food quality outside of the potential movement of agrichemicals through the food chain. Risks have been associated with food production; however, the linkage between practices and nutritional value has not been fully evaluated. Value of the raw products being produced by the agricultural system may become an important consideration. Protein value of the grain has been recognized as a variable which responds to management. Currently, there is no market incentive to produce this type of grain. Nutritional value of forages for feed is recognized for efficient livestock production, but this variable is not measured as part of the farming systems nor is there an attempt to include nutritional value as part of the comparison studies among systems. This may change as new studies are instituted and there is a broadening of the scope of the studies to include an assessment of the product quality rather than total yield.

INTERACTIONS OF THE PUBLIC AND PRIVATE SECTOR

Public and private sector involvement in agricultural research will change in the 21st century, if the current trends continue. There is increased interaction between the agricultural industry and the research community at the intellectual level as opposed to the recent past when product testing was the major product of many cooperative agreements. on-farm demonstrations and research will become increasingly used as a means of developing systems and concepts at the field and farm scale. This also suggests that farmers will be more actively involved in the research planning and evaluation stages than in the past. As discussed earlier, there are several examples in which the farmers have become successful participants in the research program and are willing to share their records with the research community to assist with research objectives.

Interactions with other agribusiness sectors will also be enhanced over the next few years. There are several reasons to believe that there will be a closer link between agricultural industry and research. The first is the development of the Cooperative Research and Development Agreements (CRADA) of the federal government. These programs encourage cooperative ventures to be taken at the development level to encourage researchers to cooperate with industry in bringing a product through development and testing. The patent rights are protected and the researcher is free to publish the results. These programs are just beginning and are showing some success at promoting better relationships. They permit access to the research community to take advantage of the expertise contained within both university and federal research organizations and often benefit local industry. The need to incorporate equipment changes and evaluate different germplasm as cropping systems change will require that research and industrial development be linked through cooperative studies.

RESEARCH NEEDS AND CHALLENGES

There are several research challenges that must be addressed in agriculture if the goal of a sustainable agriculture is to be achieved by the beginning of the next century. Throughout this book, there are summaries given by each of the authors from their own perspective; however, there will be a need to address the problems of sustainable agriculture with an integrated, multi- and interdisciplinary approach. Benbrook (1991) summarized several research challenges for sustainable agriculture and concluded that we are at the crossroads of integrating the sciences of agriculture and ecology.

We present several research challenges as a means of promoting discussion about how they might be accomplished.

- Integrate traditional research into field and farm-based studies to provide for a more complete understanding of the principles involved in changes in agricultural practices.

- Develop an understanding of systems level research which incorporates both agricultural and ecological principles.
- Develop procedures from which comparisons can be made in terms of the effect of different farming practices on the soil, crop, or environment.
- Integrate physical, chemical, and biological processes to develop quantitative understanding of the changes which occur within the soil as a result of management changes.
- Develop improved methods for sampling soil nutrients within the soil profile which are related to plant needs.
- Incorporate germplasm studies into sustainable agriculture research to improve the genetic response to different farming practices.
- Evaluate the changes which occur within the field ecosystem as a result of management changes: for example, weed population changes as a result of residue management; bird and field wildlife populations as affected by management practices and insect control procedures.
- Conduct long-term field research across a series of climates and soils to evaluate the potential limitations to systems development.
- Incorporate economic, mathematical, and systems research into the research infrastructure for development of transfer models.

The "Research and Science" chapter in *Alternative Agriculture* lists several research areas which the committee felt necessary to improve our knowledge base on sustainable agriculture (National Research Council, 1989). This list covers all of the aspects of science associated with the agricultural sector. There have been several lists developed for what needs to be completed. The limitation to achieving the goals set forth in several of the books is the effective development of an infrastructure which will allow these aspects to be incorporated into a research program.

The challenges for the 21st century are simply stated: integrate the available information into a scale which the farming community can use and be confident in for their decision-making process; identify the research gaps and fill them in a context of agroecology; incorporate the user into the research team; and satisfy the societal and environmental needs. The challenges are large but not unreachable. Agriculture must develop a strategy and a plan of how we are going to complete these tasks. This will be good for all people and an effort in which we can all claim credit.

REFERENCES

Anderson, M. D. and W. Lockeretz. 1991. On-farm research techniques. Report on a workshop. Institute of Alternate Agriculture, Greenbelt, MD.

Anderson, M. D. and W. Lockeretz. 1992. Sustainable agricultural research in the ideal and in the field. *J. Soil Water Conserv.* 47:100–104.

Atlas, E. and C. S. Giam. 1988. Ambient concentrations and precipitation scavenging of atmospheric organic pollutants. *Water, Air Soil Pollut.* 38:19–36.

Becherer, J. 1991. Crop residue management: what CTIC surveys show. In: Crop Residue Management for Conservation, Proceedings of a National Conference, August 8 to 9, 1991, Lexington, KY, Soil and Water Conservation Society, Ankeny, IA, pp. 11–12.

Benbrook, C. M. 1991. Introduction. In: National Research Council. 1991. *Sustainable Agriculture Research and Education in the Field*. National Academy Press. Washington, D.C., pp. 1–10.

Berry, E. C. and D. L. Karlen. 1993. Comparison of alternative farming systems. II. Earthworm population density and species diversity. *Am. J. Altern. Agric.* 8:21–26.

Bromley, D. W. 1991. Technology, technical change, and public policy: the need for collective decisions. *Choices.* 2nd quarter. pp. 5–13.

Bues, C. E. and R. E. Dunlap. 1990. Conventional versus alternative agriculture: The paradigmatic roots of the debate. *Rural Sociology* 55:590–616.

Busby, F. 1990. Sustainable agriculture: who will lead? *J. Soil Water Conserv.* 45:89–91.

Cook, R. J. 1991. Challenges and rewards of sustainable agriculture research and education. In: National Research Council. 1991. *Sustainable Agriculture Research and Education in the Field*. National Academy Press. Washington, D.C., pp. 32–76.

Council of Agricultural Science and Technology (CAST). 1985. Agriculture and Groundwater Quality. Rep. No. 103. Council of Agricultural Science and Technology, Ames, IA.

Crosson, P. 1991. Sustainable agriculture in North America: issues and challenges. *Can. J. Agric. Econ.* 39:553–565.

Doll, J. D. and C. E. Francis. 1992. Participatory research and extension strategies for sustainable agricultural research. *Weed Technol.* 6:473–482.

Fedkiw, J. 1991. Nitrate occurrence in U.S. waters (and related questions). A reference summary of published sources from an agricultural perspective. U.S. Department of Agriculture. Washington, D.C.

Francis, C. A. and G. Youngberg. 1990. Sustainable Agriculture: an overview. In: C. A. Francis, C. B. Flora, and L. D. King (Eds.) *Sustainable Agriculture in Temperate Zones*. John Wiley & Sons, New York, pp. 1–23.

Francis, C., J. King, J. DeWitt, J. Bushnell, and L. Lucas. 1990. Participatory strategies for information exchange. *Am. J. Altern. Agric.* 5:153–160.

Franzluebbers, A. J. and C. A. Francis. 1991. Farmer participation in research and extension: N fertilizer response in crop rotations. *J. Sustain. Agric.* 2:9–30.

Goolsby, D. A., R. C. Coupe, and D. J. Markovchick. 1991. Distribution of selected herbicides and nitrate in the Mississippi River and its major tributaries, April through June 1991. Water-Resources Investigations Rep. 91–4163, U.S. Geological Survey, Denver, CO.

Gray, R. 1991. Economic measures of sustainability. *Can. J. Agric. Econ.* 39:627–635.

Hallberg, G. R. 1989a. Pesticide pollution of groundwater in humid United States. *Agric. Ecosystems Environ.* 26:299–367.

Hallberg, G. R. 1989b. Nitrate in the ground water of the United States. In: R. F. Follett (Ed.) *Nitrogen Management and Ground Water Protection.* Elsevier, New York, pp. 35–74.

Hallberg, G. R., B. C. Kross, R. D. Libra, L. F. Burmeister, L. M. B. Weih, C. F. Lynch, D. R. Bruner, M. Q. Lewis, K. L. Cherryholmes, J. K. Johnson, and M. A. Culp. 1990. The Iowa State-Wide Rural Well-Water Survey Design Report. A systematic sample of domestic drinking water quality. Geological Survey Bureau. Tech. Inf. Ser. No. 17. Iowa Department of Natural Resources. Des Moines, IA.

Harwood, R. R. 1990. A history of sustainable agriculture. In: Edwards, C. A., R. Lal, P. Madden, R. H. Miller, and G. House. (Eds.) *Sustainable Agriculture Systems.* Soil and Water Conservation Society, Ankeny, IA, pp. 3–19.

Hillel, D. J. 1991. *Out of the Earth: Civilization and the Life of the Soil.* Free Press. New York.

Hitzhusen, F. J. 1991. The economics of sustainable agriculture: Adding a downstream perspective. *J. Sustain. Agric.* 2(2):75–89.

Holden, L. R. and J. A. Graham. 1992. Results of the national alachlor well water survey. *Environ. Sci. Technol.* 26:935–943.

Ikerd, J. 1990. Agriculture's search for sustainability and profitability. *J. Soil Water Conserv.* 45:18–23.

Karlen, D. L. and T. S. Colvin. 1992. Alternative farming system effects on profile nitrogen concentrations on two Iowa farms. *Soil Sci. Soc. Am. J.* 56:1249–1256.

Keeney, D. R. 1986. Sources of nitrate to groundwater. *Crit. Rev. Environ. Control* 16(3):257–303.

Keeney, D. R. 1989. Toward a sustainable agriculture: Need for clarification of concepts and terminology. *Am. J. Altern. Agric.* 4:101–105.

Keeney, D. R. 1990. Sustainable agriculture: definition and concepts. *J. Prod. Agric.* 3:281–285.

Keeney, D. R. 1992. Research to meet natural resource needs in the 1990's. Presented to the First International Crop Science Congress, Ames, IA. July 1993.

Keeney, D. R. and T. H. Deluca. 1993. Des Moines River nitrate in relation to watershed agriculture practices: 1945 versus 1980s. 22:101–105.

Kross, B. C., G. R. Hallberg, D. R. Bruner et al. 1990. The Iowa State-Wide Rural Well-Water Survey, Water Quality Data: Initial Analyses. Geological Survey Bureau. Tech. Inf. Ser. No. 19. Iowa Department of Natural Resources, Des Moines, IA.

Lal, R., D. J. Eckert, N. R. Fausey, and W. M. Edwards. 1990. Conservation tillage in sustainable agriculture. In: Edwards, C. A., R. Lal, P. Madden, R. H. Miller, and G. House. (Eds.) *Sustainable Agriculture Systems.* Soil and Water Conservation Society, Ankeny, IA, pp. 203–225.

Leopold, A. 1949. *Sand County Almanac* (first printing). Oxford University Press, New York.

Lockeretz, W. 1988. Open questions in sustainable agriculture. *Am. J. Altern. Agric.* 3:174–181.

Lockeretz, W. 1991. Information requirements of reduced-chemical production methods. *Am. J. Altern. Agric.* 6:97–103.

Meisinger, J. J., W. L. Hargrove, R. L. Mikkelsen, J. R. Williams, and V. W. Benson. 1991. Effects of cover crops on groundwater quality. In: W. L. Hargrove (Ed.) *Cover Crops for Clean Water*. Soil and Water Conservation Society, Ankeny, IA, pp. 57–68.

National Research Council. 1989. *Alternative Agriculture*. National Academy Press. Washington, D.C., 448 pp.

National Research Council. 1991. *Sustainable Agriculture Research and Education in the Field*. National Academy Press. Washington, D.C.

Nations, B. K. and G. R. Hallberg. 1992. Pesticides in Iowa precipitation. *J. Environ. Qual.* 21:486–492.

Northwest Area Foundation. 1992. Which Row to Hoe? A regional perspective on alternative directions in commercial agriculture. An interim report of the Northwest Area Foundation, St. Paul, MN.

Nowak, P. 1992. Why farmers adopt production technology. *J. Soil Water Conserv.* 42:14–16.

Offutt, S. 1991. Agriculture's role in protecting water quality. *J. Soil Water Conserv.* 45:94–96.

Onstad, C. A., M. R. Burkart, and G. D. Bubenzer. 1991. Agricultural research to improve water quality. *J. Soil Water Conserv.* 46:184–188.

Papendick, R. I. 1991. STEEP: A model for conservation and environmental research and education. In: National Research Council. 1991. *Sustainable Agriculture Research and Education in the Field*. National Academy Press. Washington, D.C., pp. 133–144.

Pierce, F. J. 1991. Erosion productivity impact prediction. In: R. Lal and F. J. Pierce (Eds.) *Soil Management for Sustainability*. Soil and Water Conservation Society, Ankeny, IA, pp. 35–52.

Regnier, E. E. and R. R. Janke. 1990. Evolving strategies for managing weeds. In: Edwards, C. A., R. Lal, P. Madden, R. H. Miller, and G. House. (Eds.) *Sustainable Agriculture Systems*. Soil and Water Conservation Society, Ankeny, IA, pp. 174–202.

Rosmiller, G. E. 1992. Six problems that affect agricultural policy. *Choices*. 1st quarter. pp. 14–17.

Schaller, F. 1991. An agenda for research on the impacts of sustainable agriculture. Assessment and recommendations of a panel of social scientists. Institute of Alternative Agriculture, Greenbelt, MD.

Shennan, C., L. E. Drinkwater, A. H. C. van Bruggen, D. K. Letourneau, and F. Workneh. 1991. Comparative study of organic and conventional tomato production systems: An approach to on-farm systems studies. In: National Research Council. 1991. *Sustainable Agriculture Research and Education in the Field*. National Academy Press. Washington, D.C., pp. 109–132.

Smith, S. 1992. "Farming" It's declining in the U.S. *Choices*. 1st quarter. pp. 8–10.

Smolik, J. D. and T. L. Dobbs. 1991. Crop yields and economic returns accompanying the transition to alternative farming systems. *J. Prod. Agric.* 4:153–161.

Thomassin, P. J., J. C. Henning, and L. Baker. 1991. Old paradigms revisited and new directions. *Can. J. Agric. Econ.* 39:689–698.

Weil, R. R. 1990. Defining and using the concept of sustainable agriculture. *J. Agron. Educ.* 19:126–130.

Wilken, G. C. 1991. Sustainable agriculture is the solution, but what is the problem? Occas. Pap. No. 14, Board for International Food and Agricultural Development and Economic Cooperation, AID, Washington, D.C. April 1991.

Zimdahl, R. L. 1991. Weed Science. A plea for thought. USDA-CSRS. Washington, D.C.

Index